CURRENT ISSUES IN MOLECULAR VIROLOGY - VIRAL GENETICS AND BIOTECHNOLOGICAL APPLICATIONS

Edited by **Victor Romanowski**

Current Issues in Molecular Virology - Viral Genetics and Biotechnological Applications
http://dx.doi.org/10.5772/50089
Edited by Victor Romanowski

CBS, Edition **2016**

Contributors

Marcelo Berretta, Alessandra Tenório Costa, Rodrigo Kazuo Makiyama, Alessandra Vasconcellos Nunes, Juliana Pereira Bravo, Ivan De Godoy Maia, Ewan P Plant, Saiz, Martin Bisaillon, Victor Romanowski, Santiago Haase, Maria Leticia Ferrelli, Matias Luis Pidre, Hongxuan He, Amauri Alcindo Alfieri, Michele Lunardi, Alice F. Alfieri, Rodrigo Otonel, Maria R. Albiach-Marti, Claudia Herenu, Andrea Soledad Pereyra

Published by InTech

Janeza Trdine 9, 51000 Rijeka, Croatia

Notice

Publishing Process Manager Sandra Bakic

Technical Editor InTech DTP team

Cover InTech Design team

Additional hard copies can be obtained from orders@intechopen.com

Current Issues in Molecular Virology - Viral Genetics and Biotechnological Applications, Edited by Victor Romanowski
p. cm.
ISBN 978-953-51-1207-5

Contents

Preface

Viruses are essentially itinerating genomes conveniently packaged in protein shell structures, sometimes surrounded by lipid membranes derived from the host cells they highjack in order to reproduce. In fact, being able to introduce a discrete number of genes in a cell, viruses made their enormous contribution to the exploration of mechanisms of genome replication and gene expression long before the advent of recombinant DNA technology. Studies on phages that infect bacteria built a significant body of knowledge in the early years of molecular biology. In parallel, active research was oriented at viruses that cause diseases in humans, other animals and plants (one chapter of this book deals with citrus tristeza virus). In particular, medical virology has been by far the preeminent area of interest. Efforts have been made aimed at curing and preventing viral diseases that caused large numbers of victims long before their viral etiology was recognized. In Europe 400,000 people died annually of smallpox in the 18th century. Based upon observation and empirical trials a vaccination procedure was devised to prevent this dreadful infection that caused high morbidity and mortality. Vaccination was likely practiced in Africa, India, and China long before the 18th century, when it was introduced to Europe. Edward Jenner's work represented the first scientific attempt to control an infectious disease by the deliberate use of vaccination. Strictly speaking, he did not discover vaccination but was the first person to confer scientific status on the procedure and to pursue its scientific investigation.

Biological sciences have gone a long way since then. Molecular biology of viruses (after all, viruses are more or less complex associations of macromolecules) and studies on virus-host interactions have provided a wealth of knowledge that helps designing different prevention strategies aimed at innate and adaptive immune responses. Several chapters of this book focus on viral protein complexes, gene expression, nucleotide sequences and genetic constelations in viral populations related with the design and production of new immunogens, and establishment of vaccination schemes to prevent viral diseases.

Drifting away from the human victims of viral diseases, it is worth mentioning that the prosperous silk industry in China drove attention to an economically relevant viral infection of the silk worm (Bombyx mori) starting a chase for the pathogen that affected the silk production. Three chapters in this book deal with viruses that belong to the family responsible for the economical losses of the silk industry: Baculoviridae. The studies on baculoviruses range from pathogenesis to viral genomics and gene expression, and most of the members are regarded more as friends than threats; the biology of baculoviruses has been harnessed for diverse applications such as microbial pest control, protein expression and gene transduction. Alternative gene transduction strategies are dealt with in the chapter summarizing different gene delivery systems. Advances in molecular virology have paved the way to alternative novel vaccine and gene therapy strategies.

In summary, this book is only a small collection of chapters dealing with examples of RNA and DNA viruses, and issues such as how these "gene packages" have learnt to take advantage of their hosts, molecular recognition events that hosts may use to counterattack the viruses, and how researchers have developed strategies to use viruses or their parts as tools for different purposes.

Dr. Víctor Romanowski
Instituto de Biotecnología y Biología Molecular
(IBBM, UNLP-CONICET)
Facultad de Ciencias Exactas, Universidad Nacional de La Plata
Argentina

Chapter 1

The Complex Genetics of *Citrus tristeza* virus

Maria R. Albiach-Marti

Additional information is available at the end of the chapter

http://dx.doi.org/10.5772/56122

1. Introduction

The 2000 x 11 nm long bipolar flexuous filamentous particles of *Citrus tristeza virus* (CTV) (genus *Closterovirus*, family *Closteroviridae*) (Figure 1) contain a single-stranded positive-sense RNA genome of 19.3 kb, which is encapsidated in two different capsid proteins that coat the opposite ends of the virions [1, 2]. CTV is the largest identified RNA virus infecting plants and the second largest worldwide after the animal *Coronaviruses*. The virus is phloem limited and it is transmitted by aphids (*Hemiptera: Aphididae*) (Figure 1), and mechanically by graft propagation of virus-infected plant tissues. CTV isolates from different hosts and areas display great variability either biologically or genetically. There are wild CTV isolates that consist basically of a main genotype and its quasispecies, but others could contain a mixture of strains (groups of viral variants with similar sequence) that differ in symptomology and in viral transmission efficiency by aphids. These CTV strains could bear divergent CTV genotypes. Additionally, wild isolates are also composed by a population of defective RNAs (D-RNAs) that could change by aphid or graft transmission or by host passage [3].

The *Tristeza* syndrome, induced by CTV, has devastated entire commercial citrus industries around the world, since it has caused the death of hundred million trees worldwide. In point of fact, this virus is present in most of the citrus producing areas infecting nearly all species, cultivars and hybrids of *Citrus* spp. and related genera. Phenotypically, CTV induces different grade and wide range of symptoms in *Citrus* species. In effect, depending on the virus isolate and the variety/rootstock combination, CTV strains can cause different syndromes in the field like 'decline' (QD) or 'stem pitting' (SP). Some CTV isolates induce a third syndrome, in glasshouse conditions, that is referred as 'seedling yellows' (SY). Furthermore, CTV causes a myriad of different symptom combinations in indicator plants depending on the CTV strain, or the mixture of strains, present in the plant host indexed. Remarkably, there are mild CTV strains that cause a complete lack of symptoms in almost all species and varieties of citrus, including those present in the citrus orchards, even though these mild viruses multiply to high titers [4, 5].

The study of the CTV genetics and the virus-host interactions have been hampered during long time as a consequence of the difficulties of experimenting with a virus with a large RNA genome, assembled in fragile particles and present in reduced amounts in a tree, where CTV could take long time to colonize the entire plant and to induce symptomatology. For that reason, CTV was for decades a virus complicated to isolate and characterize. Moreover, the elevated diversity of CTV populations impeded the separation of the sequence variants, composing a specific isolate, to analyze each of the genotypes independently in order to understand every aspect of viral infection. Likewise, the myriad of diseases induced by CTV, depending on the *Citrus* host, viral strain and environmental conditions, challenged the study of the host-plant interactions. In the last century, the study of CTV genetics was focused in generating molecular techniques to improve CTV detection and genotype differentiation [5]. However, in a decade, a remarkable progress has been achieved in developing the genetic engineering tools to overcome the challenges of examining CTV genetics. A cDNA clone (*T36-CTV9*) of the Florida isolate T36 was generated and an *in vitro* genetic system was developed to analyze CTV genotypes, D-RNAs, mutants and self-replicating constructs in *Nicotiana benthamiana* protoplasts or indexing plants [6, 7, 8, 9, 10]. The last advances in CTV genetics and the different biotechnological approaches used to study CTV are discussed in this chapter.

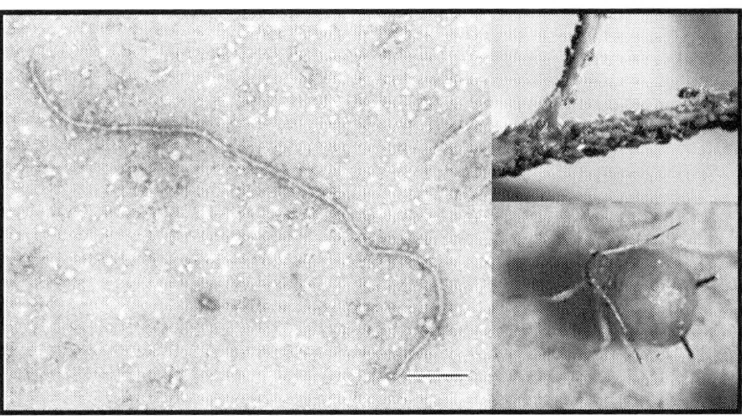

Figure 1. Left: Viral particles from *Swinglea glutinosa (Blanco) Merr.* protoplasts, transfected with CTV isolate T36, collected at 4 dpi and examined by SSEM electron microscopy. The bar indicates 200 nm. From Albiach-Marti et al. [72]. Top right: colony of *Toxoptera citricida* (Photo: Dr. A. Urbaneja). Bottom right: *Aphis gossiipi* (Photo: Dr. A. Hermoso de Mendoza)

2. *Citrus tristeza virus* genome structure, organization and gene function

The CTV RNA genome structure resembles that of *Coronaviruses*, and it is organized in twelve open reading frames (ORFs) and two non-translated regions (NTR) at the 5′ and 3′ terminus (Figure 2) [2]. The 5′ termini of the CTV genome is protected with a cap structure [2]. The 5′

NTR of around 107 nt contains the sequences necessary for both replication and particle assembly [6, 23, 24]. Remarkably, the CTV 5′NTR predicted secondary structure is similar even for divergent genotypes and folded in two stem-loops separated by a short spacer region [23, 25]. The 3′NTR (273 nt) lacks a poly-A tract and does not appear to fold in a tRNA-like structure [2] but instead consists a predicted secondary structure of minimum energy of 10 stem-loop (SL) structures [26].

The CTV genome maintains the two characteristic clusters of genes of the family *Closteroviridae* (Figure 2) [11]. The *replication gene block,* which is also conserved in the supergroup of sindbis-like viruses, comprises ORF 1a and 1b and makes up the 5′ half of the viral genome [2] (Figure 2). The ORF1a encodes a 349 kDa polyprotein with two papain-like protease domains, a type I methyltransferase-like domain, and a helicase-like domain bearing the motifs of the superfamily I helicases. The ORF1b encodes a 54 kDa protein with RNA-dependent RNA polymerase (RdRp) domains. When ORF 1 is are directly translated from the positive-strand gRNA yield a 400 kDa polyprotein [2]. The conserved *quintuple gene block* (Figure 2) is related with virion assembly and trafficking in the plant [11]. This consists of the major coat protein (CP) of 25kDa, the minor coat protein (CPm) of 27kDa and other three proteins, p61, HSP70h and p6. HSP70h is a 65 kDa protein homologue of the HSP70 plant heat-shock proteins [2], a family of plant chaperones involved in protein-protein interactions, translocation into organelles, and intracellular trafficking [12]. The p6 gene encodes a small hydrophobic protein that belongs to the single-span transmembrane proteins [2]. While CP, CPm, p61 and HSP70h are necessary for proper particle assembly, p6 is required for systemic invasion of host plant [13, 14]. The additional five ORFs located at the 3′ half of the genome (Figure 2) are the p20, an homologue of p21 of *Beet yellows virus* (BYV) (genus *Closterovirus*), and four genes encoding proteins with no homologue in other closteroviruses (p33, p18, p13 and p23) (Figure 2) [11]. The p20 protein is the main component of the CTV-induced amorphous inclusion bodies [15] and it is essential for systemic infection [14]. The multifunctional protein p23 contains a Zn finger domain that binds cooperatively both single-stranded (ss) and double-stranded (ds) RNA molecules in a non-sequence specific manner [16]. In addition, p23 controls asymmetrical accumulation of positive and negative RNA strands during viral replication, ensuring the presence of enough quantity of positive genomic RNA (gRNA) ready for virion assembly [17].

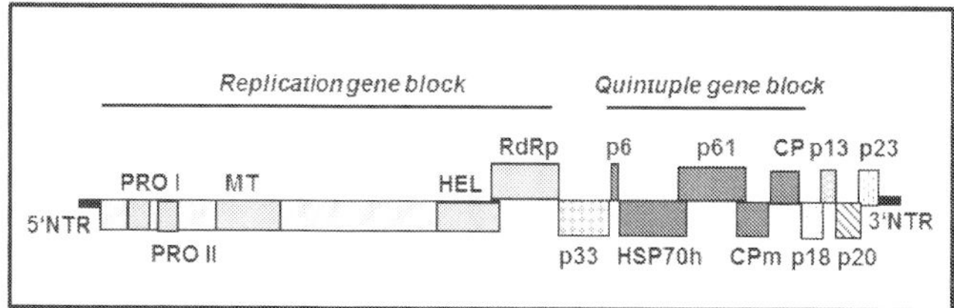

Figure 2. Scheme of CTV genome structure and organization. CTV ORFs are delimited by boxes. The acronyms PRO, MT, HEL and RdRp indicate protein domains of papain-like protease, methyltranferase, helicase and RNA-dependent RNA polymerase, respectively. HSP70h, CPm and CP indicate ORFs encoding a homologue of heat shock protein 70, the minor and the major coat proteins, respectively. From Karasev et al. [2].

In relation to host-plant interactions, CTV is a virus with a large genome and complex genetics, while the citrus host includes many species, varieties, and intergenic hybrids with which the virus could interact causing a range of physiological and biochemical responses. In fact, CTV evolved ending up with three proteins, CP, p20 and p23, which are suppressors of the plant RNA silencing mechanism in *N. benthamiana* and *N. tabacum* plants [18]. Unexpectedly, the ORFs that encode proteins p33, p18 and p13 are not required either for replication or assembly [6, 13] or for systemic infection of Mexican lime [*C. aurantifolia* (Christm.) Swing.] and *C. macrophylla* Wester plants [14]. Nevertheless, they are involved in CTV infection and movement in other citrus hosts [19].

Furthermore, several CTV genomic regions have been found to be related with viral symptom development in citrus hosts. The symptomatology determinant of SY syndrome was located at the 3′ region composed by p23 ORF and the 3′NTR [20]. Nevertheless, the p33, p18 and p13 are involved in the SP syndrome development [21], although the participation in this process of other CTV regions, undetected until the moment, has not been discarded. Mild strain cross protection has been widely applied for millions of citrus trees in Australia, Brazil and South Africa [4, 5] to protect against SP economic losses. The mechanism of this type of viral superinfection exclusion is mainly a mystery. Recently, it has been found that the lack of the functional CTV p33 protein completely eliminated the ability of the virus to exclude superinfection by the same or closely related virus [22].

3. *Citrus tristeza virus* sequence diversity

Sequencing the complete genome of CTV was the first breakthrough towards the study of CTV genetics [2]. Actually, there are twenty CTV genomic sequences available. These are T36 and T30 from Florida [2, 27]; VT from Israel [28]; SY568R from California [29, 30]; T385 and T318A from Spain [31, 32]; NuagA from Japan [33]; Qaha (AY340974) from Egypt; Mexican isolate (DQ272579); B165 form India [34]; NZ-M16, NZ-B18, NZRB-TH28, NZRB-TH30, NZRB-M12, NZRB-M17 and NZRB-G9 from New Zealand [35, 36]; HA16-5 and HA18-9 from Hawaii [37] and Kpg3 from China [38]. Genetic comparison of these CTV genomes revealed an extreme genomic divergence for genotypes of the same viral species (Figure 3). Nevertheless, these divergent CTV genotypes retained the same genomic organization [3].

Phylogenetic analysis classified the twenty CTV genomic sequences in seven main genotypes [35, 37, 38]. Six of them induce severe syndromes: (1) T36-like (T36, Qaha and Mexican); (2) the RB group plus HA18-9; (3) the VT-like (VT, NUagA, T318A, SY568 and Kpg3); (4) HA16-5; (5) B165 and NZ-B18; and the (6) NZ-M16 genomic sequences [35, 37, 38]. The group 7 consisted in the T30-like asymptomatic or mild genotypes (T30, T385). Sequence comparison of complete CTV genomes yielded nucleotide identities from 79.9%, between Qaha (a T36-like strain) and VT, to 99.3% (between T30 and T385) (Figure 3) [37]. The most conserved sequences were located in the 3′NTR region, which is almost identical in most of the cases (Figure 3). The nucleotide divergence was mostly concentrated at the 5′ half of the CTV genome and increased towards the 5′NTR region to raise, in some cases, nucleotide identities as low as 42% [27, 37,

39] (Figure 3). This pattern of genomic divergence was more evident between the T36-like genotypes and close relatives (groups 1 and 2) and the other five CTV groups [36]. However, two paths of sequence divergence were observed [39]. The sequence divergence between CTV genotype groups 3 to 7, although slightly increased in the 5′NTR region, was relatively constant in proportion and distribution along the genome [37, 39]. On the other hand, the T36-like genotypes and close relatives showed considerable genetic distance to the other five main CTV genotypes [36]. Actually, the comparison of the genomic sequences of T30 and T36 diverged from 5% in the 3′ NTR to as high as 58% in the 5′NTR (Figure 3) [27]. Based in these two paths of sequence divergence detected between CTV genomic sequences [39], it was speculate that the T36 genotype and relatives evolved from a recombinant of a CTV genome and an unknown virus millions of years ago in Asia [28].

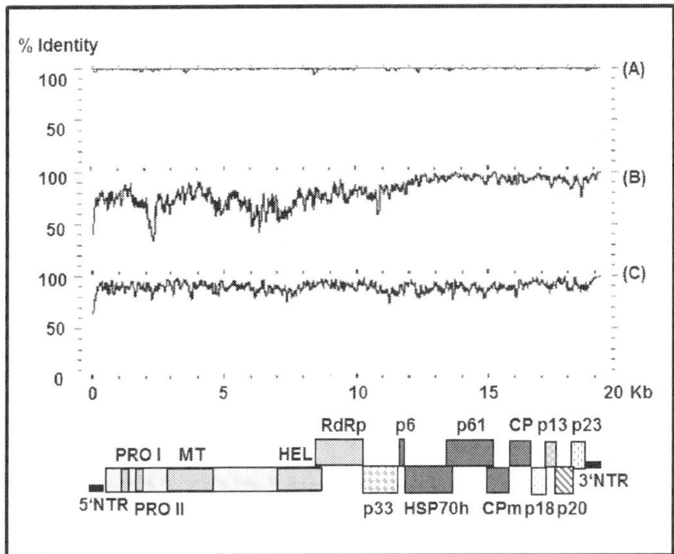

Figure 3. Graphic of the nucleotide identity along the CTV genome when comparing T30 genomic sequence with the sequences of (A) T385 (B) T36 and (C) VT genomes. From Albiach-Marti et al. [27].

Comparison of each of the CTV regions pointed to an unevenly distributed sequence variation along the CTV genome, likely reflecting different selective pressures along the genomic RNA [26, 37, 39]. Analysis of nucleotide diversity in some coding regions between CTV strains yield values higher than 0.13. However, most of the nucleotide exchanges were reported at the third codon position, indicating the preservation of the protein sequence among divergent genotypes. Actually, the ratio between non-synonymous and synonymous substitutions assessed for CTV coding regions was below the value 1, thus suggesting selective pressure for amino acid conservation [40]. In addition, analysis of the CTV genomic and D-RNAs sequences indicate homologous and non-homologous recombination events among different genotypes

[30, 32, 41, 42, 43], possibly as a result of mixed infections on trees that are recurrently inoculated by aphid transmission.

Conversely, in spite of this genetic variability, sequence comparisons of some CTV genomes revealed a remarkable viral genetic stasis as the genomes of some CTV strains, separated geographically and in time, were found essentially identical [27]. This genetic stability has been explained as a consequence of strong selection and competition between the mutants that arise in each replication cycle, which creates equilibrium in the viral quasispecies distribution [27]. In this context, there is a hypothesis to explain the high sequence variability found in the wild CTV isolates [3, 27]. In a fist stage, each of the main genotypes evolved separately in different *Citrus* species at their point of origin in Asia. This was followed by the dispersal of the main CTV genotypes to different environments around the world with the advent of the modern citrus industry in the XIX century. After that, RNA virus mutation, due the error-prone nature of RNA-dependent RNA polymerases, in addition to recombination events between diverged sequence variants, plus selection, genetic drift and gene flow could have been promoted rapid evolution [3, 5, 27].

4. *Citrus tristeza virus* replication and gene expression

CTV replication is an extraordinary process that generates at least 35 different species of viral RNA in CTV-infected cells (Figure 4) [44] plus a myriad of D-RNAs [45, 46, 47] (Figure 5). The viral genomic sequences necessary for CTV replication are the *replication gene block* plus the 3′ and 5′NTRs, which contain the cis-acting elements indispensable for this process (Figure 4). In fact, a T36 CTV replicon consisting in only these genomic regions is able to self-replicate in protoplasts of *N. benthamiana* [6]. As indicated previously, the CTV 5′NTR predicted secondary structure folded into two SL separated by a short spacer region [25]. Directed mutations disrupting this predicted secondary structure were shown to abolish replication, whereas compensatory mutations resumed replication, suggesting that the secondary structure of the 5′ NTR is more important than the primary structure for CTV replication [23]. Conversely, the basic function of the 3′ NTR (273 nt) is minus-strand initiation for the CTV gRNA and the subgenomic (sg) RNAs [26]. The 3′NTR consists in a predicted secondary structure of 10 SL structures. While the core of the 3′ replication signal was located in the primary structure of three of the central stem-loops (SL4, SL6 and SL8), the secondary structure of the other stem-loops (SL3, SL5, SL7 and SL9) proved dispensable but required for efficient replication [26]. In addition, all CTV genomes retain a CCA triplet at the 3′ termini necessary to initiate replication [26].

Wild CTV populations could be composed by divergent genotypes [3]. In the case of mixed infections in the same plant cell, it is essential to determine whether a specific replicase complex is able to recognize the cis–acting elements of the 3′or 5′ NTR of other genomic variants. The exchange of 5′NTR and 3′NTR sequences, from different main genotypes, into the *T36-CTV9* infectious clone decreases replication as the degree of sequence divergence increases. Therefore, indicating partial compatibility of the T36 replicase complex with diverged 5′ and 3′ cis-acting elements, thus suggesting limited heterologous replication in mixed viral infections [6].

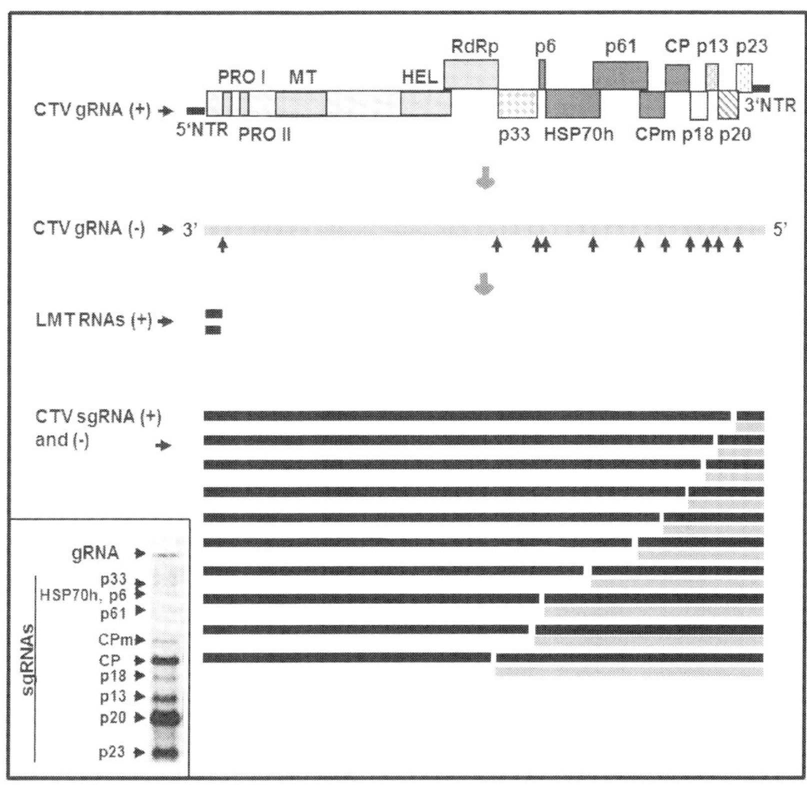

Figure 4. RNA species produced *in cis* during CTV replication. Main panel: Scheme of the different CTV RNA species. Black lines: single-stranded positive-sense RNAs. Grey lines: single-stranded negative-sense RNAs. The acronyms gRNA, sgRNA and LMT RNA indicate genomic, subgenomic and *low molecular-weight tristeza* RNAs, respectively. The signs (+) and (-) specify plus and minus-strand RNAs, respectively. Black arrows on the CTV gRNA(-) line designate the approximate position of the CTV controller elements in the CTV genome. Small left panel: Accumulation of T36 double strain (ds) RNAs in citrus plants showing the 3´coterminal sgRNAs produced during replication and expression of the ten 3´half ORFs of the CTV genome. Northern-blot hybridization performed using a single-stranded negative-sense riboprobes specific to the 3' end of T36 genomic sequence. From Karasev et al. [2, 50], Gowda et al. [44, 51, 52] and Ayllon et al. [49, 48].

In the first step of the CTV genome replication, the viral replicase uses the single stranded positive-sense gRNA (CTV gRNA (+)) from the uncoated viral particles, as template to generate a homologous single-stranded negative –sense CTV gRNA (CTV gRNA (-)) (Figure 4). The CTV gRNA(-) molecules will function as basis for the synthesis of the CTV progeny of positive-strand gRNAs. The CTVgRNA(+) molecules would act as RNA messenger for expression of viral proteins or as a pool of CTV gRNAs ready to be incorporated into virions to produce newly infectious viral particles. The new CTV gRNA(+) could also serve as template for the synthesis of fresh CTVgRNA(-) molecules to start all over the process [44].

Another function of the CTV gRNA (-) is to serve as template to produce high quantities of single and double strain sgRNAs during the expression mechanism of the ten ORFs situated at the 3'half of the genome (Figure 4 and 5) [44, 48, 49, 50]. Unlike the large animal viruses of the *Nidovirales*, the 3' sgRNAs of CTV do not share a common 5' terminus and the sgRNA transcription mechanism resembles the transcriptional mechanism of other Sindbis-like viruses [50]. The synthesis of each 3' coterminal sgRNA is controlled by its corresponding cis-acting element (controller element (CE)) (Figure 4). Probably each CTV CE could act as promoter or terminator of the CTV RNAs during the replication process [44, 48, 49, 50]. However, if the CEs function as internal promoters for the generation of positive-strand sgRNAs, using as template the CTVgRNA(-) molecules (Figure 4), or act as terminators for the synthesis of negative-strand gRNAs (by premature termination at the CE site), or both, is still unclear [44]. In addition to the plus and minus- sense 3'coterminal sgRNAs, the CEs corresponding to each of the ten 3'ORFs produce a reduced amount of a set of 5'coterminal positive-strand sgRNAs (Figure 4), probably due to premature termination during the synthesis of the CTV gRNA(+) [44]. Moreover, CTV generates significant amounts of *low molecular-weight tristeza* (LMT1 and LMT2), two positive-strand 5'co-terminal sgRNAs population with heterogeneous 3'termini at nt 842-854 and 744-746, respectively (Figure 4 and 5) [46, 47]. LMT 1 and LMT 2 are generated and accumulated differently [51, 52]. LMT1 is likely created by premature termination during CTV gRNA(+) synthesis at a 5' CE situated in the PRO I domain of the replicase (Figure 4). This 5' CE acts as a strong promoter when placed immediately upstream of the ORFs near the 3' terminus [51]. In contrast of the 3' CEs, which are able to generate plus and minus-strand sgRNAs, the 5' CE of the LMT 1 only promoted the synthesis of positive-strand sgRNAs (Figure 4) [51]. In fact, the RNA termination and initiation sites of the 5' CE, compared to those of 3' CEs, occur at opposite ends of the corresponding minimal active CE site [49]. Therefore, as a result of the replication process, CTV produces high amounts of viral RNA species in the infected cell (Figure 4). The total (gRNAs plus sgRNAs) positive to negative-strand RNA ratio (approximately 40 to 50:1) falls within the range of the genomic RNAs of most positive-strand RNA viruses, particularly the more similar alphavirus super-group and large complex viruses of the *Nidovirales* [26]. However, during CTV replication, only the positive-strand gRNA accumulates approximately 10 to 20 times more than their negative-strand gRNA homologues, a rather lower ratio compared to those generated during other RNA viruses replication [17].

The expression of the CTV genome, which potentially yields at least nineteen protein products, resembles that of *Coronaviruses* [50]. This remarkable process includes at least three different RNA expression mechanisms widely used by positive-strand RNA viruses: proteolytic processing of the polyprotein precursor, translational frameshifting and the generation of a nested set of ten 3'-coterminal sgRNAs [50]. Therefore, the ORFs 1a and 1b are directly translated, from the positive-strand gRNA, to yield a 400 kDa polyprotein that is later proteolytically processed in, at least, nine protein products. The ORF1b encodes a 54 kDa protein with RdRp domains that is occasionally translated after ORF 1a by a +1 ribosomal frameshifting [2]. Additionally, as indicated above, the 10 ORFs located at the 3' half of the CTV genome are expressed by the synthesis of ten 3' co-terminal sgRNAs (Figure 4). Each 3' sgRNA serve as RNA messenger for the translation of its 5' proximal ORF [13, 46] and the expression of each of the ten 3′ proximal ORFs is regulated independently both in amount and timing [46, 47].

5. *Citrus tristeza virus* defective RNAs

In addition to the 35 different species of RNA created during replication, CTV could accumulate considerable amounts of D-RNAs in infected cells (Figure 5) [46]. CTV D-RNAs vary in size, abundance and sequence [41, 45, 46] and could be encapsidated into particles and could be transmitted by aphids [45].

Generally, D-RNAs bear a genome from 2.0 to 5.0 kb and are composed by variable portions of the 3′ and 5′ termini of CTV genomic RNA with large internal deletions (Figure 5). Nevertheless, some D-RNAs comprising the two termini and a non-contiguous internal sequence or a non-viral sequence, plus large D-RNAs of 10-12 kb including in their 5′ proximal region the ORFs 1a and 1b, or with a 3′ region homologous to the ten CTV 3′ terminal ORFs, have been described [41, 43, 46, 55]. These large D-RNAs resembled the RNAs 1 and 2 distinctive of the bipartite *Criniviruses,* also included in the *Closteroviridae* family. Moreover, the D-RNA containing the complete CTV replicase constitutes a novel class of large self-replicating D-RNAs [47].

CTV D-RNAs characteristic genomic structure suggests an origin in the recombination events during viral replication. In this way, some large D-RNA bear a 5′ termini identical o slightly larger than the 5′ sgRNA generated by the CE of the p33 ORF [47], and the small ones usually contain a 3′ termini identical to 3′ sgRNA of p23 ORF [43]. Additionally, a repeated 4-5 nt, (corresponding to two CTV genomic regions) was reported flanking the D-RNA 3′ and 5′ termini junction sites indicating that D-RNAs are probably created during the generation of the positive-strand sgRNA or gRNA by a template-switching mechanism [41, 43].

D-RNAs require the viral machinery for their survival. The D-RNA replication *in trans* was examined using infectious D-RNAs and the *in vitro* genetic system of *T36/CTV9* [10, 6]. The minimal D-RNA sequence required for replication are a 5′ proximal region of 1kb and a 3′ termini limited to the CTV 3′NTR. In addition, efficient replication of D-RNAs involves some spacing between these terminal cis-acting signals and a continuous ORF through most of the 5′ proximal regions of the D-RNA sequence [10, 56]. CTV field isolates are composed by viral populations of divergent genotypes. In this case, an important point is to understand the dynamics of generation and accumulation of D-RNAs in a specific plant cell infected with distinct CTV genotypes. Mawassi et al., [56] demonstrate that some wild-type populations of CTV are capable of supporting the replication of synthetic divergent D-RNAs. However, replacement of 5′ region (which is the most variable among CTV strains) of a particular synthetic D-RNA, with the corresponding sequence from different main CTV genotypes, resulted in chimeric D-RNAs that were replicated to detectable levels by some CTV genotypes but not with the others. Consequently, differential specificities of distinct CTV replicase complexes with divergent D-RNA replication signals are possibly affecting the maintenance of D-RNA population structures in the infected plant cell.

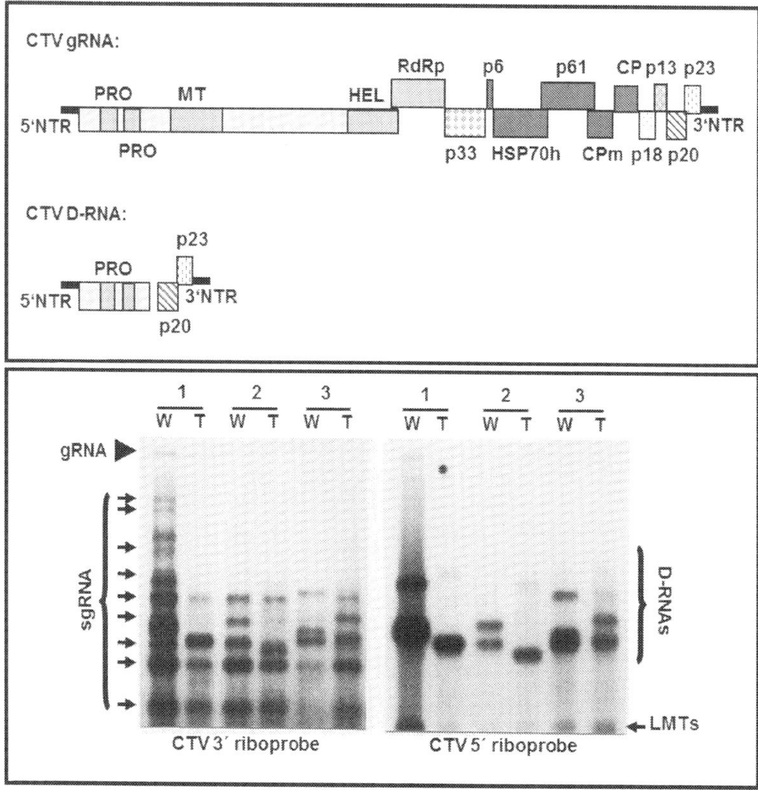

Figure 5. Different species of D-RNAs in CTV populations. Top panel: graphic representation of a usual small CTV D-RNA compared with the CTV genome. Bottom panel: accumulation of CTV RNAs in bark extracts from sweet orange plants infected with three field isolates (1 to 3 lines), before (lines w) and after (lines T) aphid transmission. Northern-blot hybridization performed using a single-stranded minus-sense riboprobes specific to the 3′ end (left panel) and the 5′ end (right panel) of the T36 genome. From Mawassi et al. [46] and Albiach-Marti et al. [45]

In other viral pathosystems, D-RNAs have the capacity of interfering with the viral replication process of their helper virus (named defective interfering (DI) RNAs), but those function was not reported for CTV [10]. Although their biological role is presently unknown, at least in one case, the presence of D-RNAs was suggested to modulate symptom development either increasing or decreasing CTV symptom expression [57]. Most of the CTV D-RNAs contain a complete region p23 and the 3′NTR [43] that is associated with SY development [20], thus they could have a role in symptom modulation. Therefore, it will be necessary to promote further research to elucidate the role of the D-RNAs (or DI-RNAs) in CTV replication or in modulation of pathogenic responses in the infected plant host.

6. *Citrus tristeza virus* encapsidation

Differently of most elongated viruses, CTV particles are encapsidated by two different capsid proteins that coat the opposite ends of the virion [24]. About the 97% of the CTV genome is coated by CP, while the remainder 3% is coated by CPm resulting in viral particles with the emblematic tail of the members of the *Closteroviridae* family [11, 24]. The coordinate action of HSP70h and p61, in addition to the CP and CPm coat proteins, are required for proper assembly of CTV particles [13]. The previously described CTV 5′NTR conserved SL structures contains the origin of assembly of CPm, overlapping the sequences that function as a cis-acting element required for gRNA synthesis [24]. During CTV assembly, CPm begins coating the gRNA at the 5′ NTR to about nt 630. However, in the absence of HSP70h or p61, CPm may coat larger segments or even the complete gRNA. Probably, HSP70h or p61 bind to the transition zone between CP and CPm (around 630 nt) and restrict CPm to the virion tail [24]. The protein homologous to HSP70h and p61 in BYV are coordinately assembled with CPm in the virion structure and remain attached to the viral particles [58]. Although the assembly of HSP70h and p61 is not directly confirmed for CTV, RNA transcripts lacking one or both of these ORFs were unable to produce infective CTV virions [24]. Several strains with divergent genotypes could coexist in the same citrus cell. Analysis of the encapsidation of heterologous CPms in absence of HSP70h and p61 indicated reduction or lack of CPm assembly. Nevertheless, the presence of HSP70h and p61 restored CPm assembly to wild-type levels. This indicated that the HSP70h and p61 could play an important role facilitating heterologous CPm assembly in mixed infections [59].

The region coated by CPm also overlaps the LMT2 5′ sgRNA (650 nts). Actually, LMT2 production is correlated to virion assembly since mutations in the 5′ NTR that abolish encapsidation also eliminate accumulation of LMT2. Although this represents the first evidence of a viral RNA processed by the assembly mechanism, the exact function of LMT2 in CTV assembly is unknown at the moment [52].

7. *Citrus tristeza virus* genetics and plant-host interactions

In order to infect a plant, CTV needs to enter in the cell, and to overcome the constitutive and/or inducible plant defences, to re-program the plant cellular machinery for its viral multiplication. The infection process will continue with the assembly of new viral particles that will move cell to cell through the plasmodesmata. This process will be completed with the viral long distance movement through the plant vascular structure to colonize systemically the plant. Each CTV gene product seems to have a primary genetic function required for the survival of the virus. However, there are secondary genetic interactions, which cause or trigger resistance or pathogenic responses in the citrus host [60, 61]. *Citrus* genus contains multitude of species, cultivars and intergenic hybrids, with which CTV could interact causing a range of physiological and biochemical responses. These could be from either pathogenic or asymptomatic phenotypes to limited or complete plant resistance [5]. Although most of these mechanisms are still a mystery, new discoveries towards the understanding of the genetics of CTV movement in the plant, host-range, host resistance and pathogenicity have been reported recently [19, 20, 21, 62, 63, 64].

7.1. *Citrus tristeza virus* host range and plant systemic infection

Citrus tristeza virus natural plant hosts belong to the order *Geraniales*, family *Rutaceae*, subfamily *Aurantoidea*. There are also non-rutaceous hosts that have been experimentally infected with CTV strains like *Passiflora gracilis* or *Passiflora coerulea*. Some citrus hosts are usually susceptible to CTV infection like Mexican lime or *C. macrophylla*. Other citrus host are tolerant to some CTV strains like sweet orange [*C. sinensis* (L.) Osb.] and grapefruit (*C. paradisi* Macf.), or tolerant to almost all known CTV strains as mandarins (*C. reticulata* Blanco). Finally, pummelos [*C. grandis* (L.) Osb.], sour orange (*C. aurantium* L.) and the hybrid rootstock Swingle citrumelo exhibit a differential degree of resistance depending on the CTV strain. In addition, some *Citrus* relatives within subfamily *Aurantioideae*, like *Poncirus trifoliata* (L.) Raf., as well as *P. trifoliata* intergenic hybrids remain resistant or immune to most of the CTV strains [4, 5]. Consequently, these data highlight an elevated complexity in the CTV systemic infection and host range genetics.

Several CTV genes are related with systemic infection of citrus plants [14, 62]. Viral mutants with a deletion in the p6 and p20 ORFs failed to infect citrus plants systemically, suggesting their possible roles in virus translocation or infection of the whole plant. Likewise, the p6 homologue in BYV is a movement protein [65], and similarly to homologous proteins function in BYV [11], CP, CPm, HSP70h and p61 probably participate in the viral movement. CTV genome has several ORFs that are non-conserved in the family *Closteroviridae*, thus unique for CTV. Unexpectedly, three of these ORFs (p33, p18 and p13) neither are required for replication and assembly [6, 13] nor for systemic infection of Mexican lime and *C. macrophylla* [14]. However, p33, p18 and p13 were demonstrated to be CTV host range determinants. The p33 ORF is necessary for the systemic infection of sour orange and lemon trees. Likewise, either p33 or p18 ORF is enough for systemic infection of grapefruit trees. Similarly, p33 or p13 ORF is sufficient to invade whole calamondin (*C. mitis*) plants. As a result of the acquisition of multiple non-conserved genes (p33, p18, and p13), probably CTV increased the possibilities to interact with multiple hosts, thus extending its host range during the course of its evolution [19].

7.2. *Citrus tristeza virus* suppressing genes of plant silencing mechanism

The plant constitutive defence consists of the RNA mediated post-transcriptional silencing mechanism (PTGS) that implies the specific degradation of the viral dsRNA in small interfering RNAs (siRNAs), which guides a specific plant ribonuclease to disintegrate the viral genomes in the cytoplasm. Besides the antiviral role, the plant silencing mechanism has important functions in regulating plant gene expression (miRNA metabolism) [60]. In order to infect plants, viruses developed a strategy to block this silencing mechanism: the suppressing genes. This strategy allows viral replication but interfere with host gene expression, thus inducing disease [60, 66]. As indicated previously, CTV evolved ending up with three proteins that are suppressors of the plant RNA silencing mechanism in *N. benthamiana* and *N. tabacum* plants. The p23 inhibits intercellular RNA silencing, while CP impedes intracellular RNA silencing and p20 limits both inter and intracellular RNA silencing [18]. Although, CP, p20 and p23 have not been yet reported as suppressors of the citrus silencing mechanism, their presence in the CTV genome is in concordance with wide host range among citrus species and hybrids, previously described, and consequently, with the complexity of CTV-citrus interactions. In fact, in spite of the existence of these three silencing suppressors, accumulation of siRNAs in CTV-infected susceptible hosts is 50% of the total RNAs in the plant [64]. The CTV siRNAs accumulation in infected plants is directly proportional to the virus accumulation and varies depending on the citrus host. Deep sequencing analysis of these siRNAs, from CTV-infected plants, indicated that they mainly consisted in small RNAs of 21-22 nt derived essentially from the CTV genome [64].

7.3. Genetic determinants of the *Citrus tristeza virus* pathogenic syndromes

Viruses possess the potential to disrupt host physiology either by usurpation of substantial amount of plant metabolic resources or by the interaction of a specific viral product with the host components [60]. CTV induces three hallmark syndromes, plus a myriad different symptom patterns in indexing plants. *Tristeza* disease or QD syndrome consists in overgrowth of the scion at the bud union, loss of root mass, and therefore death of citrus commercial varieties grafted on sour orange rootstock [5]. The SP syndrome consists in deep pits in the wood under depressed areas of bark in commercial varieties of sweet orange and grapefruit trees grafted on any rootstock. Usually SP do not cause tree death, but severe stunting and unmarketable fruit, thus causing elevated economic damages [5]. The SY syndrome is characterized by stunting, leaf chlorosis and sometimes a complete cessation of growth on sour orange, grapefruit or lemon [*C. limon* (L.) Burm. f.] seedlings (Figure 6). Although, SY syndrome might be found at the field in top–grafted plants and it is not economically valuable, it could be examined in the greenhouse in a timely manner and has a substantial diagnostic value for CTV pathotype differentiation [5]. On the contrary, the development of QD and SP extends over 10 to 40 years at the field [1], a period too long to screen the CTV isolates. Although SP pathotype could be likely examined in glasshouse conditions, there are no reliable methods to reproduce the QD in those conditions [5]. Therefore, the degree of severity of a specific CTV isolate, strain or genotype usually is assessed by using indexing plants (Mexican lime, *C. macrophylla*, sour orange, sweet orange and Duncan grapefruit) [60]. In this case, the degree of CTV symptomology ranges from the mild phenotypes, which are almost asymptomatic, to the highly virulent CTV isolates that could generate vein clearing, leaf cupping, dwarfing, stem pitting and the plant death [5]. This diversity and grade of symptom responses to CTV infection suggests the possible presence of more than one mechanism of pathogenicity taking place during the CTV-*Citrus* interactions.

CTV multiplication generates great quantities of viral products like, at least, 19 viral proteins, 35 RNA species (gRNAs, sgRNAs and LMTs) and D-RNAs along with a complicated process of replication, gene expression, assembly and movement, where the interaction with host factors is essential. Consequently, during the CTV-*Citrus* interaction there are multiple opportunities to generate disease. In fact, analysis of Mexican lime transcriptome using microarrays, after infection with a severe CTV isolate, showed altered expression of 334 genes and about half of them without significant similarity with other known sequences [63]. In this context, identifying a specific genetic determinant that is responsible for a specific disease symptom under field or glasshouse conditions could be a real challenge [20].

Although serological or molecular markers were correlated with some CTV pathotypes [5], direct linkage of these markers to symptom development has not been established. Nevertheless, the CP, p20 and p23, reported as suppressors of the plant silencing mechanism [18], could be candidates for symptom determinants since they could potentially disrupt the miRNA metabolism, thus possibly inducing disease. Indeed, several viral suppressors of RNA-mediated gene silencing have been identified as pathogenicity determinants [66]. Actually, when p23 is ectopically expressed in transgenic limes or transgenic sour orange plants induces virus-like symptoms. However, the symptomatology pattern developed in these transgenic plants is different than those induced by natural virus infection. Additionally, the grade of symptom severity observed in these p23 transgenic plants is directly proportional to the p23 production level, and independent of the viral source or sequence of the p23 gene [67, 68]. Nevertheless, the symptom intensity in wild virus-infected limes or sour oranges is radically different between severe and mild isolates of virus. Yet, the different response in transgenic plants could be related to the fact that, in this case, the p23 protein is produced constitutively in most cells, while the expression of p23 is limited to phloem-associated cells in nature [20].

As previously described, a distinctive phenotype of some isolates of CTV is the ability to induce *Seedling yellows* in sour orange, lemon and grapefruit seedlings (Figure 6). To delimit the viral sequences associated with the SY syndrome, T36/T30 hybrids were generated by substituting severe sequences, located in the 3′ moiety of the T36-CTV9 infectious clone, for homologous asymptomatic sequences from the T30 genome. The T36/T30 hybrids were analyzed in *N. benthamiana* and citrus plants [20]. The SY determinant was mapped to the region encompassing the p23 gene and the 3′ NTR (nt 18394-19296) (Figure 6) [20]. The 3′NTR has been used to generate transgenic plants resistant to CTV [69] and it has also been related to symptom development for other virus [70]. Likewise, the p23 is an obvious candidate for SY symptom determinant since it is one of the most highly expressed CTV proteins [54], a RNA-binding protein responsible for asymmetric replication [16, 17], and it is a viral suppressor of RNA-mediated gene silencing mechanism [18]. Additionally, p23 ORF has been used to produce transgenic plants searching for resistance to CTV [71].

The study of the devastating QD syndrome is especially important under de economical point of view. In this case, the extremely difficult task of reproducing this syndrome in glasshouse conditions hinders the study of the QD genetic determinants. However, since a strong correlation between SY and QD has been observed in the biological evaluation of a wide range of CTV isolates [5], it could be possible, but not yet confirmed, that determinant(s) for the decline disease map similarly to that of SY. Therefore, the CTV hybrids, used for evaluation of the SY genetics determinant [20], have been directly assessed in decline-susceptible grafted combinations of scion and rootstock in field conditions. In addition, since the hybrids are made by recombinant DNA technologies, these assays require special permits from the plant protection and environmental safety authorities [20].

Figure 6. Development of seedling yellows syndrome (SY) in CTV infected plants. Top panel: SY symptoms in (A) sour orange and (B) Duncan grapefruit seedlings compared to (C) a healthy sour orange plant. Bottom panel: SY symptoms in (1) T36/T30 hybrid [P23-3´NTR], (2) isolate T30 (3) healthy (4) T36 infectious clone CTV9 and (5) T36/T30 hybrid [HSP70h-P61] sour orange seedlings. From Albiach-Marti et al. [20].

In order to map the stem pitting determinants, the effect on symptom development in *C. macrophylla* of deletions in p33, p18, and p13 ORFs were evaluated [21]. Although the T36 full-length construct (*T36-CTV9*) causes only very mild SP symptoms in this host, certain deletion combinations (p33 and p18 and/or p13) greatly increased SP symptoms, while other combinations (p13 or p13 plus p18) resulted in reduced SP [21]. Remarkably, the stem-pitting phenotype seems to be induced as result of a balance between the expressions of different viral genes.

7.4. Host resistance to *Citrus tristeza virus* infection

There are different *Citrus* species and relatives that exhibit total or limited resistance to CTV infection. Pummelos, sour orange and the rootstock Swingle citrumelo display a differential degree of resistance depending on the CTV strain. However, some *Citrus* relatives, within subfamily *Aurantioideae*, like *P. trifoliata*, *Swinglea glutinosa* (Blanco) Merr., and *Severinia buxifolia* (Poir) Ten, as well as *P. trifoliata* intergenic hybrids like citranges (sweet orange × *P. trifoliata*), remain resistant or immune to most of the CTV strains [4, 5].

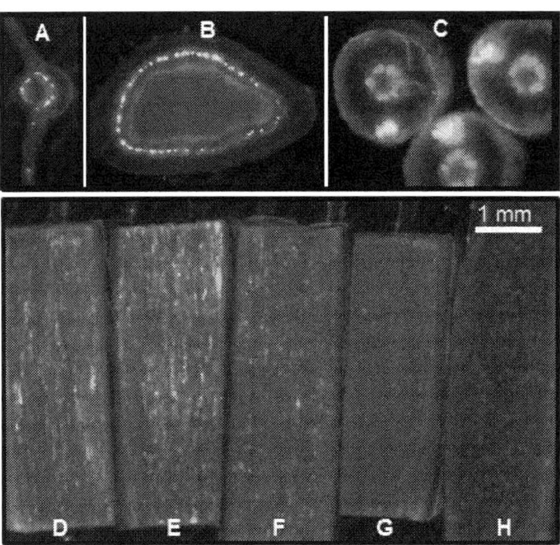

Figure 7. Systemic infection of CTV in different citrus host. Top panel: localization of construct *CTV-BCN5-GFP*, derived from the recombinant virus *T36-CTV9, in* (A) leaf, (B) shoot and (C) roots of tolerant host *C. macrophylla*. Bottom panel: localization of *BCN5-GFP* in a bark flap of (D) Mexican lime, (E) *C. macrophylla*, (F) sweet orange, (G) sour orange and (H) Duncan grapefruit. Pictures were taken in a confocal microscope under UV light. Pictures from Folimonov et al. [9] and Folimonova et al. [62].

Resistance of plants to viruses results from blockage of a basic step in the virus life cycle. This blockage can result from the lack of a factor(s) in the plant that is necessary for virus multiplication and movement (passive resistance) or activation of a defense mechanism (active resistance) [60]. One of the most effective methods of characterizing resistance mechanisms is to determine whether the resistance is expressed at the single-cell level. Albiach-Martí et al., [72] studied the nature of this CTV resistance mechanism and reported efficient multiplication of CTV in resistant *P. trifoliata* and its hybrids (Carrizo citrange, US119 and Swingle citrumelo) and *S. buxifolia* and *S. glutinosa* protoplasts (Figure 1). Thus, the resistance mechanism in these plant species affects a viral step subsequent to replication and assembly of viral particles, probably preventing CTV movement. Similar results were obtained in CTV inoculation experiments of resistant pummelo and sour orange protoplasts (Albiach-Martí, unpublished data). Likewise, the CTV systemic infection of Duncan grapefruit (a descent of pummelo) and sour orange plants was examined using a stable virus-based vector *CTV-BC5/GFP*, which was generated from the *T36-CTV9* recombinant virus (Figure 7) [9]. The susceptible host *C. macrophylla* and Mexican lime and the tolerant host sweet orange were used as controls [62]. CTV infection sites, after cell to cell movement, consisted of clusters of 3 to 12 cells in the susceptible species, while in Duncan grapefruit and sour orange there were fewer CTV infection sites and they were usually single cells, indicating absence of cell to cell movement in both cases (Figure 7) [62]. However, the long-distance movement mechanism of CTV appears to be inefficient in some extend, since the majority of phloem-associated cells in the bark flaps have not been infected, even for *C. macrophylla* and Mexican lime susceptible hosts (Figure 7) [62]. In these experiments, accumulation of T36 seemed related to host susceptibility. Actually, the hypothesis points to plant silencing as a probable cause of this resistance mechanism [62]. However, inadequate interactions of the CTV host range determinants (p33, p18 and p13) [19] with the host factors, which allow viral movement, have not been discarded.

8. *Citrus tristeza virus* genetic determinants related with aphid transmission

While CTV dispersal between new areas or countries occurs by graft propagation of virus-infected plant tissues, aphid transmission is responsible of local spread [1]. Viruliferous aphids of *Toxoptera citricida* (Kirkaldy) and *Aphis gossypii* (Glover) species are able to transmit CTV in a semipersistent manner [1] (Figure 1). However, *A. spiraecola* (Patch) and *T. aurantii* (Boyer de Fonscolombe) have also been reported as CTV vectors, although with less efficiency than *A. gossypii*. The aphid *T. citricida* is the most effective transmitting CTV and the most efficient and fast in the spatial and temporal viral spreading in citrus orchards. Moreover, when *T. citricida* appears in a new citrus area, the interaction between CTV and *T. citricida* seems to shift a specific mild or QD viral population to severe SP one [5]. This incidence suggests that *T. citricida* is more effective transmitting the minor virulent SP populations than the endemic mild or QD CTV genotypes. Citrus is the primary host of *T. citricida*, while *A. gossypii* populations build up in other crops. Probably *T. citricida* evolved with citrus and CTV and this could explain its high efficiency transmitting this virus [3].

The CTV genes or sequences related with aphid transmission are mostly unknown. However, usually for viral transmission, a *helper component* or the CTV virion has to interact with the mouthparts and the foregut of the aphids. Therefore, the protein components of the CTV particles (CP, CPm, HSP70h, p61) are candidates for aphid transmission determinants. In fact, CPm, which composes the particle tail structure of *Lettuce infectious yellows virus* (LIYV) (genus *Crinivirus*, family *Closteroviridae*), a close relative to CTV, is involved in viral transmission by *Bemisia tabaci* [75]. Similarly, the CTV CPm is suspected to affect aphid transmission [73, 74]. Comparison of CPm protein sequences from transmissible and non- transmissible CTV strains yield five mutations that appear to be conserved in transmissible CTV strains. These ones could affect aphid transmission efficiency by altering the conformation of the protein or masking motifs, which could be involved in the interaction between CPm and aphid stylet [76]. Although the special abilities of *T. citricida* are partially explained by its high efficiency in viral transmission [3], it seems that could be distinct interaction of this aphid with the coat proteins corresponding to different CTV genotypes. Additionally, the transmission mechanism of CTV by *A. gossypii* may possibly be, to some extent, different to the one by *T. citricida*.

9. Conclusions

Citrus tristeza virus research continues pushing the molecular virology technology to further limits. Molecular tools have been developed to study CTV gene expression, replication, assembly, systemic infection, viral movement, and plant-host interactions. The scientific results reveal a virus with a complex genetics that has become a model for molecular virology studies and viral biotechnology development. However, in spite of the CTV complicated genetics, further efforts need to be applied to engineer viral-based vectors, or additional biotechnological approaches, with the aim of understanding the mechanisms of viral movement, pathogenesis, resistance and aphid transmission, and the role of the D-RNAs in the CTV infection or the pathogenesis process. This valuable information will be applied to implement biotechnological strategies in order to control the devastating CTV epidemics.

Acknowledgements

The author thanks W.O. Dawson, S. Gowda, B. Belliure-Ferrer and B. Sabater for critically reviewing this manuscript. Likewise, the author is grateful to Drs A. Urbaneja, A. Hermoso de Mendoza and S. Foliminova for kindly provide the pictures included in Figures 1 and 7. This book chapter was financed by ValGenetics.

Author details

Maria R. Albiach-Marti

Address all correspondence to: maria@valgenetics.com

Molecular Phytopathology Unit, ValGenetics, The University of Valencia Science Park, C/ Catedrático Agustín Escardino, Paterna, Valencia, Spain

References

[1] Bar-joseph, M, Marcus, R, & Lee, R. F. (1989). The continuous challenge of citrus tristeza virus control. *Annual Review Phytopathology*, , 27, 291-316.

[2] Karasev, A. V, Boyko, V. P, Gowda, S, Nikolaeva, O. V, Hilf, M. E, Koonin, E. V, Niblett, C. L, Cline, K, Gumpf, D. J, Lee, R. F, Garnsey, S. M, Lewandowski, D. J, & Dawson, W. O. (1995). Complete sequence of the citrus tristeza virus RNA genome. *Virology*, , 208, 511-520.

[3] Albiach-marti, M. R. Molecular Virology and Pathogenicity Determinants of Citrus Tristeza Virus. In Garcia ML. Romanowski V (ed). Viral Genomes. Rijeka: InTech; (2012). Available from www.intechopen.com, 275-302.

[4] Bar-joseph, M, & Dawson, W. O. (2008). *Citrus tristeza virus*. In *Encyclopedia of Virology*, Third edition evolutionary biology of viruses. Elsevier Ltd. , 1, 161-184.

[5] Moreno, P, Ambrós, S, Albiach-martí, M. R, Guerri, J, & Peña, L. (2008). *Citrus tristeza virus*: a pathogen that changed the course of the citrus industry *Molecular Plant Patholology*, , 9, 251-268.

[6] Satyanayanana, T, Gowda, S, Boyko, V. P, Albiach-martí, M. R, Mawassi, M, Navas-castillo, J, Karasev, A. V, Dolja, V, & Hilf, M. E. Lew&owsky, D.J., Moreno, P., Bar-Joseph, M., Garnsey S. M. & Dawson W.O. ((1999). An engineered closterovirus RNA replicon & analysis of heterologous terminal sequences for replication. *Proc. Natl. Acad. Sci. USA*, , 96, 7433-7438.

[7] Satyanayanana, T, Bar-joseph, M, Mawassi, M, Albiach-martí, M. R, Ayllón, M. A, Gowda, S, Hilf, M. E, Moreno, P, Garnsey, S. M, & Dawson, W. O. (2001). Amplification of *Citrus tristeza virus* from a cDNA clone & infection of citrus trees. *Virology*, , 280, 87-96.

[8] Gowda, S, Satyanarayana, T, Robertson, C. J, Garnsey, S. M, & Dawson, W. O. (2005). Infection of citrus plants with virions generated in *Nicotiana benthamiana* plants agro-infiltrated with binary vector based *Citrus tristeza virus*. In *Proceedings of the 16th Conference of the International Organization of Citrus Virologists* (Hilf, M.E., Duran-Vila, N. & Rocha-Peña, M.A., eds). Riverside, CA: IOCV, , 23-33.

[9] Folimonov, A. S, Folimonova, S. Y, Bar-joseph, M, & Dawson, W. O. (2007). A stable RNA virus-based vector for citrus trees. *Virology*, , 368, 205-216.

[10] Mawassi, M, Satyanayanana, T, Albiach-martí, M. R, Gowda, S, Ayllón, M. A, Robertson, C, & Dawson, W. O. (2000). The fitness of *Citrus tristeza virus* defective RNA is affected by the length of their 5′ and 3′ termini and by coding capacity. *Virology*, , 275, 42-56.

[11] Dolja, V. V, Kreuze, J. F, & Valkonen, J. P. T. (2006). Comparative & functional genomics of closteroviruses. *Virus Research*, , 117, 38-51.

[12] Bukau, B, & Horwich, A. L. (1998). The Hsp70 & Hsp60 chaperone machines. *Cell*, , 92, 351-366.

[13] Satyanayanana, T, Gowda, S, Mawassi, M, Albiach-martí, M. R, Ayllón, M. A, Robertson, C, Garnsey, S. M, & Dawson, W. O. (2000). Closterovirus encoded HSP70 homolog & p61 in addition to both coat proteins function in efficient virion assembly. *Virology*, , 278, 253-265.

[14] Tatineni, S, Robertson, C, Garnsey, S. M, Bar-joseph, M, Gowda, S, & Dawson, W. O. (2008). Three genes of *Citrus tristeza virus* are dispensable for infection and movement throughout some varieties of citrus trees. *Virology.*, 376(2), 297-307.

[15] Gowda, S, Satyanayanana, T, Davis, C. L, Navas-castillo, J, Albiach-martí, M. R, Mawassi, M, Valkov, N, Bar-joseph, M, Moreno, P, & Dawson, W. O. gene product of *Citrus tristeza virus* accumulates in the amorphous inclusion bodies. *Virology,*, 274, 246-254.

[16] López, C, Navas-castillo, J, Gowda, S, Moreno, P, & Flores, R. (2000). The 23-kDa protein coded by the 3′-terminal gene of citrus tristeza virus is an RNA-binding protein. *Virology,*, 269, 462-470.

[17] Satyanarayana, T, Gowda, S, Ayllón, M. A, Albiach-martí, M. R, Rabindram, R, & Dawson, W. O. p23 protein of *Citrus tristeza virus* controls asymmetrical RNA accumulation. *Journal Virology,*, 76, 473-483.

[18] Lu, R, Folimonov, A, Shintaku, M, Li, W. X, Falk, B. W, Dawson, W. O, & Ding, S. W. (2004). Three distinct suppressors of RNA silencing encoded by a 20-Kb viral RNA genome. *Proc. Natl. Acad. Sci. USA,*, 101, 15742-15747.

[19] Tatineni, S, Gowda, S, & Dawson, W. O. Heterologous minor coat proteins of Citrus tristeza virus strains affect encapsidation, but the coexpression of HSP70h and restores encapsidation to wild-type levels. Virology (2010)., 61.

[20] Albiach-marti, M. R, Robertson, C, Gowda, S, Tatineni, S, Belliure, B, Garnsey, S. M, Folimonova, S. Y, Moreno, P, & Dawson, W. O. (2010). The pathogenicity determinant of *Citrus tristeza virus* causing the seedling yellows syndrome maps at the 3′-terminal region of the viral genome. *Molecular Plant Pathology,*, 11, 55-67.

[21] Tatineni, S, & Dawson, W. O. Enhancement or attenuation of disease by deletion of genes from citrus tristeza virus. Journal of Virology (2012)., 86(15), 7850-7857.

[22] Folimonova, S. Y. Superinfection exclusion is an active virus-controlled function that requires a specific viral protein. Journal of Virology. (2012)., 2012(86), 10-5554.

[23] Gowda, S, Satyanayanana, T, Ayllón, M. A, Moreno, P, Flores, R, & Dawson, W. O. (2003b). The conserved structures of the 5′nontranslated region of *Citrus tristeza virus* are involved in replication & virion assembly. *Virology,*, 317, 50-64.

[24] Satyanayanana, T, Gowda, S, Ayllón, M. A, & Dawson, W. O. (2004). Closterovirus bipolar virion: evidence for initiation of assembly by minor coat protein and its restriction to the genomic RNA 5′region. *Proc. Natl. Acad. Sci. USA,*, 101, 799-804.

[25] López, C, Ayllón, M. A, Navas-castillo, J, Guerri, J, Moreno, P, & Flores, R. (1998). Molecular variability of the 5′ & 3′ terminal regions of citrus tristeza virus RNA. *Phytopathology,*, 88, 685-691.

[26] Satyanarayana, T, Gowda, S, Ayllón, M. A, Albiach-martí, M. R, & Dawson, W. O. (2002a). Mutational analysis of the replication signals in the 3′-non translated region of *Citrus tristeza virus. Virology, ,* 300, 140-152.

[27] Albiach-martí, M. R, Mawassi, M, Gowda, S, Satyanarayana, T, Hilf, M. E, Shanker, S, Almira, E. C, Vives, M. C, López, C, Guerri, J, Flores, R, Moreno, P, Garnsey, S. M, & Dawson, W. O. (2000b). Sequences of *Citrus tristeza virus* separated in time and space are essentially identical. *Journal of Virology, ,* 74, 6856-6865.

[28] Mawassi, M, Mietkiewska, E, Gofman, R, Yang, G, & Bar-joseph, M. (1996). Unusual sequence relationships between two isolates of citrus tristeza virus. *Journal General Virology, ,* 77, 2359-2364.

[29] Yang, Z. N, Mathews, D. M, Dodds, J. A, & Mirkov, T. E. (1999). Molecular characterization of an isolate of citrus tristeza virus that causes severe symptoms in sweet orange. *Virus Genes, ,* 19, 11-142.

[30] Vives, M. C, Rubio, L, & Sambade, A. Mirkov, Moreno, P. & Guerri, J. ((2005). Evidence of multiple recombination events between two RNA sequence variants within a *Citrus tristeza virus* isolate. *Virology, ,* 331, 232-237.

[31] Ruiz-ruiz, S, Moreno, P, Guerri, J, & Ambrós, S. (2006). The complete nucleotide sequence of a severe stem pitting isolate of *Citrus tristeza virus* from Spain: comparison with isolates from different origins. *Archives of Virology, ,* 151, 387-398.

[32] Vives, M. C, Rubio, L, López, C, Navas-castillo, J, Albiach-martí, M. R, Dawson, W. O, Guerri, J, Flores, R, & Moreno, P. (1999). The complete genome sequence of the major component of a mild citrus tristeza virus isolate. *Journal General Virology, ,* 80, 811-816.

[33] Suastika, G, Natsuaki, T, Terui, H, Kano, T, Ieki, H, & Okuda, S. (2001). Nucleotide Sequence of *Citrus tristeza virus* seeding yellows isolate. *Journal General Plant Pathology, ,* 67, 73-77.

[34] Roy, A, & Brlansky, R. H. (2010). Genome analysis of an orange stem pitting *Citrus Tristeza Virus* isolate reveals a novel recombinant genotype. *Virus Research,. ,* 151, 118-130.

[35] Harper, S. J, Dawson, T. E, & Pearson, M. N. (2009). Complete genome sequences of two distinct and diverse Citrus tristeza virus isolates from New Zealand. *Archives of Virolology, ,* 154, 1505-1510.

[36] Harper, S. J, Dawson, T. E, & Pearson, M. N. (2010). Isolates of *Citrus tristeza virus* that overcome *Poncirus trifoliata* resistance comprise a novel strain. *Archives of Virolology, ,* 155, 471-480.

[37] Melzer, M. J, Borth, W. B, Sether, D. M, Ferreira, S, Gonsalves, D, & Hu, J. S. (2010). Genetic diversity and evidence for recent modular recombination in Hawaiian *Citrus tristeza virus. Virus Genes, ,* 40(1), 111-118.

[38] Biswas, K. K, Tarafdar, A, & Sharma, S. K. Complete genome sequence of mandarin decline Citrus tristeza virus of the Northeastern Himalayan hill region of India: comparative analyses determine recombinant. Archives of Virology (2012). , 2012(157), 3-579.

[39] Hilf, M. E, Karasev, A. V, Albiach-martí, M. R, Dawson, W. O, & Garnsey, S. M. (1999). Two paths of sequence divergence in the citrus tristeza virus complex. *Phytopathology*, , 89, 336-342.

[40] Rubio, L, Ayllón, M. A, Kong, P, Fernández, A, Polek, M. L, Guerri, J, Moreno, P, & Falk, B. W. (2001). Genetic variation of Citrus tristeza virus isolates from California & Spain: evidence for mixed infections & recombination. *Journal of Virology*, , 75, 8054-8062.

[41] Ayllón, M. A, López, C, Navas-castillo, J, Mawassi, M, & Dawson, W. O. (1999a). New defective RNAs from citrus tristeza virus: evidence for a replicase driven template switching mechanism in their generation. *Journal General Virology*, , 80, 871-882.

[42] Weng, Z, Barthelson, R, Gowda, S, Hilf, M. E, Dawson, W. O, Galbraith, D. W, & Xiong, Z. (2007). Persistent infection & promiscuous recombination of multiple genotypes of an RNA virus within a single host generate extensive diversity. *PLoS ONE*, e917., 2(9)

[43] Yang, G, Mawassi, M, Gofman, R, Gafny, R, & Bar-joseph, M. (1997). Involvement of a subgenomic mRNA in the generation of a variable population of defective citrus tristeza virus molecules. *Journal of Virology*, , 71, 9800-9802.

[44] Gowda, S, Satyanayanana, T, Ayllón, M. A, Albiach-martí, M. R, Mawassi, M, Rabindran, S, & Dawson, W. O. (2001). Characterization of the cis-acting elements controlling subgenomic mRNAs of *Citrus tristeza virus*: production of positive-and negative-stranded 3'-terminal and positive-stranded 5' terminal RNAs. Virology, , 286, 134-151.

[45] Albiach-martí, M. R, & Guerri, J. Hermoso de Mendoza, A., Laigret, F., Ballester-Olmos, J.F. & Moreno, P. ((2000a). Aphid transmission alters the genomic and defective RNA populations of citrus tristeza virus. *Phytopathology*, , 90, 134-138.

[46] Mawassi, M, Karasev, A. V, Mietkiewska, E, Gafny, R, Lee, R. F, Dawson, W. O, & Bar-joseph, M. (1995a). Defective RNA molecules associated with citrus tristeza virus. *Virology*, , 208, 383-387.

[47] Che, X, Piestum, D, Mawassi, M, Satyanayanana, T, Gowda, S, Dawson, W. O, & Bar-joseph, M. (2001). coterminal subgenonic RNAs in citrus tristeza virus-infected cells. *Virology*, , 283, 374-381.

[48] Ayllón, M. A, Satyanarayana, T, Gowda, S, & Dawson, W. O. (2005). An atypical 3'-controller element mediates low-level transcription of the p6 subgenomic mRNA of *Citrus tristeza virus*. *Molecular Plant Pathology*, , 6(2), 165-176.

[49] Ayllón, M. A, Gowda, S, Satyanarayana, T, & Dawson, W. O. (2004). Cis-acting elements at opposite ends of the Citrus tristeza virus genome differ in initiation & termination of subgenomic RNAs. Virology, , 322, 41-50.

[50] Karasev, A. V, Hilf, M. E, Garnsey, S. M, & Dawson, W. O. (1997). Transcriptional Strategy of *Closteroviruses*: Maping the 5'termini of the citrus tristeza virus subgenomic RNAs. *Journal of Virology*, , 71, 6233-6236.

[51] Gowda, S, Ayllón, M. A, Satyanayanana, T, Bar-joseph, M, & Dawson, W. O. (2003a). Transcription strategy in a *Closterovirus*: a novel 5'-proximal controler element of *Citrus tristeza virus* produces 5'- & 3'-terminal subgenomic RNAs & differs from 3'open reading frame controler elements. *Journal of Virology*, , 77, 340-352.

[52] Gowda, S, Tatineni, S, Folimonova, S. Y, Hilf, M. E, & Dawson, W. O. (2009). Accumulation of a 5' proximal subgenomic RNA of *Citrus tristeza virus* is correlated with encapsidation by the minor coat protein. *Virology*, , 389, 122-131.

[53] Hilf, M. E, Karasev, A. V, Pappu, H. R, Gumpf, D. J, Niblett, C. L, & Garnsey, S. M. (1995). Characterization of citrus tristeza virus subgenomic RNAs in infected tissue. *Virology*, , 208, 576-582.

[54] Navas-castillo, J, Albiach-martí, M. R, Gowda, S, & Hilf, M. E. Garnsey S.M & Dawson W.O. ((1997). Kinetics of accumulation of *Citrus tristeza virus* RNAs. *Virology*, , 228, 92-97.

[55] Che, X, Dawson, W. O, & Bar-joseph, M. (2003). Defective RNAs of Citrus tristeza virus analogous to Crinivirus genomic RNAs. *Virology*, , 310, 298-309.

[56] Mawassi, M, Satyanayanana, T, Gowda, S, Albiach-martí, M. R, Robertson, C, & Dawson, W. O. (2000b). Replication of heterologous combinations of helper & defective RNA of Citrus tristeza virus. Virology, , 267, 360-369.

[57] Yang, G, Che, X, & Gofman, R. Ben Shalom, Y., Piestun, D., Gafny, R., Mawassi, M., Bar-Joseph, M. ((1999). D-RNA molecules associated with subisolates of the VT strain of citrus tristeza virus which induce different seedling-yellows reactions. *Virus Genes*, , 19, 5-13.

[58] Alzhanova, D. V, Prokhnevsky, A. I, Peremyslov, V. V, & Dolja, V. V. (2007). Virion tails of beet yellows virus: coordinated assembly by three structural proteins. Virology, , 359, 220-226.

[59] Tatineni, S, Gowda, S, & Dawson, W. (2010). Heterologous minor coat proteins of Citrus tristeza virus strains affect encapsidation, but the coexpression of HSP70h and p61 restores encapsidation to wild-type levels. *Virology*, , 402, 262-270.

[60] Culver, J. N, & Padmanabhan, M. S. (2007). Virus-Induced Disease: Altering Host Physiology One Interaction at a Time. *Ann. Rev. Phytopathol.*, , 45, 221-243.

[61] Voinnet, O. (2005). Induction & suppression of RNA silencing: insights from viral infections. *Nature Gen. Rev.* , 6, 206-220.

[62] Folimonova, S. Y, Folimonov, A. S, Tatineni, S, & Dawson, W. O. (2008). *Citrus tristeza virus*: survival at the edge of the movement continuum. *Journal of Virology,* , 82, 6546-6556.

[63] Gandia, M, Conesa, A, Ancillo, G, Gadea, J, Forment, J, Pallás, V, Flores, R, Duranvila, N, Moreno, P, & Guerri, J. (2007). Transcriptional response of Citrus aurantifolia to infection by *Citrus tristeza virus*. *Virology,* , 367, 298-306.

[64] Ruiz-ruiz, S, Navarro, B, Gisel, A, Peña, L, Navarro, L, & Moreno, P. Di Serio, F. & Flores, R. ((2011). *Citrus tristeza virus* infection induces the accumulation of viral small RNAs (21-24-nt) mapping preferentially at the 3′-terminal region of the genomic RNA and affects the host small RNA profile. *Plant molecular biology,* , 75, 607-619.

[65] Peremyslov, V. V, Pan, Y. W, & Dolja, V. V. (2004). Movement protein of a closterovirus is a type III integral transmembrane protein localized to the endoplasmic reticulum. *J. Virol.* , 78, 3704-3709.

[66] Qu, F, & Morris, J. (2005). Suppressors of RNA silencing encoded by plant viruses and their role in viral infections. *FEBS Letters.* , 579, 5958-5964.

[67] Ghorbel, R, López, C, Moreno, P, Navarro, L, Flores, R, & Peña, L. (2001). Transgenic citrus plants expressing the Citrus tristeza virus p23 protein exhibit viral-like symptoms. *Molecular Plant Pathology,* , 2, 27-36.

[68] Fagoaga, C, & López, C. Hermoso de Mendoza, A.H., Moreno, P., Navarro, L., Flores R. & Peña L. ((2006). Post-transcriptional gene silencing of the p23 silencing suppressor of *Citrus tristeza virus* confers resistance to the virus in transgenic Mexican lime. *Plant Molecular Biology,* , 66, 153-165.

[69] Lopez, C, Cervera, M, Fagoaga, C, Moreno, P, Navarro, L, Flores, R, & Peña, L. (2010). Accumulation of transgene-derived siRNAs is not sufficient for RNAi-mediated protection against Citrus tristeza virus in transgenic Mexican lime. Molecular Plant Pathology, 11(1), 33-41

[70] Rodríguez- Cerezo, E. Gamble Klein, P. & Shaw, J.G ((1991). A determinant of disease symptom severity is located in the 3`-terminal noncoding regions of the RNA of a plant virus. *Proc. Natl. Acad. Sci. USA.* , 88, 9863-9867.

[71] Fagoaga, C, López, C, Moreno, P, Navarro, L, Flores, R, & Peña, L. (2005). Viral-like symptoms induced by the ectopic expresión of the p23 of *Citrus tristeza virus* are citrus specific & do not correlate with the patogenicity of the virus strain. *Molecular Plant Microbe Interaction,* , 18, 435-445.

[72] Albiach-martí, M. R, Grosser, J. W, Gowda, S, Mawassi, M, Satyanarayana, T, Garnsey, S. M, & Dawson, W. O. (2004). Citrus tristeza virus replicates and forms infec-

tious virions in protoplast of resistant citrus relatives. *Molecular Breeding, , 14,* 117-128.

[73] Febres, V. J, Ashoulin, L, Mawassi, M, Frank, A, Bar-joseph, M, Manjunath, K. L, Lee, R. F, & Niblett, C. L. protein is present at one end of citrus tristeza virus particles. *Phytopathology, , 86,* 1331-1335.

[74] Ng, J. C. K, & Falk, B. W. (2006). Virus-vector interactions mediating nonpersistent and semipersistent plant virus transmission. *Ann. Rev. Phytopathol. , 44,* 183-212.

[75] Stewart, L. R, Medina, V, Tian, T, Turina, M, Falk, B. W, & Ng, J. C. K. (2010). A mutation in the *Lettuce infectious yellows virus* minor coat protein disrupts whitefly transmission but not in planta systemic movement. *Journal of Virology, , 84,* 12165-12173.

[76] Barzegar, A, Rahimian, H, & Sohi, H. H. (2009). Comparison of the minor coat protein gene sequences of aphid-transmissible and-nontransmissible isolates of *Citrus tristeza virus. Journal of General Plant Pathology, , 76(2),* 143-151.

Chapter 2

RNA 5′-end Maturation: A Crucial Step in the Replication of Viral Genomes

Frédéric Picard-Jean, Maude Tremblay-Létourneau,
Elizabeth Serra, Christina Dimech, Helene Schulz,
Mathilde Anselin, Vincent Dutilly and
Martin Bisaillon

Additional information is available at the end of the chapter

http://dx.doi.org/10.5772/56166

1. Introduction

Viruses are a vastly diverse group of infectious particles with many different structures, mechanisms of function and ingenious strategies of invading host organisms for their own proliferation. One of the key features that ties viruses together as an inclusive group, is the reliance on living cells for replication and propagation. On their own, viruses lack the cellular machinery necessary for many life-sustaining functions including protein translation and metabolism. Regardless of the organization of a viral genome or the type of nucleic acid, infection of a host cell and viral propagation is dependent on the transcription of viral mRNA and, in turn, the translation of viral proteins as well as genome replication. Because viruses are dependent on host cell machinery for most of these processes, they have driven an outstanding virus-host co-evolution. Viruses that rely on the replication machinery of the host cell become cell-cycle dependent in their own replication. Furthermore, just as viruses have evolved ways to hijack necessary cellular proteins, cells have evolved complex mechanisms for fighting infection by detection and degradation of foreign mRNA. In order for viral mRNA to utilize host cell machinery, begin translation and remain both stable and undetected in the cytoplasm, it must contain the post-translational modifications of a host cell mRNA including, but not limited to, a 5′ cap structure. By disguising viral mRNA with the same structural elements found in host mRNA, the cellular defense mechanism can be evaded and protein translation may occur. The significance of the cap structure can be seen through the diversity of cap-synthesis pathways across vastly different viral families that all lead to the formation

of a ubiquitous RNA 5′-cap. The 5′→ 3′ direction of nucleotide triphosphate (NTP) polymerization during RNA synthesis creates a nascent mRNA molecule with a 5′-triphosphate moiety
resulting from the initial NTP on the 5′-end. Through the processes involved in cap synthesis,
the pppRNA structure is transformed into a basic, cap-0 RNA structure (^{m7}GpppN). Further
2′-O-methylations of the first and second nucleotides of the RNA may occur.

In this chapter, a number of processes used by viruses to synthesize, acquire or mimic a 5′ cap
are explored to highlight the similarities and differences in the enzymatic mechanisms that
lead to the maturation of a 5′cap on viral RNA and its importance in viral genome replication
within a host cell.

2. Description of the RNA cap structure

To understand the importance of an RNA cap structure for viruses, it is crucial to first
understand why this structure is essential to their eukaryotic hosts. Prokaryotic RNA transcription and protein translation are coupled due to the spatial proximity between DNA and
ribosomes. In eukaryotic cells however, newly synthesized RNA transcripts undergo several
nuclear post-transcriptional modifications, known as RNA processing, before they are
exported and translated in the cytoplasm. These eukaryotic pre-mRNA modifications include
the addition of a cap structure at the 5′-end, the splicing out of introns, the editing of nucleo-
bases and the addition of a poly(A) tail at the 3′-end. RNA capping is a co-transcriptional
process that occurs when an RNA molecule is 20-30 nucleotides in length. The cap structure
consists of a guanosine residue, harboring a methylation in the N-7 position, which is bound
to the terminal 5′-end nucleotide with a peculiar 5′-5′ triphosphate bridge (Fig. 1). This inverted
link between the two nucleotides prevents RNA degradation by 5′-3′ exonucleases. The second
important feature of the cap structure is the presence of the methyl group on the guanosine,
which confers a positive charge that plays an important role in its specific recognition by
specialized proteins. The cap structure fulfills many roles which ultimately lead to mRNA
translation. In the nucleus for instance, the cap structure of pre-mRNAs is recognized by the
cap binding proteins (CBP20 and CBP80). This cap binding complex (CBC) protects mRNA
from degradation and assists RNA transport from the nucleus to the cytoplasm. Once in the
cytoplasm, ribosomes and translation factors must be recruited for translation of mRNAs into
proteins. The eukaryotic translation initiation factor 4E (eIF4E) specifically binds to the RNA
cap structure [1]. This association is mediated through stacking interactions between two
aromatic residues of the eIF4E protein; the mRNA binding is further stabilized by specific
hydrogen bonds between the positive charge of the 7-methylguanosine and an acidic residue
[2]. Upon cap binding, eIF4E assembles with eIF4G (a scaffold protein) and eIF4A (an RNA
helicase) into the eIF4F complex [3]. The scaffolding protein eIF4G recruits the small 40S
ribosomal subunit through the eIF3 complex [4]. The translation initiation complex then scans
the mRNA for the start codon before recruiting the larger subunit of the ribosome, and
translation of the open reading frame (ORF) takes place [2]. Taken together, the roles fulfilled
by the RNA cap structure are crucial for RNA stability and translation. Because of this, many

eukaryotic viruses require strategies, such as RNA cap synthesis, in order to protect, replicate and translate their genomes in eukaryotic hosts.

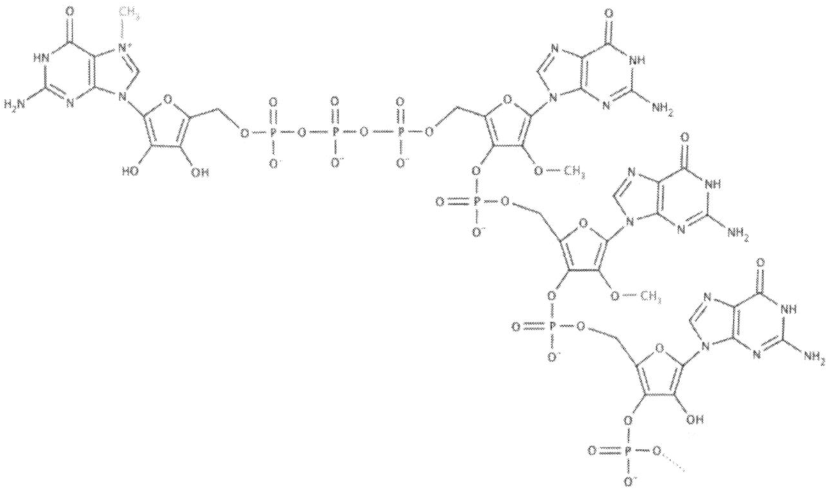

Figure 1. RNA 5'-cap structure. The RNA 5'cap structure is composed of a 7-methylguanosine (blue) linked to the RNA (black) through a 5'-5' triphosphate bridge (blue and black). The N7 methylation of the guanosine (green) confers a positive charge to the cap structure. Additional 2'O-methylation (red) can be found on the first few nucleotides.

3. Conventional and unconventional 5' RNA cap synthesis mechanism

3.1. Canonical cap synthesis by different viruses

The importance of the cap structure in eukaryote metabolism has resulted in an evolutionary pressure for viruses to adopt a similar cap structure. A series of enzymatic reactions is required to synthesize a cap structure at the 5'-end of RNA. The most pervasive enzymatic pathway, also termed "conventional capping", consists of three sequential enzymatic activities that are required to generate a functional 7-methylguanosine 5'-5'-triphosphate bridged cap structure. As a result of the directional 5' to 3' polymerization of nucleotide triphosphates (NTP) during RNA synthesis, nascent RNA bear at their 5'-end a triphosphate moiety (originating from the initial NTP). This 5'-triphosphate end of the RNA is first converted into a 5'-diphosphate end by hydrolysis of the terminal phosphate, or γ-phosphate, by an RNA triphosphatase (RTPase). This is followed by a two-step reaction catalyzed by an RNA guanylyltransferase (GTase). The enzyme first specifically binds and hydrolyzes a GTP molecule to form a covalent enzyme-GMP intermediate, which then catalyzes the transfer of the GMP moiety onto the 5'-end of a diphosphorylated acceptor RNA (ppRNA) in the second step of GTase reaction. Lastly, an

RNA (guanine-N-7)-methyltransferase (N7MTase) uses S-Adenosyl methionine (SAM) as a methyl group donor in order to methylate the guanosine residue of the cap structure at the N7 position. This sequence of enzymatic modifications yields the minimal RNA cap-0 structure (^{m7}GpppN). Subsequent methylation of the 2'-hydroxyl group of the first few nucleotides of the RNA can be catalyzed by a (nucleoside-2'-O)-methyltransferase (2'OMTase) again using a SAM molecule as a methyl-donor (Fig. 2). Further methylations on the caps proximal nucleotides convert a cap-0 structure into a cap-1 (^{m7}GpppN$_m$) or cap-2 (^{m7}GpppN$_m$N$_m$) structure.

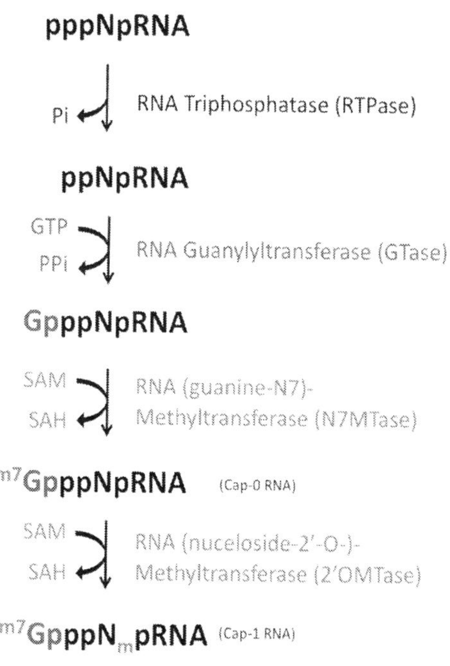

Figure 2. Canonical 5' RNA cap synthesis pathway. RNA cap-1 structures are conventionally synthesized by the sequential γ-phosphate hydrolysis by an RTPase, GMP transfer by a GTase, N7-methylation by an N7MTase, and 2'O-methylation by a 2'OMTase. The contribution of each substrate to the formation of the final 5' RNA cap structure is highlighted by a color code: pppRNA (black), GTP (blue) and SAM (green and red).

The conventional RNA 5' cap synthesis mechanism is used by a majority of viruses in order to acquire a cap structure. Most DNA viruses together with the RNA viruses from the *Bornaviridae* and *Retroviridae* families use the host RNA polymerase II (RNA Pol II) to transcribe their mRNAs. As a result, the majority of DNA virus transcripts are co-transcriptionally capped using the cellular capping apparatus. Alternatively, many RNA viruses with a cytoplasmic replication cycle, do not have access to the host RNA Pol II and therefore have evolved their own capping machinery. Over time, a wide diversity of enzyme structures and mechanisms of action have evolved to generate the same highly conserved RNA cap structure (Fig. 3). The following paragraphs describe the enzymes supporting the RTPase, GTase, N7MTase and 2'OMTase activity.

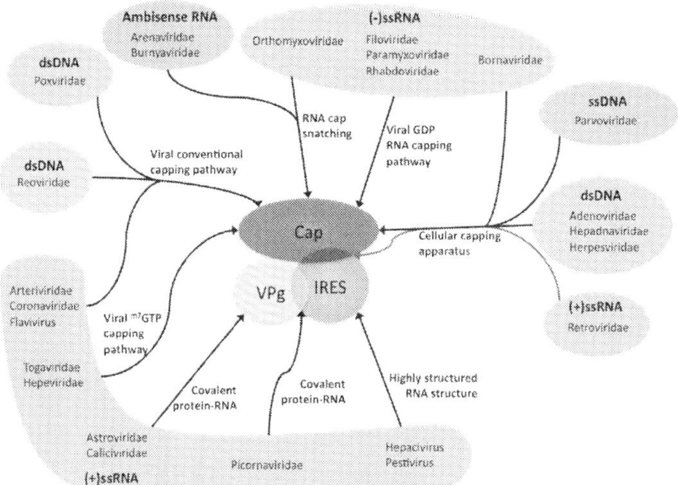

Figure 3. Viral RNA 5'-end structure and maturation. Nearly all mammalian viruses modify their RNA 5'-end through the covalent addition of a cap structure (majority) or a VPg protein (minority). Although the widely acquired RNA cap structure is chemically identical (^{m7}GpppN), viruses have evolved a large variety of mechanisms to synthesize or acquire this crucial structure. The mechanisms of 5'-end maturation vary among order and families of viruses as presented in the above schematic (for clarity, only a few viral families are presented). Within the *Flaviviridae* family, the Flavivirus synthesize a typical cap structure while the Hepacivirus and Pestivirus are the only mammalian viruses representative to harbor an unmodified 5'-triphosphate end. Their RNA 5' UTR instead folds into a highly structured three-dimensional conformation termed IRES. Of notice, the Retroviridae RNA harbor both, a cap structure and an IRES, as well as the *Picornaviridae* RNA that harbors both an IRES and a 5'-VPg-linked protein. Adapted from Decroly and al. (2012).

3.2. RNA triphosphatases

The RTPase activity is the first of the three enzymatic reactions required to synthesize a cap structure. The RTPase hydrolyzes the γ-β-phosphoanhydride bond at the 5'-end of an RNA to yield an RNA 5'-diphosphate and inorganic phosphate (Pi). Viruses have evolved a wide variety of enzyme structures and mechanisms of action to fulfill the RTPase activity, a greater diversity than is seen with any other enzymatic capping activity. RTPases are classified as either belonging to the metal-dependent family or the metal independent family based on their cofactor requirements. As indicated by its name, the first family requires a divalent cation cofactor for its activity. This metal requirement is usually satisfied by Mg^{2+}, although Mn^{2+} is also able to support the RTPase activity [5]. This family of enzymes also shares the ability to hydrolyze free NTPs, again in the presence of a metal cofactor [5, 6]. The lack of substrate specificity is speculated to be a result of the chemical similarity between an NTP and the RNA 5′-triphosphate end. The metal dependent RTPase family is further subdivided into three distinct structural groups, namely the triphosphate tunnel metalloenzyme (TTM), histidine triad-like (HIT-like) and helicase-like RTPase (Fig. 4).

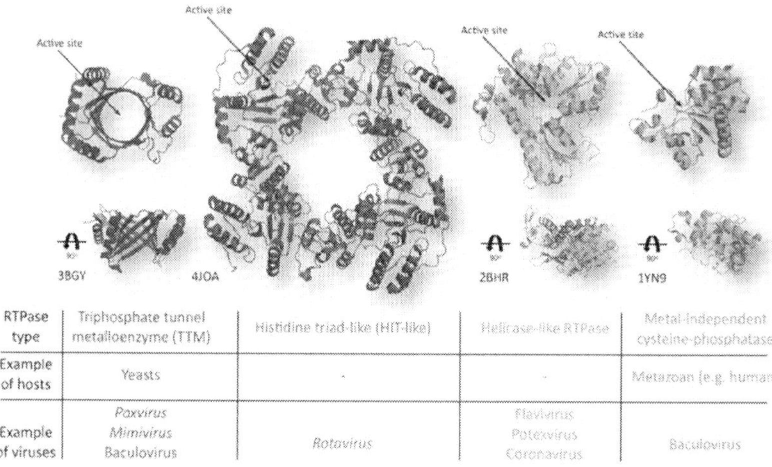

RTPase type	Triphosphate tunnel metalloenzyme (TTM)	Histidine triad-like (HIT-like)	Helicase-like RTPase	Metal-independent cysteine-phosphatase
Example of hosts	Yeasts			Metazoan (e.g. human)
Example of viruses	*Poxvirus* *Mimivirus* *Baculovirus*	*Rotavirus*	*Flavivirus* *Potexvirus* *Coronavirus*	*Baculovirus*

Figure 4. Diversity in RNA triphosphatase structure and mechanism of action. The RTPase activity can be catalyzed by various mechanisms of action, each associated with a characteristic structure. They are indicated as follows: TTM (blue), HIT-like (multimeric: dark pink, monomeric: red), Helicase-like (cyan) and metal-independent (green). The location of the active site is indicated by and arrow, and examples of hosts and viruses utilizing those enzymes are given.

The TTM enzymes are found in chlorella virus, poxviruses, baculoviruses, mimiviruses and lower eukaryotes. All TTM RTPases fold in a specific, characteristic structure. An assembly of eight antiparallel β-strands to form a tunnel scaffold surrounding the active site (Fig. 4). The interior of the tunnel is dominated by hydrophilic amino acid side chains oriented toward the center of the tunnel creating a network of interactions for the triphosphate moiety of the substrate [7]. Glutamate residues, within this amino acid network are also responsible for the coordination of the crucial cation cofactor [6]. The recognition of the RNA substrate, primarily through its triphosphate moiety, could explain the activity of the TTM RTPase against NTP substrates. Interestingly, this NTP hydrolysis is not supported by Mg^{2+}, but is rather dependent on Mn^{2+} or Co^{2+} [6]. The coordinated metal ion, in conjunction with basic lysine and arginine, activates the γ-phosphate and stabilizes the pentacoordinate phosphorane transition state. A glutamate serves as a general base catalyst to activate the nucleophilic water for the attack on the γ-phosphorus according to a one-step in-line mechanism [8]. TTM RTPases have been acquired by large DNA viruses from their hosts [7]. Interestingly, modern *Poxviridae* infect higher eukaryotes that lack TTM RTPase, underlying their evolution from viral ancestors that replicated in unicellular eukarya, from which they likely acquired a TTM RTPase.

The HIT-like RTPase is so far only represented by the NSP2 enzyme of rotaviruses (dsRNA virus). The name of this family is based on the structural resemblance between the NSP2 C-terminal domain (CTD) and the ubiquitous cellular histidine triad nucleotidyl hydrolases (HIT). The NSP2 protein associates into an octamer to form a doughnut-shaped quaternary structure (Fig. 4) [9, 10]. RNA binding grooves are found at the surface of the doughnut-shape while the active site is buried deep in an electro-positive cleft on each monomer. Despite structural similarity with HIT, NSP2 appears to be catalytically distinct. The catalytic histidine triad requires a Mg^{2+} cofactor to hydrolyze the γ-β-phosphoanhydride and form a covalent phosphate-histidine intermediate [11]. The enzyme harbours similar catalytic rates toward both NTP and pppRNA substrates. Increased affinity for RNA, conferred by the RNA binding grooves, is speculated to stimulate RTPase activity over NTPase activity *in vivo* [10]. Despite the structural similarity with HIT, currently no evidence indicates that HIT-like RTPase could have evolved from their cellular counterpart, and rather a convergent evolution is more probable [9].

The helicase-like RTPases are found in a variety of ss(+) RNA viruses of the *flavivirus, coronavirus, potexvirus* and *alphavirus* genera and the dsRNA viruses of the *Reoviridae* family. These enzymes are active NTPase-helicases and belong to the large helicase superfamilies SF1 and SF2. The NTPase activity fuels the energy-consuming strand displacement of the helicase activity. The common NTPase-RTPase catalytic site is located in a cleft formed from the junction of two RecA-like subdomains (Fig. 4). As with many nucleotide-binding proteins, the active site of helicase-like RTPases harbour both a Walker A and Walker B motif [12, 13]. The Walker A motif (GxxxxGK(T/S)), or phosphate-binding loop (P-loop), is responsible for contacting the γ-phosphate through its highly conserved arginine. The aspartate of the Walker B motif (DExD) coordinates the crucial Mg^{2+}, which stabilizes the γ and β-phosphates, while the glutamate activates the water molecule for the hydrolysis reaction [14]. The addition of the RTPase activity to an NTPase-helicase ancestor appears to result form only a minor evolutionary progression as the ancestor enzyme already displayed the key RTPase features, namely, a nucleic acid binding domain, a triphosphate binding active site and a terminal phosphate hydrolysis activity.

The second family of RTPases is the metal-independent group. Higher eukaryotic viruses that rely on capping apparatus of the cell use the host metal-independent RTPase. Moreover, baculovirus also expresses such a metal-independent RTPase. Two striking differences between this enzyme family and the metal-dependent family, are its cation-independent mechanism of action and its inability to hydrolyze free NTP [15]. Metal-independent RTPases are members of the cysteine phosphatase superfamily, sharing their signature HCxxxxxR(S/T) P-loop motif located in a deep positively charged pocket. The catalytic cysteine is located at the bottom triphosphate binding cleft formed by the characteristic α/β-fold ternary structure (Fig. 4) [15, 16]. The catalytic cycle fits a two-step phosphoryl-transfer reaction. First, the pppRNA γ-phosphate is attacked by the catalytic cysteine to form a covalent protein-cysteinyl-S-phosphate intermediate which results in the release of the ppRNA product. Next, a water molecule attacks the phosphocysteine to expel the inorganic phosphate and regenerate the enzyme [15]. The metal-independent RTPase presumably evolved from the cysteine phosphatase ubiquitously found in higher eukaryotes and was later acquired by *baculovirus* from their hosts. Interestingly, baculovirus also encodes a second TTM RTPase fulfilling the same role. This unconventional carrying of two distinct enzymes having the same activity is speculated to be an evolutionary snapshot of an RTPase transition from the lower eukaryote TTM RTPase to the higher eukaryote metal-independent RTPase.

3.3. RNA guanylyltransferase

The second step of the capping sequence is the GTase activity. GTase catalyzes the rate-limiting transfer of a GMP moiety from a GTP substrate to an acceptor ppRNA to yield an unmethylated cap structure (GpppN). GTases are members of the covalent nucleotidyltransferases super-family which also includes the ATP- and NAD$^+$-dependent DNA ligases and the ATP-dependent RNA ligases [17]. This superfamily's ternary structure is composed of the N-terminal of the nucleotidyltransferase (NT) domain fused to an oligobinding fold (OB-fold) domain in the C-terminal. These flexible proteins are able to undergo large conformational changes during their catalytic cycle. GTases share highly conserved structures and motifs, of which the hallmark KxDG(I/L) motif is present in nearly all GTases [18]. The catalytic cycle of the GTase is a complex two-step ping-pong reaction involving multiple conformational changes. First, a GTase in a conformation where the OB-fold domain is distant from the NT domain (open conformation) specifically binds a GTP molecule. This is followed by the closure of the OB-fold domain toward the NT domain (closed conformation) which is stabilized by interactions between the bound nucleotide and residues from both NT and OB fold domains. This conformational change also creates a Mg^{2+} cofactor binding site, thus the closed confor-mation represents the catalytically active form of the enzyme [19, 20]. Upon Mg^{2+} binding, the α-phosphate of the GTP is sandwiched between the catalytic lysine (form the KxDG) and the metal cofactor. Deprotonation of the lysine leads to the attack on the α-phosphate of the GTP to form a enzyme-(lysyl-N)-GMP intermediate (EpG), concomitant with the hydrolysis of a pyrophosphate molecule [20]. Following the catalysis, interactions between the bound guanylate and the OB fold domain are disrupted, leading to the reopening of the enzyme and the release of pyrophosphate. The reopening of the guanylylated enzyme allows for accom-modation of the ppRNA, which is likely followed by the closure of the OB-fold domain. Closing of the OB-fold domain returns the enzyme to its catalytically active form, which promotes the transfer of the GMP to the acceptor RNA. A final reopening allows for unmethylated capped RNA to be released and the apo-protein to be regenerated (Fig. 5) [19]. The active sites of the GTase are highly conserved, potentially due to their fairly complex catalytic cycle. Most viruses encode GTases that are, with respect to the active site, nearly identical to their eukaryotic host GTase, favouring the hypothesis of ancestral viral acquisition of the host GTase.

While nearly all GTases are highly conserved, a few recently discovered viral GTases are different. Little is currently known about those atypical GTases lacking the catalytic KxDG motif. Some segmented dsRNA viruses of the *Reoviridae* family encode for a large multiprotein capsid harbouring nucleic acid maturation functions, including GTase activity. The *Reoviri-dae* GTase is structurally different from the conventional GTase. While they lack the conserved KxDG motif, they still maintain the capacity to form an enzyme-(lysyl-N)-GMP intermediate.

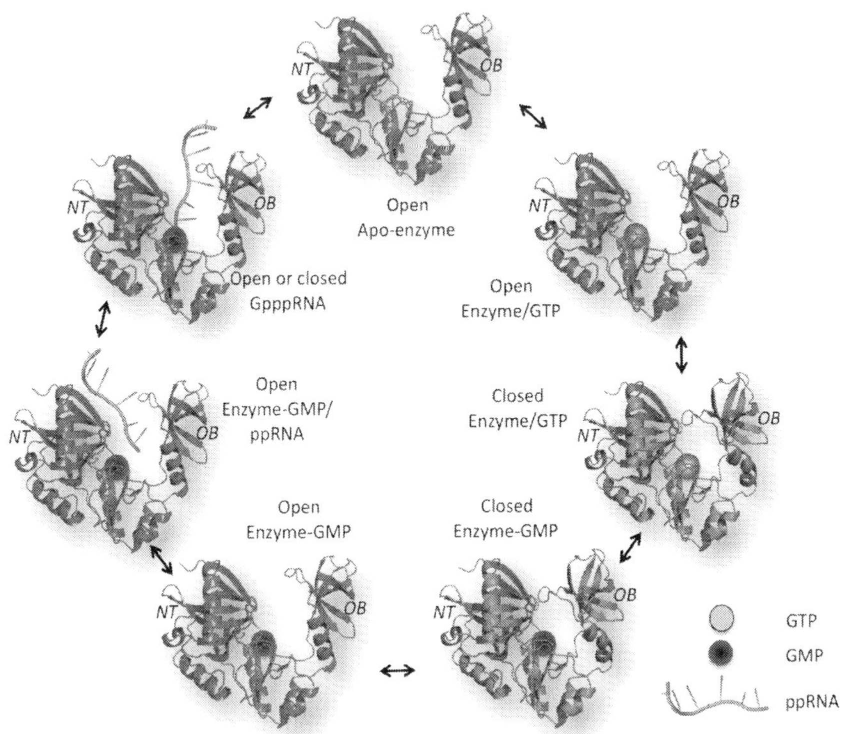

Figure 5. Structural and mechanistic pathway used by GTase. The apo-enzyme in open conformation (blue) binds a GTP substrate (gray), closes (red) and proceeds to hydrolysis thereby generating an enzyme-GMP (black) intermediate. GTase reopening (blue) allows for RNA binding (orange), the enzyme either stays open or closed to allow GMP transfer onto the RNA. Finally, the open enzyme releases the GpppN product and the apo-enzyme is regenerated.

The *flavirivus* GTases are also atypical. Their activities are found on the N-terminal portion of the RDRP-MTase peptide. They are structurally distinct from both the conventional and the *Reoviridae* GTase but they still mediate RNA guanylation through a two-step mechanism involving an EpG intermediate [21, 22]. The precise amino acid involved in the guanylate-enzyme complex formation is also speculated to be a lysine, but a histidine or an arginine residue may also play this role. Progress in the field of atypical viral capping enzymes will eventually shed light on those imprecisions.

3.4. RNA methyltransferase

The third step of the RNA 5′-end cap synthesis is the methylation of the cap guanosine by a N7MTase. An N7MTase adds a methyl group to the guanine at the N7 position in order to convert the GpppN into a functional ^{m7}GpppN cap-0 structure. The conversion of S-Adenosyl methionine (SAM) into S-Adenosyl homocysteine (SAH) provides the methyl group. N7MTases are members of the large SAM-dependent MTase family, which shares a low sequence identity but a structurally conserved SAM binding core. This SAM binding pocket, composed of a seven-stranded β-sheet flanked by six α-helices, ensures specific and proper positioning of the SAM molecule, while other structural determinants provide specificity for a range of methyl acceptors [23, 24]. For the N7MTase, those structural determinants are a positively charged RNA-accommodating groove and a GpppN binding pocket that forms extensive electrostatic interactions with the cap guanine, thereby ensuring specificity [25]. Despite a broad network of interactions with both substrates (GpppN and SAM), no direct contact is made between the N7MTase and their substrate reacting group: the guanine N7 nitrogen (methyl acceptor) and the SAM CH_3 (methyl donor). The methyl transfer is instead mediated by a direct in-line nucleophilic attack of the SAM methyl moiety by the guanine N7 nitrogen. N7MTases are not directly implicated in the transition state stabilization, but are rather optimizing the proximity and the spatial orientation between both ligands reacting groups. In addition, a favourable electrostatic environment further stimulates the catalysis [25]. The degree of conservation among N7MTases is very high and most viral and eukaryotic N7MTases only differ in their accessory domain. A rare exception is the poxvirus N7MTase, which appears to bind SAM in a slightly different conformation. Moreover, some poxviruses, such as vaccinia virus, have evolved a heterodimer N7MTase. The vaccinia virus N7MTase D1 for example relies on its association with the accessory protein D12 to be fully active [26]. The degree of conservation among N7MTases points toward a common eukaryotic ancestor acquired by viruses.

Lastly, some viruses infecting higher eukaryotes, such as *flavirirux*, *reovirus* and *poxvirus*, can further modify their RNA 5'-end through 2'-O-methylation in order to more accurately mimic their host mRNA modifications. This last modification is not required for viruses infecting lower eukaryotes as their host harbours cap-0 mRNA. The 2'OMTase methylates the first nucleotide 2'-hydroxyl group(s) of the RNA, allowing for the conversion of a $^{m7}GpppN$ (cap-0) into a $^{m7}GpppN_m$ (cap-1). The 2'OMTases are also members of the large SAM-dependent MTase family. When compared to the N7MTase, 2'OMTase harbours an additional highly conserved catalytic lysine-asparagine-lysine-glutamine tetrad [27]. These amino acids are not consecutive in the primary sequence, but they cluster together once the protein adopts its three-dimensional structure. The exact catalytic pathway is still controversial, but relies on the conserved asparagine and arginine to lower the pKa of the catalytic lysine, which is responsible for the 2'-hydroxyl group activation. Two mechanisms are proposed for this substrate activation. The first involves the lysine deprotonating the 2'-OH to form a nucleophilic 2'-oxanion. The second implicates the lysine in the formation of a non-deprotonating hydrogen bound with a 2'-hydroxyl proton, which freezes the 2'-OH rotation in an angle where the 2'-oxygen electron lone pair is steered toward the SAM methyl group. In both cases, the nucleophilic 2'-oxygen attacks the electrophilic SAM methyl group according to an in-line Sn2 mechanism [28-31]. The pentavalent methyl intermediate of the transition state is stabilized by the asparagine. Despite the structural homology with the N7MTase, the 2'OMTase harbours a distinct mechanism of methyl transfer. Interestingly, some viruses, such as the *flavivirus*, have evolved both N7MTase and 2'OMTase activities within the same enzyme [22, 32]. These dual MTases share the same SAM binding site and accessory domain but not the same mechanism of methyl transfer. The classical N7MTase and 2'OMTase mechanisms are instead present but independent. It is, for example, possible to abolish the 2'OMTase activity through disruption of the lysine-asparagine-lysine-glutamine tetrad while maintaining the N7MTase activity [22, 32]. It is important to note that the *flavivirus* dual MTase accomplishes a sequential methylation, starting with the N7 guanine methylation and followed by a repositioning of the cap structure and finally, the 2'-hydroxyl methylation [22, 32]. This sequence is virus specific and can be inverted, as exemplified by the vesicular stomatitis virus (VSV), a member of the *Rhabdoviridae* family. The VSV also encodes a dual MTase, but the 2'OMTase takes place first and is followed by the N7MTase [33]. These dual MTases have likely evolved their second MTase activity out of their initial MTase fold.

3.5. Gene organization of viral capping enzymes

In order to support viral replication and fitness, both the catalytic activity of viral enzymes involved in RNA capping as well as their localization within the cell, are crucial. Viral capping enzymes required for RNA capping have to be recruited at the site of RNA synthesis. Recruitment of the capping enzyme can be mediated by protein-protein interactions with either the RNA polymerase or a scaffold protein. While recruitment of the three distinct enzymatic activities is required in order to synthesise a cap-0 structure, the available surface for protein interactions at the RNA synthesis site is limited. Viruses have evolved multiple solutions to overcome this problem including the fusion of multiple enzymatic activities to the same polypeptide as well as protein-protein interactions between two capping enzymes to form a hetero-multimer (Fig. 6). A good example of protein-protien interaction is seen in *Paramecium bursaria Chlorella virus*, which encodes the RTPase, GTase and N7MTase activities on three different peptides [19, 34, 35]. The RTPase enzyme is likely to interact with the GTase, in a manner that is reminiscent of the lower eukaryotic capping machinery in which the Pol II co-transcriptionally recruits the RTPase-GTase heterodimer and the N7MTase separetely [36, 37]. Alternatively, viruses such as *Baculovirus* and Infectious Spleen and Kidney Necrosis virus benefit from the fusion of the RTPase and the GTase activities in a single polypeptide, thus facilitating the recruitment of the capping apparatus to the viral RNA polymerase transcription site [38, 39]. In this instance, the organization of the viral capping enzymes is most analogous to that of higher eukaryotes in which the RTPase and GTase enzymes are fused together. In this case, interaction with the GTase domain is solely responsible for RTPase-GTase recruitment to the RNA Pol II while the N7MTase is recruited separately [40]. The fusion of sequential enzymatic activities to the same multi-domain protein appears to be more robust than the heterodimer formation. Because of this, selective pressures have driven the fusion of the capping gene in a wide variety of viruses. *Alphavirus*, for example, encodes a single protein that is able to add a N7 methylated guanosine to a ppRNA, while the RTPase activity is located on a different peptide [41]. The *flavivirus* represent an even more striking example of gene organization optimization. The RTPase in this group shares a catalytic site with the NTPase/helicase (also implicated in RNA synthesis) on one protein while the GTase and the dual (N7 and 2'OMTase are fused to the RNA dependent RNA polymerase (RDRP) on a second protein.

In this example, *flavivirus* managed to pack, within two polypeptides, six different enzymatic activities, all of which are involved in RNA synthesis and maturation [21, 22, 32].

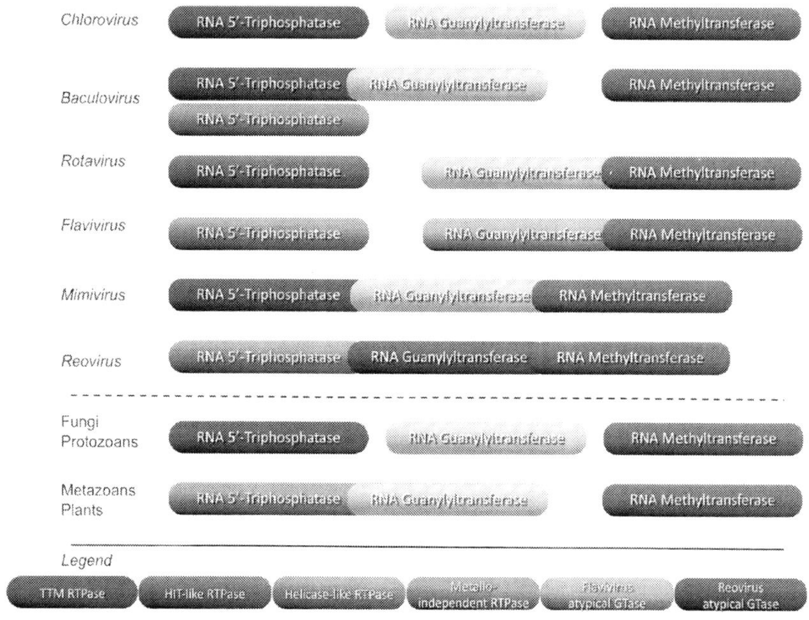

Figure 6. Gene organization of the canonical capping enzymes. Schematic representation of the genetic organization of the canonical enzymatic activity required to synthesize a cap-0 structure. Examples of organisms associated with each gene organization is indicated on the left. The color code is representative of structural and mechanistic enzymatic conservation and is detailed at the bottom of the figure.

Some viruses have even evolved a highly efficient capping enzyme, fusing together all three or four enzymatic functions required for cap synthesis into what can be described as an RNA-capping assembly line. *Mimivirus* and African swine fever virus encode a large, single protein inclusively harbouring the RTPase, GTase and N7MTase activities. This allows these viruses to efficiently modify their RNA to generate a cap-0 structure [7, 42]. The conventional cap synthesis pathway is a directional succession of enzymatic activities such that RTPase→GTase→N7MTase. Interestingly, the order of the catalytic domains within the primary sequence of these triple-activity capping enzymes follows the required capping activity sequence (NH_2-RTPase-GTase-N7MTase-COOH). As a result, they not only co-localize all capping activity to the RNA 5'-end, but also optimize the progression of the RNA through the capping activity sequence. *Poxvirus*, typified by the vaccinia virus (VV), also display a nice example of a multi-capping enzyme. The VV multi-capping enzyme, D1, possesses all three RTPase, GTase and N7MTase activities. The first two are constitutive while the N7MTase requires association with the D12 stimulatory subunit. Together this complex is able to modify an RNA 5′-end up to a cap-0 structure. It is also interesting to note that the structure of the D12 stimulatory subunit indicates that it used to be a 2′OMTase but that function is now inactive. Instead, the 2′OMTase activity is now taken over by the dedicated VP39 2′OMTase [23]. This raises the possibility of an ancestor *poxvirus* RNA-capping assembly line composed of a D1-D12-like complex that could process a 5′-triphosphate RNA into a cap-1 RNA. Such an enzymatic conveyor can currently be found in mammalian reovirus and bluetongue virus. These two viruses are members of the segmented dsRNA *Reoviridae* family and transcribe their plus-strand messenger RNA within an internal capsid particle containing the RDRP and the capping apparatus. A single protein packs together all four enzymatic activities required to synthesize a cap-1 structure (RTPase, GTase, N7MTase and 2′OMTase), although the putative RTPase activity is yet to be confirmed [43-45]. Once again, these activities are presented into a directional layout that channels the mRNA through successive enzymatic modifications with the goal of converting its 5′-triphosphate end into a cap-1 end. Moreover, this RNA capping assembly line is in direct contact with the polymerase, ensuring optimal recruitment of the nascent mRNA to the capping apparatus [46]. The λ2 and VP4 capping proteins from *reovirus* and bluetongue virus are slightly different in regard to their quaternary structure. *Reovirus* λ2, which is overall linearly shaped, associates into a pentamer to form a hollow cylinder with each active site facing the interior of the cavity, or the turret. This barrel is perpendicular to the spherical internal capsid particle and creates a channel for the nascent mRNA to exit the internal capsid particle while undergoing complete type-1 mRNA capping [44]. It is interesting that a diversity of viruses, ranging from dsDNA virus such as *Mimivirus*, African swine fever virus and *poxvirus*, to segmented dsRNA viruses including members of the *Reoviridae* family, have evolved such a complex but highly effective RNA-capping assembly line. The convergent evolution of these systems highlights the critical importance of proper RNA capping for viral genome replication and overall viral fitness.

3.6. Unconventional 5′ RNA cap synthesis mechanism evolved by different viruses

The capacity to properly cap RNA confers a distinct advantage to many eukaryotic viruses. Consequently, the selective pressure to maintain this structure is high, which is reflected by the degree of conservation among the viral capping proteins. Interestingly, this selective pressure is not directed toward the capping proteins themselves (RTPase, GTase and N7MTase), but rather toward their final product, the cap structure. Because of this, many viruses have evolved diverse biosynthetic strategies, divergent from the canonical RTPase→GTase→N7MTase pathway, allowing them to synthesize or acquire the final cap structure. This cap structure is in every aspect identical to the canonically synthesized one; only the enzymatic pathway varies. Many viruses families include members that use an unconventional 5′ RNA cap synthesis pathway. As of today, three unconventional 5′ RNA cap synthesis mechanism have been described.

3.7. The ^{m7}GTP RNA capping pathway

The ^{m7}GTP RNA capping pathway, also termed the *alphavirus*-like pathway, is found in a number of (+)ssRNA viruses of the alphavirus (Semliki Forest virus and Sindbis virus), potexvirus (Bamboo mosaic virus), tobamovirus (Tobacco mosaic virus), *Togaviridae* (Rubella virus and Chikungunya virus) and *Hepeviridae* (Hepatitis E virus) families [5, 47]. These viruses encode unique capping machinery capable of synthesizing a cap-0 structure in three sequential enzymatic reactions. The initial step is quite similar to the conventional capping mechanism in which an RTPase (nsP2 protein of Semliki Forest virus for example) hydrolyzes the γ-β-phosphoanhydride bond at the 5′-end of the RNA yielding a ppRNA [48]. Next a GTP molecule in methylated in position N7 by an atypical N7MTase (nsP1 protein of Semliki Forest virus for example). This ^{m7}GTP is then recognized as a substrate by an atypical GTase (also nsP1 of protein of Semliki Forest virus for example). The reaction results in the formation of a characteristic ^{m7}GMP-enzyme covalent complex upon the hydrolysis of a pyrophosphate group. This ^{m7}GMP group is finally transferred onto the 5′-end of the acceptor ppRNA, to yield a typical ^{m7}GpppN cap-0 structure [41, 49-52]. The overall capping reaction is then RTPase→atypical N7MTase→atypical GTase (Fig. 7). It is worth mentioning, however, that not only the order of chemical modifications differs, but also the protein mechanisms of action. The atypical N7MTase has fundamental similarities to the standard N7MTase, including the presence of a SAM binding domain, but its substrate recognition is vastly different. Atypical N7MTase proteins are unable to methylate GpppN as the canonical N7MTase does, and instead they specifically methylate GTP (and GDP to some extent) [51]. The atypical GTases are mechanistically different from their GTase counterpart in that they lack the KxDG conserved motif and mediate their ^{m7}GMP-enzyme intermediate through a conserved histidine instead of a lysine [41]. These proteins have no activity with GTP, but specifically require ^{m7}GTP to form a covalently bound enzyme complex. Therefore, the conversion from GTP to ^{m7}GTP is necessary prior to the N7-methyl-gunanylyltransferase activity [49].

Of all known eukaryotes and viruses, the ^{m7}GTP RNA capping pathway is only used by members of the (+)ssRNA viruses, which points toward a eukaryote-independent emergence of this unconventional cap synthesis mechanism. In addition, the conservation of this capping pathway throughout distantly related viruses harbouring a broad spectrum of hosts, ranging from plants to animals, suggests an evolution from a common (+)ssRNA virus ancestor.

3.8. The GDP RNA capping pathway

The GDP RNA capping pathway, also termed the *Rhabdoviridae*-like pathway, is found in representatives of many (-)ssRNA viruses of the *Rhabdoviridae* (vesicular stomatitis virus (VSV) and Rabies virus), *paramyxoviridae* (Human respiratory syncytial virus and Measles virus), *Bornaviridae* (bornavirus), and *Filoviridae* (Ebola virus and Marburg virus) families [5, 47]. These viruses encode unconventional capping machinery that catalyzes the formation of a cap-1 structure. These viruses, exemplified by VSV, encode a large L protein harbouring the RNA dependent RNA polymerase RDRP activity as well as the RNA capping activity. The latter requires a sequence of four enzymatic activities that differ from the conventional pathway, in order to generate a cap-1 structure. First, the NTPase activity is responsible for the hydrolysis of a GTP molecule into a GDP molecule. Then, an RNA GDP polyribonucleotidyl transferase (PRNTase) catalyzes a two-step reaction. The L protein hydrolyzes the (alpha-beta) phosphoanhydride bond of the pppRNA triphosphate moiety releasing a molecule of pyrophosphate and creating a covalent enzyme-pRNA intermediate. The pRNA moiety is then transferred onto the

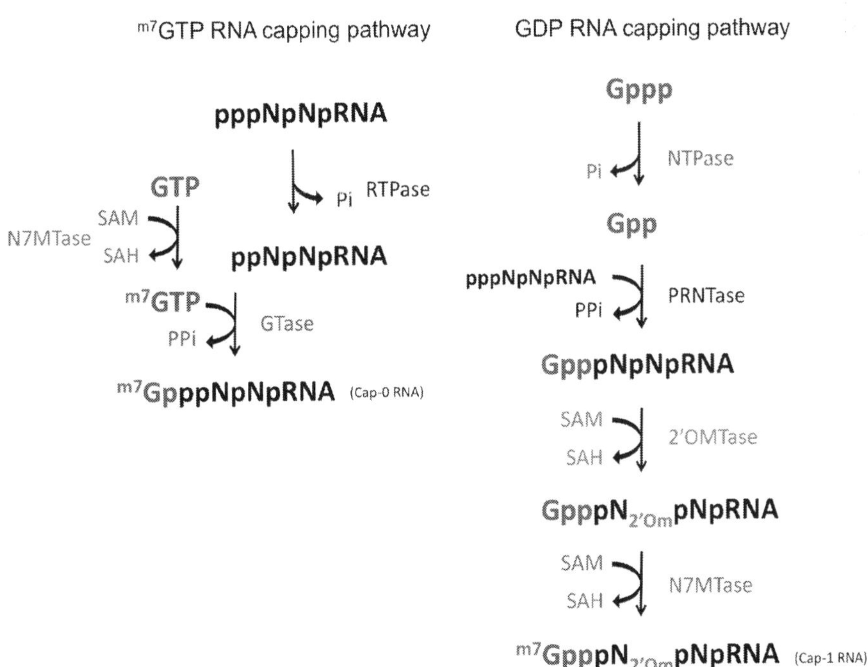

Figure 7. Unconventional 5' RNA cap synthesis mechanisms. The m7GTP capping pathway involves the hydrolysis of the RNA γ-phosphate by an RTPase, the methylation of a GTP by a N7MTase and the transfer of this m7GTP onto the diphosphorylated RNA. The GDP capping pathway is initiated by the hydrolysis of GTP to GDP by an NTPase. A PRNTase then hydrolyzes the γ-and β-phosphates of the RNA to form a covalent enzyme-pRNA intermediate. The pRNA is then transferred onto the GDP. Further methylation by the N7MTase and 2'OMTase complete the cap-1 structure. The contribution of each substrate to the formation of the final 5' RNA cap structure is highlighted by a color code: pppRNA (black), GTP (blue) and SAM (green and red).

GDP to form a GpppN block RNA. In this case, only the α-phosphate originates from the RNA whereas both the β and γ-phosphates are contributed by the GDP. Finally, synthesis of the cap-1 structure is completed by two successive methylations; the first being methylation of the first nucleotide of the 2'OH and the second being methylation of the guanine N7 nitrogen [33, 53-57]. When compared to the canonical capping reaction, this unconventional capping pathway reverses the phosphate contribution from the GTP and the RNA. The covalent enzyme-monophosphate-nucleotide intermediate is formed with the RNA instead of the GTP in an enzyme-pRNA complex instead of an enzyme-GMP complex. Similarly to the conventional capping pathway, the diphosphate cosubstrate is pre-emptively hydrolysed from its triphosphate precursor, but this time it is GDP instead of ppRNA that is generated. The PRNTase mechanism of action is also distinct from the GTase one in that the KxDG motif is replace by an HR motif and the histidine, not the lysine, is responsible for the enzyme-pRNA phosphoamide bond [55, 56]. Both the N7 and 2'OMTase activities are also present on the L protein and share the same SAM binding site. The typical lysine-asparagine-lysine-glutamine tetrad is also predicted to be at the MTase active site. The 2'O position of the GpppN is methylated prior to the guanine N7 position, which is the opposite order when compared to most canonical cap-1 methylation events [33, 53]. The overall GDP RNA capping sequence can be summarized as NTPase→PRNTase→2'OMTase→N7MTase (Fig. 7). It is very likely that an ancestral (+)ssRNA virus polymerase has evolved a PRNTase activity independently from its eukaryotic host. Both N7 and 2'OMTase, however, have likely been acquired from a eukaryotic host.

3.9. The RNA cap snatching

Some viruses, unable to synthesize their own cap structures, have evolved a clever way to acquire this important entity: steeling it from their host. This method of cap acquisition, termed RNA cap snatching, is used by representatives of the *Orthomyxoviridae* (e.g. Influenza virus, Thogoto virus), the *Arenaviridae* (e.g. Lassa virus, Machupo virus) and the *Bunyaviridae* (Hantaan virus, La Crosse virus, Tomato Spotted Wilt virus) families [5, 58]. These (-)ssRNA viruses acquire their cap structure from their hosts capped mRNA. They bind the cap structure, cleave the RNA a few nucleotides downstream and finally use this short capped RNA to prime their RDRP [59]. The *Arenaviridae* and *Bunyaviridae* express a large monomeric polymerase where the *Orthomyxoviridae* expresses an heterotrimeric polymerase (e.g. PB1, PB2 and PA protein of influenza virus) harbouring all the activities required for cap snatching. The PB2 protein of the Influenza virus, the most studied cap snatching virus, specifically binds the host mRNA cap structure. The specificity of the binding is crucial and is mediated by the aromatic stacking of the methylated gunanine coupled to a base-specific interaction with a conserved acidic residue [60]. While the mode of cap binding is similar between PB2 and other cap-binding proteins (e.g. eIF4E, nuclear cap binding complex, Vaccinia VP39) its overall fold is completely different [60]. Once the host mRNA is bound by the cap-binding PB2, the viral PA subunit cleaves the mRNA a few nucleotides downstream from the cap structure. The length of the primer RNA generated is virus-dependent, and typically ranges from 10-13 nucleotides for Influenza virus, but can be as short as 1-2 nucleotides as is seen in the Thogoto virus [59, 61, 62]. The PA endonuclease domain shares a high homology with the type II restriction enzyme, including the active site conserved $(P)Dx_n(D/E)xK$ signature motif [63]. The PA active site coordinates two Mn^{2+} cations and is believed to catalyze endonucleolytic cleavage through a common two-metal dependent mechanism [61, 64]. The short capped oligomers are next used by the PB1 RDRP as primer to initiate the transcription of the viral mRNAs [58]. PB1 also specifically binds the viral RNA (vRNA) 3' and 5'-end through a ribonucleoprotein 1-like motif ((R/K)G(F/Y)(G/A)(F/Y)Vx(F/Y)) [65]. The vRNA serves as a template for the 3' elongation of the cellular 10-13 nucleotide-capped primer. The overall cap snatching process results in the transcription of a chimeric full-length vRNA with a 5'-extension of 10-13 cellular nucleotides and a cap-2 structure (Fig. 8). Cap snatching enables viruses to acquire their hosts cap structure, which not only promotes viral replication but also impairs cellular mRNA translation, as translation of decapped cellular mRNA is impeded and the mRNA is targeted for degradation. Another consequence of cap snatching is the dependency on a pool of host mRNA molecules in order to support viral replication. (-)ssRNA viruses that utilize cap snatching have evolved ways to maintain the precious pool of eukaryotic mRNA. First, the cap binding and endonu-clease activity of the trimeric polymerase are only activated upon vRNA binding, limiting the waste (induced by the cleavage and downstream degradation) of mRNA when the vRNA are not loaded on the RDRP [66]. Secondly, some nucleocapsid proteins, first demonstrated by Hantavirus, are able to bind and protect capped mRNA from degradation in the processing bodies (P-bodies) [67]. Thus, converting the P-bodies function from mRNA decapping and decay into cellular cap storage foci. The cap snatching is only observed in segmented (-)ssRNA viruses; such a unique molecular mechanism supports the hypothesis of a common (-)ssRNA virus ancestor of today's virus, despite their tropism now ranging from plants to animals.

The incredible diversity of RNA capping pathways, protein folding and enzymatic mechanisms of action that have been evolved by viruses all lead to the synthesis of the same ubiquitous structure is a testimony to the importance of the cap structure for viral genome replication and global viral fitness.

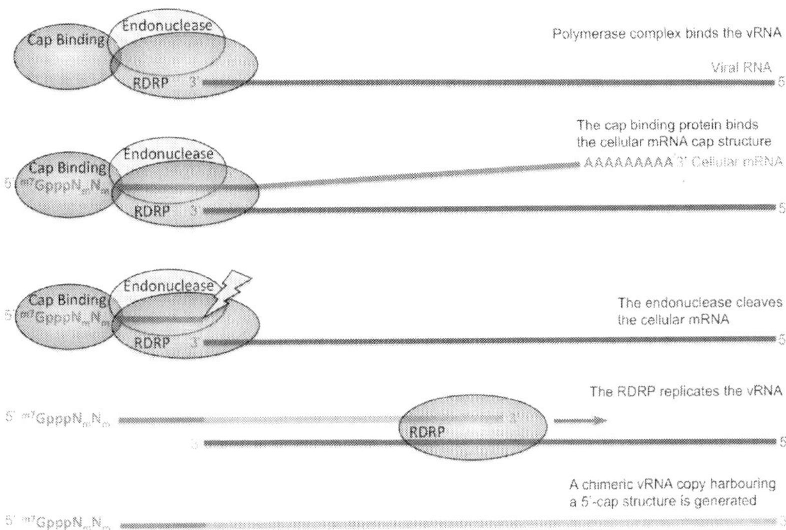

Figure 8. RNA cap snatching. The viral polymerase complex is activated upon viral RNA binding (dark blue) and specifically binds the cellular mRNA cap structure (red) via its cap binding activity. The endonuclease activity cleaves the bound cellular mRNA 10-13 nucleotides downstream of the cap structure. This short capped oligomer is then used to prime the RDRP and initiate genome replication, resulting in a chimeric (red and light blue) RNA copy harbouring the host cap structure (in this example a cap-2 structure).

4. Viral alternatives to cap structures

Most viruses harbour a cap structure at the 5′-end of their RNA. Mutations preventing the proper capping of their RNA result in infection or replication deficient viruses. This is a strong proof of the crucial importance of the cap structure for viral RNA stability and translation. Yet not all viruses harbour capped RNA, which raises the question about the mechanism they evolved to overcome this cap dependency? To answer this query it's important to ask whether it is the cap structure itself or its function that is essential. In fact, the cap structure is important for a number of different cellular processes related to mRNA metabolism. For instance, the cap structure protects the RNA from $5'\rightarrow3'$ exonucleases, preventing their degradation. The RNA cap structure also represents a definite molecular structure that is specifically recognized by the eukaryotic initiation factor 4E (eIF4E), which, together with the scaffold protein eIF4G, the RNA helicase eIF4A and the ribosome binding protein eIF3, promote RNA translation initiation. While most viruses use a cap structure to fulfill these important roles, some viruses have evolved cap-independent strategies to ensure the stability and translation of their RNA.

4.1. Viral proteins as substitutes for the cap structure

Viruses of the *Picornaviridae* (e.g. Poliovirus, Hepatitis A virus), *Potyviridae* and *Caliciviridae* (e.g. Norwalk virus, Feline calicivirus) families bear a special type of RNA 5'-end modification. The RNA 5'end of these (+)ssRNA viruses is covalently linked to a viral protein [68]. This viral genome-linked protein (VPg) is not added to the viral genome upon replication, like a regular cap structure, but is instead directly used by the RDRP as a primer to initiate RNA polymerisation. VPg is a representative of the class II nucleic acid-protein complex and does not catalyze its own covalent complex formation (like GTase or PRNTase could do) [69]. The VPg-RNA formation is instead catalyzed by a second protein, the viral RDRP, which synthesizes the primer in a template-dependent matter, resulting in a virus specific initiating primer, VPg-pUpU for *Picornaviridae* and VPg-pGpU for *Calicivirus* [70]. VPg is covalently linked to the first RNA nucleotide via a phosphodiester bond between the RNA α-phosphate and the tyrosine hydroxyl group situated in the conserved motif (E/D)EYDE(Y/W/F)[71]. The VPg protein protects the vRNA 5'-end from the cellular $5'\rightarrow3'$ exounucleases, thus limiting the vRNA degradation. Furthermore, the VPg is used to initiate the RNA polymerisation instead of being added once the RNA is synthesised. This prevents the formation of 5'-triphosphate vRNA and limits the cellular anti-viral response, which will be described later [68]. In addition to their protective role against RNA degradation, some VPg can fulfill a second important role of the cap structure, promoting the vRNA translation initiation. This is the case of the *Caliciviridae* and *Potyviridae* 15 kDA VPg that is essential for vRNA translation initiation. This VPg directly interacts with eIF4E (the cap-binding protein) and the eIF3 complex (the 40S binding complex), which promotes the assembly of the translation initiation complex to the 5'-end of the vRNA (Fig. 9) [68, 72-75]. This allows VPg-vRNA to bypass the requirement for a direct eIF4E-cap interaction in order to initiate translation. This property is not conserved among all VPg, the *Picornaviridae* VPg is much smaller (2.5 kDA) and is not involved in the vRNA translation initiation [68]. These viruses instead rely on a highly structured RNA sequence called an internal ribosome entry site (IRES) to ensure their translation (this will also be described in more detail later on). All the (+)ssRNA viruses encoding a VPg benefit from its protective effect on the viral genome, but the *Caliciviridae* and *Potyviridae* VPg have evolved an additional function, promoting vRNA translation initiation. This VPs is a striking example of a cap substitute as it fulfills two critical functions of the cap structure, namely ensuring vRNA stability and promoting translation initiation.

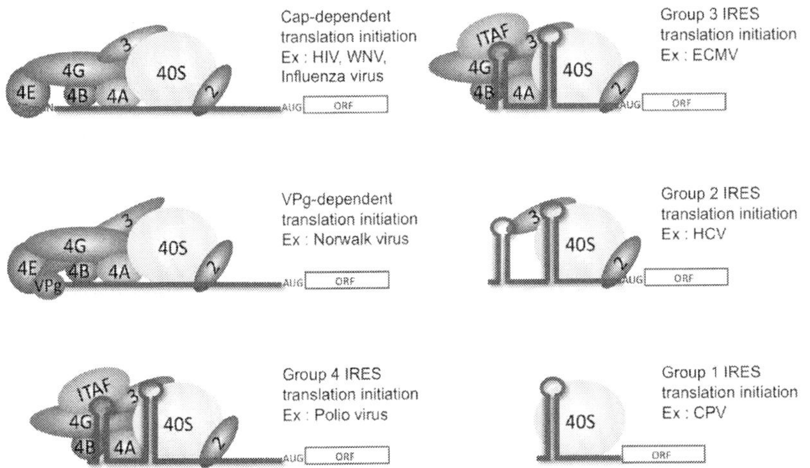

Figure 9. Mechanisms of viral translation initiation. The cap structure of RNA is specifically recognized by eIF4E, which recruits the cap-dependent translation initiation complex to the 5′-end of capped RNA. Alternatively, the viral VPg (covalently linked at the 5′-end of vRNA) can directly recruit eIF4E. The conserved viral RNA structure, located within the 5′-UTR, can directly promote translation initiation. These internal ribosome entry sites (IRES) are categorized into four different groups, each able to directly recruit a subset of initiation factors and the 40S ribosomal subunit in order to initiate translation. Of notice, some mechanisms of initiation require RNA scanning (left panel) by the initiation complex to reach the ORF while others position the ribosome directly adjacent to the transcription initiation site (right panel). Abbreviation : Human Immunodeficiency virus (HIV), West Nile virus (WNV), Encephalomyocarditis virus (ECMV), Hepatitis C virus (HCV) and Cricket paralysis virus (CPV), Open reading frame (ORF).

4.2. Highly structured 5′ RNA structure as an alternative to the cap structure

The ribonucleic acid (RNA) is a macromolecule which, according to the central dogma of molecular biology, is a transient messenger carrying the genetic information required to pilot the protein synthesis. In addition to this canonical role, RNA, given its high chemical complexity, can fulfill additional roles including genome support, ordered three-dimensional structure and even catalytic activity [76]. Many viruses have exploited this capacity of RNA to form complex structure in order to promote viral replication. Some viruses, lacking enzymatic activity to synthesize or acquire a cap structure at the 5′-end of their vRNA, have instead selected a high-order structural RNA element upstream of their coding region. This peculiar RNA sequence can fold precisely and repeatedly into a definite three-dimensional structure. This ordered structure has numerous functions including binding to other macromolecule partners. Those viruses use this *cis*-acting structure to bind directly or indirectly to ribosomal components in order to assemble the translation initiation complex at the beginning of their open reading frame (ORF). This promotes the cap-independent translation of viral genes. Such RNA structures bypassing the cap-dependency for translation initiation are called internal ribosome entry site (IRES). Many RNA virus families (e.g. *Dicistroviridae*, *Picornaviridae* and some *Flaviviridae*) use this structure to promote viral protein production. The diversity of viruses that have evolved distinct IRES structures can be divided into four categories that differ in their structure, length, mechanism of ribosome recruitment and robustness (Fig.9). The first group of IRES, which is the smallest and simplest, is encoded into the *Dicistroviridae* (e.g. Cricket paralysis virus) genome. This IRES consists of a 180 nt structure that is able to directly bind and recruit the 40S ribosomal subunit to the translation initiation site, and does not require any initiation factors nor methionyl-tRNA to initiate translation (Fig. 9) [77, 78]. The second group of IRES is similar to the first, but slightly larger with 330 nt. These include *Flaviviridae* of the *Hepacivirus* (e.g. Hepatitis C virus) and *Pestivirus* (e.g. Classical swine fever virus) genus. The second group of IRES is also able to directly bind the 40S ribosomal subunit, but requires the contribution of a limited number of initiation factors (eIF2 and eIF3) together with the methionyl-tRNA in order to initiate the vRNA translation [77-79]. Of notice, the RNA helicase eIF4A is not required for initiation of the group 1 or 2 IRES, an advantage that comes at the expense of a limited RNA unwinding capacity. Therefore the initial coding sequence of the ORF must be encoded by a non-structured RNA sequence, as an RNA structure will block translation initiation in the absence of helicase activity [78]. The *Picornaviridae* family viruses harbour IRES from the third and fourth groups and are similar in many regards. They are the largest IRES (450 nt) and the most complex. They do not directly bind the 40S ribosomal subunit and require canonical eIFs (eIF2, eIF3, eIF4A, eIF4B, eIF4G) together with additional proteins called IRES trans-activating factors (ITAFs) in order to recruit the ribosome and initiate translation [80]. The difference between these two groups lies in the positioning of the ribosome relative to the ORF. Group 3, found in the *Aphthovirus* (e.g. Foot-and-mouth disease virus) and *Cardiovirus* (e.g. Encephalomyocarditis virus) genera, recruits the ribosome at the initiating AUG codon. Group 4, found in the *Enterovirus* (e.g. poliovirus) and *Hepatovirus* (e.g. Hepatitis A virus) genera, recruits the ribosome upstream from the ORF and requires a scanning or shunting process to move along the RNA in order to reach the AUG codon and initiate translation [77, 78]. Of notice, those viral IRES (with the exception the of Hepatitis A virus IRES) are able to bypass the requirement for eIF4E, one of the limiting components of the cap-dependent translation initiation complex, to initiate their downstream ORF translation [81]. Encoding an IRES into the viral genome is an efficient mechanism evolved by viruses to fulfill a critical role of the cap structure, namely the translation initiation. The importance of this structure is exemplified by its remarkable degree of conservation. The case of the *Flaviviridae* family presents an interesting example: the members of the *Hepacivirus* and *Pestivirus* genera share a much closer homology between their IRES region than between their coding region, while members of the *Flavivirus* genus, do not have any IRES at all and synthesize a cap structure through a conventional viral RNA capping mechanism [78]. The emergence of viral alternatives to overcome the lack of a cap structure is a testimony to the crucial functions of this small structure for viral genome stability, replication and translation.

5. Recognition of the 5'-ends by the innate immune system

In humans, the RNA cap structure harbors additional methylations at the 2'-*O* site of the first and second transcribed nucleotides of the mRNAs [82]. The addition of these supplementary ribose methylations occurs via enzymatic activities located in the nucleus and cytoplasm, respectively [83, 84]. Similarly, many different viruses possess RNA 2′-O-methyltransferases in order to modify their mRNAs. The role of these methylations has however remained elusive until recently when it was demonstrated that 2′-O methylation of viral mRNAs enhances virulence through evasion of intrinsic cellular defense mechanisms [85, 86].

5.1. Innate immune response

Viral infection normally results in the generation of immunological non-self RNA species. Pattern recognition receptors are a crucial component of innate immunity that are responsible for the detection of non-self RNAs [87]. Toll-like receptors (TLRs), retinoic acid inducible gene-I (RIG-I)-like receptors (RLRs) and nucleotide oligomerization domain (Nod)-like receptors (NLRs) are important pattern recognition receptors that recognize non-self nucleic acids of pathogens [88-90]. For instance, many TLRs can detect viral nucleic acids that are found in endosomes following the release of nucleic acids from infected cells [91-95]. This eventually leads to the activation of subsequent immune reactions. In contrast, RLRs detect viral nucleic acids in the cytoplasm of the infected cells during the early phase of viral replication [96, 97]. This detection leads to the induction of interferons and inflammatory cytokines which ultimately block viral replication and promote the activation of antigen-presenting cells in order to eliminate infected cells [98].

RIG-I, MDA5, and LGP2 are important RLRs that can detect cytoplasmic viral RNAs and induce the expression of cytokines in order to establish a host antiviral state through the expression of numerous interferon-stimulated genes (ISGs) [98]. These include the protein kinase PKR and stress-inducible proteins such as IFIT1 and IFIT2 that can inhibit the protein synthesis machinery of the host cell [99-101]. What is the exact molecular signature found on viral RNAs that is detected by RLRs? Previous experiments demonstrated that RIG-I specifically recognizes 5′-triphosphate groups that can be found on some viral RNAs [102-104]. Viruses must therefore hide or modify their RNA 5′-ends in order to evade the innate immune recognition through the addition of an RNA cap structure or through the addition of alternative 5′ elements, such as viral proteins linked to the 5′end in order to hide their uncapped ends. This last strategy is used for instance by poliovirus which encodes a protein, VPg, which is covalently linked to the 5′ end of the plus-strand genomic RNA [105]. Viruses that are unable to maturate their RNA 5′-end have instead evolved immune-evasion strategies to prevent ISGs induction. For instance, the Hepacivirus protease inhibits the signal transduction resulting from RIG-I activation [106, 107].

5.2. Importance of the RNA cap 2′-O-methylation

Recent studies suggest that 2′-O-methylation of viral RNAs can enhance the replication of viruses through evasion of the innate immune response [85, 86]. For instance, coronaviruses that lack a functional 2′-O-methyltransferase activity induce a higher expression level of type I interferon [86]. Moreover, these mutant viruses can replicate efficiently in the absence of some RLRs such as MDA5 [86]. Similarly, poxvirus and coronavirus mutants that lack 2′-O-methyltransferase activities show an enhanced sensitivity to IFIT proteins. Therefore, it appears that

2'-O-methylation of cellular mRNAs has evolved as a molecular signature in order to distinguish between self and non-self RNA during viral infection, and that ribose 2'-O-methylation in the cap structure of viral RNAs plays an important role in viral escape from innate immune recognition. Not surprisingly, it has been suggested that the development of pharmacological strategies that could inhibit viral 2'-O-methyltransferases could represent a novel therapy against viruses that replicate in the cytoplasm of infected cells [85]. In fact, it was previously shown that mutations of the 2'-O-methyltransferase catalytic residues can block or attenuate replication [22, 32] and that viral inhibitors such as sinefungin can inhibit methylation and suppress the replication of certain viruses, such as West Nile virus, in cell culture [108].

6. Conclusion

This chapter explored the viral diversity of enzymatic activities and mechanistic pathways converging to the maturation of the 5' cap on viral RNA. The cap structure provides tremendous advantages to eukaryotic viruses in terms of vRNA stability, gene translation and immune evasion. Some viruses have evolved enzymatic mechanisms of action unknown to the eukaryotic domain in order to synthesize this critical structure. Other viruses have developed novel cap synthesis mechanisms that generate a 5' cap structure chemically identical to their hosts, yet formed by an entirely new process. Finally, particular viruses have also evolved unique mechanisms to steal or mimic the host cap structure. In conclusion, the incredible diversity and conservation of the mechanisms evolved by viruses to synthesize, acquire or mimic the 5' cap structure is a testimony to the importance of viral RNA capping for viral replication, fitness and infectivity.

Acknowledgements

M.B. is a 'Chercheur Boursier' Senior from the Fonds de Recherche en Santé du Québec and a member of the Centre de Recherche Clinique Étienne-Lebel.

Author details

Frédéric Picard-Jean, Maude Tremblay-Létourneau, Elizabeth Serra, Christina Dimech, Helene Schulz, Mathilde Anselin, Vincent Dutilly and Martin Bisaillon*

*Address all correspondence to: Martin.Bisaillon@USherbrooke.ca

Département de Biochimie, Faculté de Médecine et des Sciences de la Santé, Université de Sherbrooke, Sherbrooke, Québec, Canada

References

[1] Fischer, P. M. Cap in hand: targeting eIF4E. Cell Cycle. (2009). , 8(16), 2535-41.

[2] Baron-benhamou, J, Fortes, P, Inada, T, Preiss, T, & Hentze, M. W. The interaction of the cap-binding complex (CBC) with eIF4G is dispensable for translation in yeast. RNA. (2003). , 9(6), 654-62.

[3] Rhoads, R. E. eIF4E: new family members, new binding partners, new roles. J Biol Chem. (2009). , 284(25), 16711-5.

[4] Sonenberg, N, & Eif, E. the mRNA cap-binding protein: from basic discovery to translational research. Biochem Cell Biol. (2008). , 86(2), 178-83.

[5] Decroly, E, Ferron, F, Lescar, J, & Canard, B. Conventional and unconventional mechanisms for capping viral mRNA. Nat Rev Microbiol. (2012). , 10(1), 51-65.

[6] Lima, C. D, Wang, L. K, & Shuman, S. Structure and mechanism of yeast RNA triphosphatase: an essential component of the mRNA capping apparatus. Cell. (1999). , 99(5), 533-43.

[7] Benarroch, D, Smith, P, & Shuman, S. Characterization of a trifunctional mimivirus mRNA capping enzyme and crystal structure of the RNA triphosphatase domain. Structure. (2008). , 16(4), 501-12.

[8] Bisaillon, M, & Shuman, S. Structure-function analysis of the active site tunnel of yeast RNA triphosphatase. J Biol Chem. (2001). , 276(20), 17261-6.

[9] Jayaram, H, Taraporewala, Z, Patton, J. T, & Prasad, B. V. Rotavirus protein involved in genome replication and packaging exhibits a HIT-like fold. Nature. (2002). , 417(6886), 311-5.

[10] Vasquez-Del Carpio RGonzalez-Nilo FD, Riadi G, Taraporewala ZF, Patton JT. Histidine triad-like motif of the rotavirus NSP2 octamer mediates both RTPase and NTPase activities. J Mol Biol. (2006). , 362(3), 539-54.

[11] Carpio, R. V, Gonzalez-nilo, F. D, Jayaram, H, Spencer, E, Prasad, B. V, Patton, J. T, et al. Role of the histidine triad-like motif in nucleotide hydrolysis by the rotavirus RNA-packaging protein NSP2. J Biol Chem. (2004). , 279(11), 10624-33.

[12] Luo, D, Xu, T, Hunke, C, Gruber, G, Vasudevan, S. G, & Lescar, J. Crystal structure of the NS3 protease-helicase from dengue virus. J Virol. (2008). , 82(1), 173-83.

[13] Benarroch, D, Selisko, B, Locatelli, G. A, Maga, G, Romette, J. L, & Canard, B. The RNA helicase, nucleotide 5'-triphosphatase, and RNA 5'-triphosphatase activities of Dengue virus protein NS3 are Mg2+-dependent and require a functional Walker B motif in the helicase catalytic core. Virology. (2004). , 328(2), 208-18.

[14] Hanson, P. I, & Whiteheart, S. W. AAA+ proteins: have engine, will work. Nat Rev Mol Cell Biol. (2005). , 6(7), 519-29.

[15] Changela, A, Ho, C. K, Martins, A, Shuman, S, & Mondragon, A. Structure and mechanism of the RNA triphosphatase component of mammalian mRNA capping enzyme. EMBO J. (2001). , 20(10), 2575-86.

[16] Changela, A, Martins, A, Shuman, S, & Mondragon, A. Crystal structure of baculovirus RNA triphosphatase complexed with phosphate. J Biol Chem. (2005). , 280(18), 17848-56.

[17] Shuman, S, & Lima, C. D. The polynucleotide ligase and RNA capping enzyme superfamily of covalent nucleotidyltransferases. Curr Opin Struct Biol. (2004). , 14(6), 757-64.

[18] Chu, C, Das, K, Tyminski, J. R, Bauman, J. D, Guan, R, Qiu, W, et al. Structure of the guanylyltransferase domain of human mRNA capping enzyme. Proc Natl Acad Sci U S A. (2011). , 108(25), 10104-8.

[19] Hakansson, K, Doherty, A. J, & Shuman, S. Wigley DB. X-ray crystallography reveals a large conformational change during guanyl transfer by mRNA capping enzymes. Cell. (1997). , 89(4), 545-53.

[20] Hakansson, K, & Wigley, D. B. Structure of a complex between a cap analogue and mRNA guanylyl transferase demonstrates the structural chemistry of RNA capping. Proc Natl Acad Sci U S A. (1998). , 95(4), 1505-10.

[21] Issur, M, Geiss, B. J, Bougie, I, Picard-jean, F, Despins, S, Mayette, J, et al. The flavivirus NS5 protein is a true RNA guanylyltransferase that catalyzes a two-step reaction to form the RNA cap structure. RNA. (2009). , 15(12), 2340-50.

[22] Zhou, Y, Ray, D, Zhao, Y, Dong, H, Ren, S, Li, Z, et al. Structure and function of flavivirus NS5 methyltransferase. J Virol. (2007). , 81(8), 3891-903.

[23] De La Pena, M, Kyrieleis, O. J, & Cusack, S. Structural insights into the mechanism and evolution of the vaccinia virus mRNA cap N7 methyl-transferase. EMBO J. (2007). , 26(23), 4913-25.

[24] Martin, J. L, & Mcmillan, F. M. SAM (dependent) I AM: the S-adenosylmethionine-dependent methyltransferase fold. Curr Opin Struct Biol. (2002). , 12(6), 783-93.

[25] Fabrega, C, Hausmann, S, Shen, V, Shuman, S, & Lima, C. D. Structure and mechanism of mRNA cap (guanine-N7) methyltransferase. Mol Cell. (2004). , 13(1), 77-89.

[26] Mao, X, & Shuman, S. Intrinsic RNA (guanine-7) methyltransferase activity of the vaccinia virus capping enzyme D1 subunit is stimulated by the D12 subunit. Identification of amino acid residues in the D1 protein required for subunit association and methyl group transfer. J Biol Chem. (1994). , 269(39), 24472-9.

[27] Egloff, M. P, Benarroch, D, Selisko, B, Romette, J. L, & Canard, B. An RNA cap (nucleoside-2′-O-)-methyltransferase in the flavivirus RNA polymerase NS5: crystal structure and functional characterization. EMBO J. (2002). , 21(11), 2757-68.

[28] Decroly, E, Debarnot, C, Ferron, F, Bouvet, M, Coutard, B, Imbert, I, et al. Crystal structure and functional analysis of the SARS-coronavirus RNA cap 2′-O-methyltransferase nsp10/nsp16 complex. PLoS Pathog. (2011). e1002059.

[29] Dong, H, Zhang, B, & Shi, P. Y. Flavivirus methyltransferase: a novel antiviral target. Antiviral Res. (2008). , 80(1), 1-10.

[30] Li, C, Xia, Y, Gao, X, & Gershon, P. D. Mechanism of RNA 2′-O-methylation: evidence that the catalytic lysine acts to steer rather than deprotonate the target nucleophile. Biochemistry. (2004). , 43(19), 5680-7.

[31] Hodel, A. E, Gershon, P. D, & Quiocho, F. A. Structural basis for sequence-nonspecific recognition of 5′-capped mRNA by a cap-modifying enzyme. Mol Cell. (1998). , 1(3), 443-7.

[32] Ray, D, Shah, A, Tilgner, M, Guo, Y, Zhao, Y, Dong, H, et al. West Nile virus 5′-cap structure is formed by sequential guanine N-7 and ribose 2′-O methylations by nonstructural protein 5. J Virol. (2006). , 80(17), 8362-70.

[33] Rahmeh, A. A, Li, J, Kranzusch, P. J, & Whelan, S. P. Ribose 2′-O methylation of the vesicular stomatitis virus mRNA cap precedes and facilitates subsequent guanine-N-7 methylation by the large polymerase protein. J Virol. (2009). , 83(21), 11043-50.

[34] Ho, C. K, Gong, C, & Shuman, S. RNA triphosphatase component of the mRNA capping apparatus of Paramecium bursaria Chlorella virus 1. J Virol. (2001). , 75(4), 1744-50.

[35] Ho, C. K, Van Etten, J. L, & Shuman, S. Expression and characterization of an RNA capping enzyme encoded by Chlorella virus PBCV-1. J Virol. (1996). , 70(10), 6658-64.

[36] Ho, C. K, Lehman, K, & Shuman, S. An essential surface motif (WAQKW) of yeast RNA triphosphatase mediates formation of the mRNA capping enzyme complex with RNA guanylyltransferase. Nucleic Acids Res. (1999). , 27(24), 4671-8.

[37] Gu, M, Rajashankar, K. R, & Lima, C. D. Structure of the Saccharomyces cerevisiae Cet1-Ceg1 mRNA capping apparatus. Structure. (2010). , 18(2), 216-27.

[38] Gross, C. H, & Shuman, S. RNA 5′-triphosphatase, nucleoside triphosphatase, and guanylyltransferase activities of baculovirus LEF-4 protein. J Virol. (1998). , 72(12), 10020-8.

[39] He, J. G, Deng, M, Weng, S. P, Li, Z, Zhou, S. Y, Long, Q. X, et al. Complete genome analysis of the mandarin fish infectious spleen and kidney necrosis iridovirus. Virology. (2001). , 291(1), 126-39.

[40] Ho, C. K, Sriskanda, V, Mccracken, S, Bentley, D, Schwer, B, & Shuman, S. The guanylyltransferase domain of mammalian mRNA capping enzyme binds to the phos-

phorylated carboxyl-terminal domain of RNA polymerase II. J Biol Chem. (1998). , 273(16), 9577-85.

[41] Ahola, T, Laakkonen, P, Vihinen, H, & Kaariainen, L. Critical residues of Semliki Forest virus RNA capping enzyme involved in methyltransferase and guanylyltransferase-like activities. J Virol. (1997). , 71(1), 392-7.

[42] Pena, L, Yanez, R. J, Revilla, Y, Vinuela, E, & Salas, M. L. African swine fever virus guanylyltransferase. Virology. (1993). , 193(1), 319-28.

[43] Sutton, G, Grimes, J. M, Stuart, D. I, & Roy, P. Bluetongue virus VP4 is an RNA-capping assembly line. Nat Struct Mol Biol. (2007). , 14(5), 449-51.

[44] Reinisch, K. M, Nibert, M. L, & Harrison, S. C. Structure of the reovirus core at 3.6 A resolution. Nature. (2000). , 404(6781), 960-7.

[45] Cheng, L, Sun, J, Zhang, K, Mou, Z, Huang, X, Ji, G, et al. Atomic model of a cypovirus built from cryo-EM structure provides insight into the mechanism of mRNA capping. Proc Natl Acad Sci U S A. (2011). , 108(4), 1373-8.

[46] Nason, E. L, Rothagel, R, Mukherjee, S. K, Kar, A. K, Forzan, M, Prasad, B. V, et al. Interactions between the inner and outer capsids of bluetongue virus. J Virol. (2004). , 78(15), 8059-67.

[47] Ghosh, A, & Lima, C. D. Enzymology of RNA cap synthesis. Wiley Interdisciplinary Reviews- RNA. (2010). , 1(1), 152-72.

[48] Vasiljeva, L, Merits, A, Auvinen, P, & Kaariainen, L. Identification of a novel function of the alphavirus capping apparatus. RNA 5'-triphosphatase activity of Nsp2. J Biol Chem. (2000). , 275(23), 17281-7.

[49] Ahola, T, & Kaariainen, L. Reaction in alphavirus mRNA capping: formation of a covalent complex of nonstructural protein nsP1 with 7-methyl-GMP. Proc Natl Acad Sci U S A. (1995). , 92(2), 507-11.

[50] Li, Y. I, Chen, Y. J, Hsu, Y. H, & Meng, M. Characterization of the AdoMet-dependent guanylyltransferase activity that is associated with the N terminus of bamboo mosaic virus replicase. J Virol. (2001). , 75(2), 782-8.

[51] Magden, J, Takeda, N, Li, T, Auvinen, P, Ahola, T, Miyamura, T, et al. Virus-specific mRNA capping enzyme encoded by hepatitis E virus. J Virol. (2001). , 75(14), 6249-55.

[52] Merits, A, Kettunen, R, Makinen, K, Lampio, A, Auvinen, P, Kaariainen, L, et al. Virus-specific capping of tobacco mosaic virus RNA: methylation of GTP prior to formation of covalent complex FEBS Lett. (1999). , 126-m7GMP.

[53] Li, J, Fontaine-rodriguez, E. C, & Whelan, S. P. Amino acid residues within conserved domain VI of the vesicular stomatitis virus large polymerase protein essential for mRNA cap methyltransferase activity. J Virol. (2005). , 79(21), 13373-84.

[54] Ogino, T, & Banerjee, A. K. Unconventional mechanism of mRNA capping by the RNA-dependent RNA polymerase of vesicular stomatitis virus. Mol Cell. (2007). , 25(1), 85-97.

[55] Ogino, T, & Banerjee, A. K. The HR motif in the RNA-dependent RNA polymerase L protein of Chandipura virus is required for unconventional mRNA-capping activity. J Gen Virol. (2010). Pt 5):1311-4.

[56] Ogino, T, Yadav, S. P, & Banerjee, A. K. Histidine-mediated RNA transfer to GDP for unique mRNA capping by vesicular stomatitis virus RNA polymerase. Proc Natl Acad Sci U S A. (2010). , 107(8), 3463-8.

[57] Testa, D, & Banerjee, A. K. Two methyltransferase activities in the purified virions of vesicular stomatitis virus. J Virol. (1977). , 24(3), 786-93.

[58] Ruigrok, R. W, Crepin, T, Hart, D. J, & Cusack, S. Towards an atomic resolution understanding of the influenza virus replication machinery. Curr Opin Struct Biol. (2010). , 20(1), 104-13.

[59] Lamb, R. A, & Horvath, C. M. Diversity of coding strategies in influenza viruses. Trends Genet. (1991). , 7(8), 261-6.

[60] Guilligay, D, Tarendeau, F, Resa-infante, P, Coloma, R, Crepin, T, Sehr, P, et al. The structural basis for cap binding by influenza virus polymerase subunit PB2. Nat Struct Mol Biol. (2008). , 15(5), 500-6.

[61] Dias, A, Bouvier, D, Crepin, T, Mccarthy, A. A, Hart, D. J, Baudin, F, et al. The cap-snatching endonuclease of influenza virus polymerase resides in the PA subunit. Nature. (2009). , 458(7240), 914-8.

[62] Weber, F, Haller, O, & Kochs, G. Nucleoprotein viral RNA and mRNA of Thogoto virus: a novel "cap-stealing" mechanism in tick-borne orthomyxoviruses? J Virol. (1996). , 70(12), 8361-7.

[63] Yuan, P, Bartlam, M, Lou, Z, Chen, S, Zhou, J, He, X, et al. Crystal structure of an avian influenza polymerase PA(N) reveals an endonuclease active site. Nature. (2009). , 458(7240), 909-13.

[64] Crepin, T, Dias, A, Palencia, A, Swale, C, Cusack, S, & Ruigrok, R. W. Mutational and metal binding analysis of the endonuclease domain of the influenza virus polymerase PA subunit. J Virol. (2010). , 84(18), 9096-104.

[65] Li, M. L, Ramirez, B. C, & Krug, R. M. RNA-dependent activation of primer RNA production by influenza virus polymerase: different regions of the same protein subunit constitute the two required RNA-binding sites. EMBO J. (1998). , 17(19), 5844-52.

[66] Sugiyama, K, Obayashi, E, Kawaguchi, A, Suzuki, Y, Tame, J. R, Nagata, K, et al. Structural insight into the essential PB1-PB2 subunit contact of the influenza virus RNA polymerase. EMBO J. (2009). , 28(12), 1803-11.

[67] Mir, M. A, Duran, W. A, Hjelle, B. L, Ye, C, & Panganiban, A. T. Storage of cellular 5′ mRNA caps in P bodies for viral cap-snatching. Proc Natl Acad Sci U S A. (2008). , 105(49), 19294-9.

[68] Goodfellow, I. The genome-linked protein VPg of vertebrate viruses- a multifaceted protein. Curr Opin Virol. (2011). , 1(5), 355-62.

[69] Drygin, Y. F. Natural covalent complexes of nucleic acids and proteins: some comments on practice and theoryon the path from well-known complexes to new ones. Nucleic Acids Res. (1998). , 26(21), 4791-6.

[70] Steil, B. P, & Barton, D. J. Cis-active RNA elements (CREs) and picornavirus RNA replication. Virus Res. (2009). , 139(2), 240-52.

[71] Gavryushina, E. S, Bryantseva, S. A, Nadezhdina, E. S, Zatsepin, T. S, Toropygin, I. Y, Pickl-herk, A, et al. Immunolocalization of Picornavirus RNA in infected cells with antibodies to Tyr-pUp, the covalent linkage unit between VPg and RNA. J Virol Methods. (2011). , 171(1), 206-11.

[72] Goodfellow, I, Chaudhry, Y, Gioldasi, I, Gerondopoulos, A, Natoni, A, Labrie, L, et al. Calicivirus translation initiation requires an interaction between VPg and eIF 4 E. EMBO Rep. (2005). , 6(10), 968-72.

[73] Chaudhry, Y, Nayak, A, Bordeleau, M. E, Tanaka, J, Pelletier, J, Belsham, G. J, et al. Caliciviruses differ in their functional requirements for eIF4F components. J Biol Chem. (2006). , 281(35), 25315-25.

[74] Daughenbaugh, K. F, Fraser, C. S, Hershey, J. W, & Hardy, M. E. The genome-linked protein VPg of the Norwalk virus binds eIF3, suggesting its role in translation initiation complex recruitment. EMBO J. (2003). , 22(11), 2852-9.

[75] Herbert, T. P, Brierley, I, & Brown, T. D. Identification of a protein linked to the genomic and subgenomic mRNAs of feline calicivirus and its role in translation. J Gen Virol. (1997). Pt 5):1033-40.

[76] Eddy, S. R. Non-coding RNA genes and the modern RNA world. Nat Rev Genet. (2001). , 2(12), 919-29.

[77] Kieft, J. S. Viral IRES RNA structures and ribosome interactions. Trends Biochem Sci. (2008). , 33(6), 274-83.

[78] Jackson, R. J. Alternative mechanisms of initiating translation of mammalian mRNAs. Biochem Soc Trans. (2005). Pt 6):1231-41.

[79] Martinez-salas, E, Pacheco, A, Serrano, P, & Fernandez, N. New insights into internal ribosome entry site elements relevant for viral gene expression. J Gen Virol. (2008). Pt 3):611-26.

[80] King, H. A, Cobbold, L. C, & Willis, A. E. The role of IRES trans-acting factors in regulating translation initiation. Biochem Soc Trans. (2010). , 38(6), 1581-6.

[81] Ali, I. K, Mckendrick, L, Morley, S. J, & Jackson, R. J. Activity of the hepatitis A virus IRES requires association between the cap-binding translation initiation factor (eIF4E) and eIF4G. J Virol. (2001). , 75(17), 7854-63.

[82] Furuichi, Y, & Shatkin, A. J. Viral and cellular mRNA capping: past and prospects. Adv Virus Res. (2000). , 55, 135-84.

[83] Wei, C. M, Gershowitz, A, & Moss, B. Methylated nucleotides block 5′ terminus of HeLa cell messenger RNA. Cell. (1975). , 4(4), 379-86.

[84] Langberg, S. R, & Moss, B. Post-transcriptional modifications of mRNA. Purification and characterization of cap I and cap II RNA (nucleoside-2′-)-methyltransferases from HeLa cells. J Biol Chem. (1981). , 256(19), 10054-60.

[85] Daffis, S, Szretter, K. J, Schriewer, J, Li, J, Youn, S, Errett, J, et al. O methylation of the viral mRNA cap evades host restriction by IFIT family members. Nature. (2010). , 468(7322), 452-6.

[86] Zust, R, Cervantes-barragan, L, Habjan, M, Maier, R, Neuman, B. W, Ziebuhr, J, et al. Ribose 2′-O-methylation provides a molecular signature for the distinction of self and non-self mRNA dependent on the RNA sensor Mda5. Nat Immunol. (2011). , 12(2), 137-43.

[87] Janeway, C. A. Jr. Approaching the asymptote? Evolution and revolution in immunology. Cold Spring Harb Symp Quant Biol. (1989). Pt , 1, 1-13.

[88] Takeuchi, O, & Akira, S. Pattern recognition receptors and inflammation. Cell. (2010). , 140(6), 805-20.

[89] Fujita, T. A nonself RNA pattern: tri-p to panhandle. Immunity. (2009). , 31(1), 4-5.

[90] Beutler, B, Eidenschenk, C, Crozat, K, Imler, J. L, Takeuchi, O, Hoffmann, J. A, et al. Genetic analysis of resistance to viral infection. Nat Rev Immunol. (2007). , 7(10), 753-66.

[91] Alexopoulou, L, Holt, A. C, Medzhitov, R, & Flavell, R. A. Recognition of double-stranded RNA and activation of NF-kappaB by Toll-like receptor 3. Nature. (2001). , 413(6857), 732-8.

[92] Heil, F, Hemmi, H, Hochrein, H, Ampenberger, F, Kirschning, C, Akira, S, et al. Species-specific recognition of single-stranded RNA via toll-like receptor 7 and 8. Science. (2004). , 303(5663), 1526-9.

[93] Diebold, S. S, Kaisho, T, Hemmi, H, & Akira, S. Reis e Sousa C. Innate antiviral responses by means of TLR7-mediated recognition of single-stranded RNA. Science. (2004). , 303(5663), 1529-31.

[94] Hemmi, H, Kaisho, T, Takeuchi, O, Sato, S, Sanjo, H, Hoshino, K, et al. Small antiviral compounds activate immune cells via the TLR7 MyD88-dependent signaling pathway. Nat Immunol. (2002). , 3(2), 196-200.

[95] Hemmi, H, Takeuchi, O, Kawai, T, Kaisho, T, Sato, S, Sanjo, H, et al. A Toll-like receptor recognizes bacterial DNA. Nature. (2000). , 408(6813), 740-5.

[96] Takeuchi, O, & Akira, S. Innate immunity to virus infection. Immunol Rev. (2009). , 227(1), 75-86.

[97] Loo, Y. M, & Gale, M. Jr. Viral regulation and evasion of the host response. Curr Top Microbiol Immunol. (2007). , 316, 295-313.

[98] Haller, O, & Weber, F. Pathogenic viruses: smart manipulators of the interferon system. Curr Top Microbiol Immunol. (2007). , 316, 315-34.

[99] Hui, D. J, Terenzi, F, Merrick, W. C, Sen, G. C, & Mouse, p. blocks a distinct function of eukaryotic initiation factor 3 in translation initiation. J Biol Chem. (2005). , 280(5), 3433-40.

[100] Terenzi, F, Hui, D. J, Merrick, W. C, & Sen, G. C. Distinct induction patterns and functions of two closely related interferon-inducible human genes, ISG54 and ISG56. J Biol Chem. (2006). , 281(45), 34064-71.

[101] Fensterl, V, & Sen, G. C. The ISG56/IFIT1 gene family. J Interferon Cytokine Res. (2011). , 31(1), 71-8.

[102] Hornung, V, Ellegast, J, Kim, S, Brzozka, K, Jung, A, Kato, H, et al. Triphosphate RNA is the ligand for RIG-I. Science. (2006). , 314(5801), 994-7.

[103] Pichlmair, A, Schulz, O, Tan, C. P, Naslund, T. I, Liljestrom, P, Weber, F, et al. RIG-I-mediated antiviral responses to single-stranded RNA bearing 5'-phosphates. Science. (2006). , 314(5801), 997-1001.

[104] Schlee, M, Roth, A, Hornung, V, Hagmann, C. A, Wimmenauer, V, Barchet, W, et al. Recognition of 5' triphosphate by RIG-I helicase requires short blunt double-stranded RNA as contained in panhandle of negative-strand virus. Immunity. (2009). , 31(1), 25-34.

[105] Nomoto, A, Detjen, B, Pozzatti, R, & Wimmer, E. The location of the polio genome protein in viral RNAs and its implication for RNA synthesis. Nature. (1977). , 268(5617), 208-13.

[106] Gale, M. Jr., Foy EM. Evasion of intracellular host defence by hepatitis C virus. Nature. (2005). , 436(7053), 939-45.

[107] Hiscott, J, Lin, R, Nakhaei, P, & Paz, S. MasterCARD: a priceless link to innate immunity. Trends Mol Med. (2006). , 12(2), 53-6.

[108] Dong, H, Ray, D, Ren, S, Zhang, B, Puig-basagoiti, F, Takagi, Y, et al. Distinct RNA elements confer specificity to flavivirus RNA cap methylation events. J Virol. (2007). , 81(9), 4412-21.

Chapter 3

Baculovirus Gene Expression

Marcelo F. Berretta, M. Leticia Ferrelli,
Ricardo Salvador, Alicia Sciocco and
Víctor Romanowski

Additional information is available at the end of the chapter

http://dx.doi.org/10.5772/56955

1. Introduction

Baculoviridae is a diverse family of insect viruses with large, double-stranded, circular DNA genomes packaged in rod-shaped, enveloped nucleocapsids. A characteristic feature of baculoviruses is the production of paracrystalline occlusion bodies (OBs) which surround the assembled virions at late times of infection. Baculoviruses produce lethal infections in their hosts and OBs protect the virions in the environment after death of the insect until uptake by another susceptible host. Other insect viruses also produce OBs such as entomopoxviruses (EPV) and cytoplasmic polyhedrosis viruses (CPV). EPV and CPV replicate in the cytoplasm of infected cells; in contrast, baculoviruses replicate within the nucleus. According to the size and shape of OBs, baculoviruses were traditionally classified into two genera: nucleopolyhedrovirus (NPV), which produce large OBs known as polyhedra, and granulovirus (GV), which produce small ovoid OBs or granules. The major proteins that form each class of OB are known as polyhedrin and granulin, respectively. Recently, after several baculoviral genomes have been sequenced, a new classification based on the phylogenetic relationships between species within the family was accepted [1-2]. Four genera were defined: *Alphabaculovirus* (lepidopteran NPV), *Betabaculovirus* (lepidopteran GV), *Gammabaculovirus* (hymenopteran NPV) and *Deltabaculovirus* (dipteran NPV). Baculoviruses are not infectious to vertebrates and they were known from far before they were recognized as viral entities as they produced disease outbreaks in the silkworm rearing. There are more than 600 species described in the literature [3] and some of them are widely used as bioinsecticides to control insect pests in agriculture and forestry [4]. A

feature that differentiates baculovirus from other DNA viruses that replicate in the nuclei of infected cells is that they encode a novel DNA-dependent RNA polymerase. This enzyme, of uncertain evolutionary origin, is responsible for transcription of baculovirus late and very late genes. At this time of the infection the transcription of most cellular genes is shutoff and the synthesis of the polyhedrin/granulin becomes prominent to finally account for up to 95% of total cellular protein production. This high capacity of protein synthesis has been exploited for the development of baculoviruses as vectors for expression of foreign proteins.

2. Two types of virus progeny serve at different steps of host invasion

Alphabaculoviruses and betabaculoviruses produce virions of two phenotypic classes: occlusion-derived virus (ODV) and budded-virus (BV). Nucleocapsids of both types of virus particles are assembled in the nucleus of the infected cell. During the late phase of infection, when proteins of the nucleocapsid are expressed, BVs are produced as the newly formed nucleocapsids exit the cell, acquiring their envelopes from the cell membrane during the budding process. BVs disseminate the infection within the host by entering other cells via a mechanism of receptor-mediated endocytosis. ODVs arise at very late times, when nucleocapsids are enveloped in membrane units derived from the nuclear envelope to finally be embedded in the OBs. OBs persist in the environment after liquefaction of the insect cadaver and are responsible for the horizontal transmission of the virus between hosts. Upon ingestion by a susceptible insect larva, OBs are dissolved in the alkaline environment of the midgut and ODVs are released. The ODVs move through the peritrophic membrane and nucleocapsids are delivered into midgut epithelial cells through a mechanism of membrane fusion mediated by specific viral proteins known as *per os* infectivity factors (PIFs). This primary infection in the midgut is followed by a secondary infection, consisting in the dissemination of BVs to other tissues. In contrast to this infection cycle, which is typical of lepidopteran-specific baculoviruses, infections of gammabaculoviruses and deltabaculoviruses are restricted to the midgut of their hosts.

3. Nuclear events associated with infection

Baculovirus infection causes cells to enlarge and stop dividing; the nucleus swells and forms the virogenic stroma (VS), which is the nuclear compartment where the viral DNA is replicated and the nucleocapsids of virus progeny are assembled. The host chromatin adopts a marginal distribution at 24 hours post infection (hpi), when the VS becomes evident. At this time, it was shown that histone H4 fused to fluorescent protein markers colocalizes with the chromatin in the periphery but not with the VS [5]. The new distribution of the chromatin is determined by the replication of the virus and may have effects in changes operated in the expression of host genes and the progression of the cell cycle. At late times, ODVs become occluded into OBs in the periphery of the VS.

Apparently, the nucleocapsids of NPVs enter the nucleus through the nuclear pores, whereas the genome of GVs is probably injected [6]. Virions of baculoviruses are devoid of histones, in turn the DNA in the nucleocapsid is packed in association with viral protein P6.9, a basic DNA-binding protein. This small polypeptide is rich in arginine, serine and threonine residues, a feature similar to proteins called protamines present in the nuclei of spermatids in many animals and plants. The positive arginine residues in protamines neutralize the negative charges in the DNA backbone while serine and threonine mediate interaction between protamine molecules, resulting in a high condensation of genomic DNA. Once it is uncoated into the nucleus, the DNA dissociates from P6.9 through the phosphorylation of the protein. During infection the viral DNA appears to be organized in the form of nucleosomal-like structures in association with P6.9, as suggested by experiments of micrococcal nuclease digestion of isolated nuclei [7].

4. Transcription program of baculovirus genes

In general, genes of DNA viruses are transcribed in a temporal sequence and the process is highly regulated by infection-derived mechanisms and proteins from both host and viral origin. This stepwise mode of gene expression ensures the availability of gene products required for the progression into the next phase of the infection. Baculoviruses express their genes in three successive phases designated as early, late and very late (figure 1). Early genes are transcribed by the host RNA polymerase II before virus DNA replication, while late and very late genes are transcribed by a virus-encoded RNA polymerase, after starting of viral DNA replication. Products of a number of early genes are required for virus DNA synthesis and for expression of late/very late genes, and at least one late gene product is also needed for expression of very late genes. The progression of the infection into the late phase correlates with the transcriptional shutoff of cellular and early viral genes. Genes belonging to the different temporal classes are encoded in both DNA strands without any associative distribution in the genome. A number of baculovirus genes contains promoters with sequence elements characteristic of both early and late classes. Their transcription is regulated independently at each temporal phase which ensures their expression throughout the infection [8]. Most of our knowledge of baculovirus gene regulation comes from studies in *Autographa californica* nucleopolyhedrovirus (AcMNPV), which is the type species of the family. This alphabaculovirus has a wide host range and causes productive infections in permissive insect cell-lines Sf21 (and its clonal isolate Sf9), derived from *Spodoptera frugiperda* [9] and TN368, from *Trichoplusia ni* [10]. The genome of AcMNPV encodes 154 predicted open reading frames (ORFs) [11]. Roughly, one half corresponds to late genes, according to the sequence elements present in their promoters. About 25 gene products, mostly of early genes, have functions directly or indirectly related to gene expression.

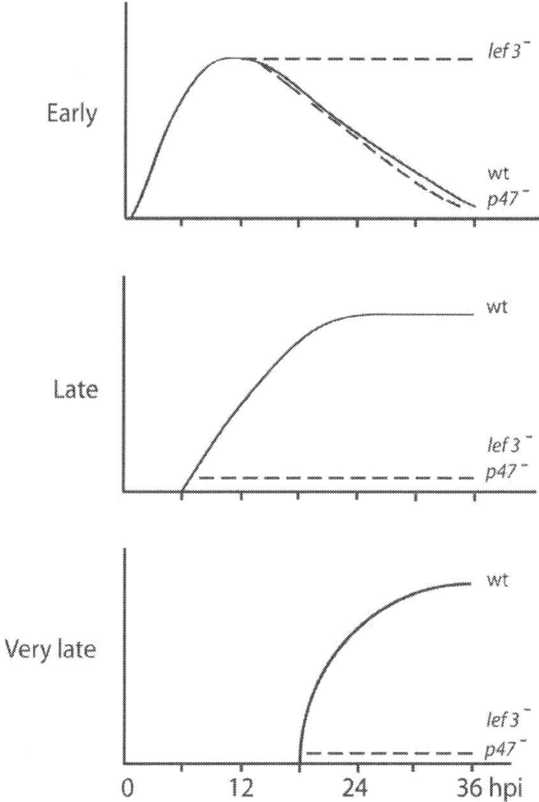

Figure 1. Kinetics of transcription of baculovirus genes. The time courses of steady-state level of transcripts (y axis) of the three AcMNPV temporal classes of genes, in Sf21 cells, are shown schematically. Solid lines represent mRNAs in wild-type (wt) infection; dotted lines represent idealized transcriptional profiles in infections in which a virus gene required for either DNA replication or late transcription was silenced by RNAi (denoted by "-"). Early genes are transcribed by the host RNA polymerase II before virus DNA replication, which initiates 6 to 9 hours post infection (hpi). Transcripts of early genes are detectable within 1 hpi, reach a maximum between 6 to 12 hpi and diminish thereafter (shutoff). Transcription of late and very late genes depends on viral DNA replication and is performed by a virus-encoded RNA polymerase. Transcription of late genes begins with the onset of DNA replication and continues mainly up to 24 hpi, while transcription of very late genes bursts around 18 hpi and continues through 72 hpi. The figure shows that late/very late gene transcription is abolish by silencing *p47*, a gene that encodes a subunit of the RNA polymerase, which is essential for the activity of the enzyme. A similar effect is shown as consequence of silencing *lef3*, a gene essential for virus DNA replication, since DNA replication is required for transcription of late/very late genes. However, the effects of silencing these genes on expression of early genes are different. When *p47* is silenced, early gene transcription declines as in wild-type infection; in contrast, silencing of *lef3* blocks the reduction of transcription, indicating that DNA replication, rather than late transcription, is a primary determinant of the shutoff of early genes (adapted from [65]).

5. Expression of early genes

At the time the virus reaches the nucleus, the template DNA for transcription as well as proteins of the virion that are carried over and may activate transcription are at their lowest levels. Therefore, the success of the infection depends on the ability of the virus to efficiently redirect the cellular system to express early genes encoded in its genome. Baculovirus DNA is infectious without any accompany protein, as proved by the infection that follows after transfection of permissive cells with viral DNA. This indicates that baculovirus early promoters are responsive to the RNA pol II and accordingly, their structural organization resembles that of the host genes which are transcribed into mRNAs.

Promoters of baculovirus early genes consist in a core region and regulatory proximal sequences that may be recognized for specific binding of transcription factors from either the host or the virus. The core promoter includes two characteristic elements: a TATA-box-like sequence and a transcription initiator (INR), although one or both are absent in some early promoters [12]. The TATA box is the site for assembly of the preinitiation transcription complex by first binding of TATA binding protein (TBP). After the RNA pol II is recruited to the complex, transcription starts about 30 nucleotides downstream of the position of the TATA-box. The INR determines the starting site for transcription (nucleotide +1 of the primary transcript), and ensures proper initiation when there is no TATA-box present. CAGT is the most conserved INR sequence motif in baculovirus early promoters. Other activating sequences can be found either upstream of the core promoter or downstream, in the 5′UTR of the regulated gene.

Besides the sequences within the promoter region that modulate expression of a gene through the binding of regulatory proteins, there are sequences that enhance transcription from promoters even if they are located at a long distance. In baculoviruses there are non-coding regions known as homologous regions (*hr*) that play the role of enhancers [13] (see below).

Baculovirus early genes can be subdivided into two categories: immediate-early (*ie*) and delayed-early genes. Expression of *ie*-genes does not require viral factors, whereas the transcription of delayed-early genes was shown to need activation by *ie*-genes in transient expression assays. The major transactivator of AcMNPV early genes is the product of *ie1*, a gene that is present in all lepidopteran baculoviruses [14].

Other *ie*-genes known to regulate the expression of early genes are *ie2* and *pe38*. *ie2* is conserved only in group I alphabaculoviruses, one the two phylogenetic lineages in which members of this genus can be separated. In transient expression experiments it was determined that *ie2* increases IE1-mediated transactivation of early promoters when *ie1* is present at low concentrations [15]

5.1. Immediate early transactivator IE1

AcMNPV IE1 is a 582 aminoacids long protein exhibiting general characteristics of transcription factors (figure 2). It has a modular organization with domains associated to different functions: dimerization, nuclear import, DNA-binding, transactivation and replication [16-17].

Dimer formation is required for nuclear localization since the protein mutated in the dimerization domain cannot be imported into the nucleus [18]. IE1 transactivates early promoters including its own. Two mechanisms are postulated by which IE1 is capable of activating an early promoter: one independent and one dependent on DNA-binding. In the first one, IE1 activates transcription by interaction with cell factors recruited to the promoter regulatory regions. In the DNA-binding-dependent mechanism, the activation depends on binding of IE1 to *hr* sequences which function as enhancers. The *hrs* contain a variable number of imperfect palindromic repeats in tandem, separated by non-palindromic sequences. The palindromes are conserved within a genome but differ widely between genomes. In AcMNPV the imperfect palindrome consensus sequence has a length of 28 nucleotides with a central *EcoRI* site. According to the current knowledge, each subunit of the IE1 dimer interacts with a corresponding hemipalindrome during binding to an *hr*. Binding of IE1 to the enhancer increases the effective concentration of IE1 molecules able to interact with cellular factors in the promoter region. It was found that *hrs* also bind cellular factors in sites overlapping palindromic repeats and within interpalindromic regions [19-20]. *Hrs* are enriched in sequence motifs similar to cAMP and TPA response elements known to interact with transcription factors of the bZIP family. This is consistent with the ability of *hrs* to enhance transcription from baculovirus early promoters even in the absence of viral factors (see [21] for a review).

It has been reported that AcMNPV IE1 down-regulates the expression of certain genes. Promoters of these genes contain a sequence motif similar to one half of a typical *hr* palindrome which still functions as a target for IE1 binding; however, IE1 bound to this sequence is no longer able to promote activation, instead it functions as a repressor [22].

IE1 is the only known baculoviral gene that is expressed as part of a product of alternative splicing designated IE0. Compared to IE1, AcMNPV IE0 contains 54 additional aminoacids at its N-terminus. Both protein species are required for an efficient infection, although each one is dispensable given the other is present [23].

6. Expression of late and very late genes

Experiments using α-amanitin, an inhibitor of RNA pol II, showed that the synthesis of virus-specific mRNA becomes resistant to the drug with the progression of the infection, indicating that a novel RNA polymerase is induced in infected cells [24-25]. In AcMNPV-infected cells this viral encoded enzyme transcribes late genes mostly from 6 to 24 hpi and very late genes between 18 and 72 hpi. These times correlate with the production of BVs and ODVs, respectively. Expression of late genes depends on viral DNA replication. When replication is blocked with the DNA synthesis inhibitor aphidicolin, transcription of late genes is also inhibited [26]. In accordance to this dependency, whose nature is not known, genes involved in DNA replication are also required for transcription of late genes, and therefore, they are considered as a subset of the factors regulating late gene expression, collectively known as late expression factors (*lefs*). The remaining *lefs*, including those encoding the multi-subunit RNA polymerase, are considered to regulate transcription-specific events. AcMNPV *lefs* were discovered using

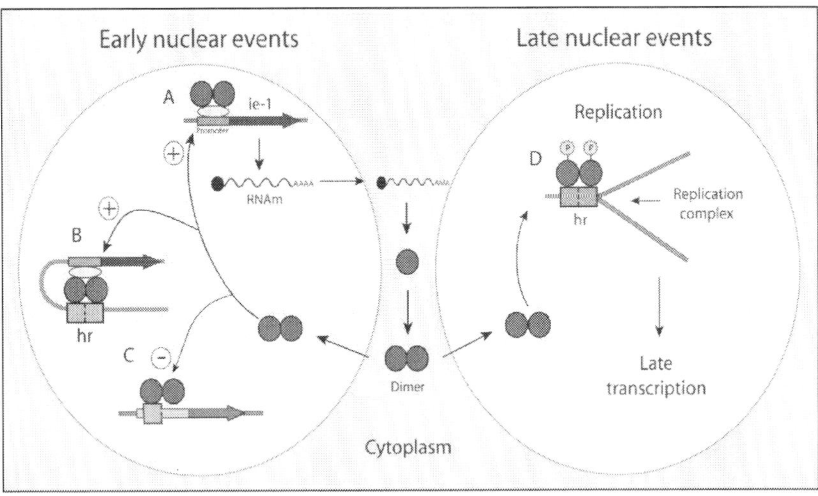

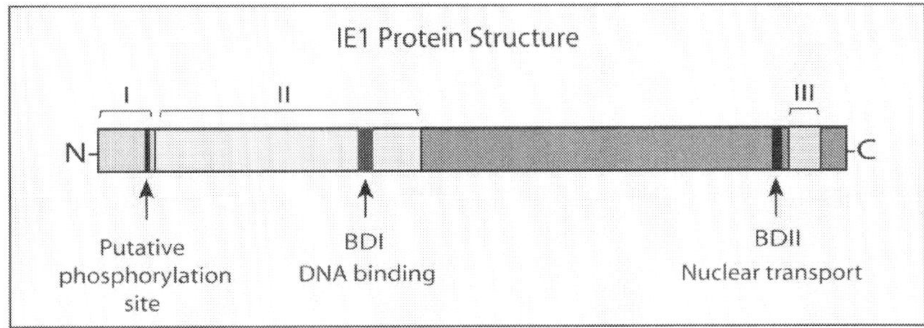

Figure 2. Functions of IE1 during the replicative cycle of AcMNPV. IE1 is targeted to the nucleus by a nuclear localization element (NLE) that becomes functional upon IE1 dimerization. The NLE is determined by a small basic domain (BDII) located at the C-terminus of the molecule adjacent to a helix-loop-helix dimerization domain. IE1 transactivates early promoters through interacting directly with transcription factors in the promoter (A) or via binding to *hr* enhancers (B). The *hr*-mediated transactivation mechanism depends on the interaction of the basic domain I (BDI) of each IE1 monomer in the dimer with a corresponding hemipalindrome of an *hr* repeat. IE1 may bind sequences within a promoter that resemble an *hr* palindrome half-site and down-regulate rather than activate transcription, as observed in transient expression assays (C). With the onset of viral DNA replication, early transcripts are down-regulated and there is a switch to the viral-encoded RNA polymerase which takes over transcription of late genes. At this stage, IE1 is required for the replication of virus DNA. *Hrs* may serve as origins of replication and it appears that binding of IE1 to *hrs* recruits requisite factors to assemble the replication complex (D). This IE1 function is associated with the N-terminal domain of the protein designated as the replication domain. This domain mediates phosphorylation of IE1 at the initiation of DNA replication. It has been proposed that this event is timely regulated and determines the functional switch of IE1 from transcription- to replication-associated activities. IE1 functional domains for replication, transactivation and dimerization, are indicated in lower panel with I, II and III, respectively (adapted from [37]

different approaches including temperature-sensitive mutations mapping and transient assays of plasmid DNA replication and late gene expression (for review see [8]). In transient expression assays, a plasmid containing a reporter gene under control of a late promoter was cotransfected into cells with an overlapping clone library representing the AcMNPV genome [27]. Genomic DNA fragments containing *lefs* were identified as a consequence of the lack of reporter activity when they were omitted in cotransfections. Gradual shortening of those fragments led to the identification of each *lef*. Nineteen AcMNPV *lefs* were identified as required for activation of the late promoter in this system: *lef1* to *lef12, ie1, ie2, dnapol, p143, p35, p47* and *39k* [28] (Table 1). In addition to the *lefs*, a gene designated *vlf1* was found necessary to support expression from a very late promoter. Some other genes influence DNA replication or late gene expression either directly or indirectly, and may be considered as *lefs* also [8]. Differences in reports on the relative contribution of specific genes appear to be consequence of different experimental approaches utilized in studies.

ORF name	ORF number	Amino acid residues	Homologs in baculovirus lineages[§]	Functional class	LEF function
lef1	14	266	core	Replication	DNA primase
lef2	6	210	core	Replication	primase accessory protein
dnapol	65	984	core	Replication	DNA polymerase
p143	95	1221	core	Replication	DNA helicase
lef11	37	112	α-I, α-II, β, γ	Replication	
lef3	67	385	α-I, α-II, β	Replication	ssDNA binding protein
ie1	147	582	α-I, α-II, β	Replication	transactivator of early genes, *hr*-binding protein
lef7	125	226	α-I, α-II*, β*	Replication	possible ssDNA binding protein
ie2	151	408	α-I	Replication	transactivator of early genes, cell-cycle control
p35	135	299	α-I*, α-II*, β*	Replication	apoptosis inhibitor
lef4	90	464	core	Transcription	RNA polymerase subunit, capping enzyme,
lef8	50	876	core	Transcription	RNA polymerase subunit
lef9	62	490	core	Transcription	RNA polymerase subunit
p47	40	401	core	Transcription	RNA polymerase subunit
lef5	99	265	core	Transcription	transcription initiation factor
lef6	28	173	α-I, α-II, β	Transcription	
39k/pp31	36	275	α-I, α-II*, β	Transcription	DNA binding protein
lef10	53A	78	α-I, α-II*, β*	Transcription	
lef12	41	181	α-I, α-II*	Transcription	

[§] Core genes have homologs in all sequenced baculoviruses. α-I, α-II: Group I and Group II of alphabaculoviruses, respectively. β: betabaculoviruses, γ: gammabaculoviruses.

*There is at least one member in the indicated group having an homolog of the corresponding gene.

Table 1. AcMNPV *lefs*

6.1. Replication lefs

Among *lefs* involved in viral DNA replication [29-30], four baculovirus core genes are essential for this process: *lef1*, *lef2*, *dnapol* and *p143*; they were required in transient assays for plasmid replication as well as for late gene expression. *lef1* is a DNA primase that interacts with *lef2*, a primase accessory protein [31]. *dnapol* encodes a DNA polymerase with 3' to 5' exonuclease activity. The sequence of *dnapol* is the most conserved among baculovirus replication *lefs*; the phylogeny of the family Baculoviridae inferred from its sequence is congruent with that resulting from the analysis of all core genes altogether, suggesting that *dnapol* is an ancestral gene [32]. P143 is a helicase that requires ATP to unwind DNA. Two other AcMNPV replication *lefs*, *ie1* and *lef3*, which are present in all lepidopteran baculoviruses, are also essential as determined in transient assays. Nevertheless, their absence in the genome of γ- and δ-baculoviruses suggests that there may be functional homologs of these genes involved in replication/gene expression in species belonging to these genera. LEF3 is a single stranded DNA-binding protein that promotes unwinding of DNA duplex and annealing of complementary strands [33]. LEF3 interacts with p143 and shuttles this molecule to the nucleus [34].

The actual function of IE1 in DNA replication is poorly understood, nonetheless it appears to depend on the *hr*-binding capacity of IE1. *Hr* regions function not only as enhancers of early genes but also serve as origins of DNA replication in plasmid replication assays [35]. It was shown that in the presence of an *hr* element, transiently expressed IE1 adopts a focal distribution within the nucleus. When LEF3 and P143 are simultaneously expressed they localize to the *hr*-induced IE1 foci [36]. This suggests that IE1 functions by recruiting viral replication factors to the *hr* origin. The switch to the replication activity of IE1 seems to be timely regulated by phosphorylation [37].

Other genes have a stimulatory effect in transient replication/late expression assays. These are *ie2*, *lef7* and *p35*. IE2 is a transactivator involved in cell cycle arrest [38] and LEF7 has sequence similarity to single stranded-DNA binding proteins. Both are present in all genomes of group I α-baculoviruses and LEF7 is also present in some other α- and β-baculoviruses. P35 blocks the apoptotic response of cells triggered by DNA replication through inhibition of effector caspases [39]. The contribution of P35 in transient assays reflects its protective effect against apoptosis, and therefore represents an indirect requirement for replication and late gene expression. Actually, it may be replaced in the assays by a member of the baculovirus *iap* (inhibitor of apoptosis) gene family with similar results. *ie2*, *lef7* and *p35* are dispensable for the infection of TN368 cells by AcMNPV [40]; hence, they are considered as host range factors able to extend the infectivity of AcMNPV towards Sf21 cells.

AcMNPV *lef11*, which was necessary for late gene expression in transient assays, was not required in transient plasmid replication assays. However, an AcMNPV *lef11*-null bacmid was unable to replicate its DNA [41]. Therefore, *lef11* is essential for DNA replication in the context of the virus infection and may be considered as a replication *lef*.

6.2. Transcription lefs

The viral RNA polymerase is a complex of the products of four baculovirus core genes: *lef4*, *lef8*, *lef9* and *p47*. LEF8 and LEF9 have motifs present in the two large subunits of RNA polymerases from prokaryotes and eukaryotes, and are supposed to participate of the catalytic domain. Their sequences are the most highly conserved among LEFs. The role of LEF4 as a capping enzyme is discussed below, while the specific function of P47 remains to be elucidated. The polymerase complex was chromatographycally isolated as a fraction that was active in *in vitro* transcription assays using template DNA containing signals specific of baculovirus late gene promoters [42]. In the complex, the four subunits are present in an equimolar ratio.

lef5 is a core gene that encodes a protein with sequence similarity to the eukaryotic transcription elongation factor TFIIS, however, *in vitro* assays evidenced the ability of LEF5 to increase the transcriptional activity of the viral RNA polymerase at the initiation step rather than to have any effect in the elongation process [43]. The remaining *lefs* have a distribution among species restricted to lepidopteran baculoviruses. *39k* (also known as *pp31*) and *lef10* are considered essential for late gene expression although their specific role in transcription is not known. 39K binds single- and double-stranded DNA and localizes to the virogenic stroma during infection. LEF10 is a small polypeptide without homology to known proteins. *lef6* and *lef12* are considered auxiliary lefs because although they were shown necessary in transient assays (*lef12* is dispensable in TN368 cells), knockout mutants for these genes sustained late gene expression with minor deviations from wild type virus [44-45].

6.3. Late and very late mRNAs synthesis and regulation

Promoters of late genes contain a TAAG sequence motif from which transcription is initiated. There are less TAAG motifs in baculovirus genomes than expected by random occurrence, suggesting that the activity of this sequence as a late promoter selects negatively its random distribution. The integrity of this motif is strictly necessary for transcription, while adjacent sequences up to eighteen nucleotides may affect the level of expression [46]. There may be more than one functional TAAG over a variable distance upstream the translational start codon of the regulated gene [47].

Late transcripts usually span more than one ORF; likewise, one specific ORF may be represented in transcripts with different 5' or 3' ends. The significance of these polycistronic messages is not known and it is generally assumed that only the leading gene in the message is translated into protein. Late genes are encoded in both DNA strands, distributed over the genome, therefore there may be opposite late transcripts with complementary stretches. It is not known if this may play any regulatory role considering that baculovirus genes are susceptible to silencing by double-stranded RNA [48]. Late transcripts are capped and polyadenylated at their 5' and 3' ends, respectively. At least two enzymatic activities required for capping reside in the LEF4 subunit of the RNA polymerase. This protein functions as RNA triphosphatase and guanylyltransferase but lacks activity of N7-methyltransferase, which is required for methylation of the cap structure in position N7 of guanine [49-50]. A gene responsible for this activity has not been identified in baculovirus. The structure of cap 1 mRNAs includes methylation of the 2'hydroxyl group of the ribose of the first transcribed

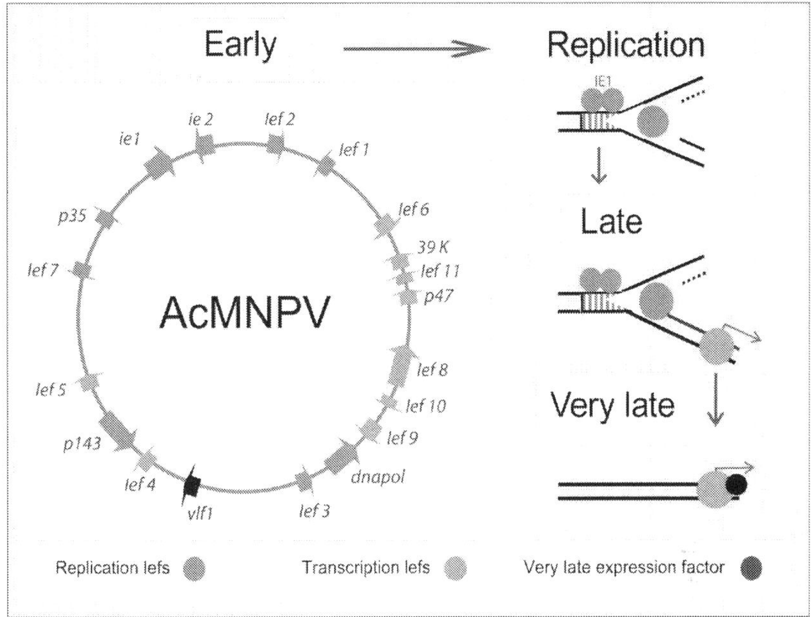

Figure 3. Cascade of baculovirus gene transcription events. Diagram of the AcMNPV genome indicating the localization of the genes encoding key proteins involved in the regulatory network of transcription and DNA replication.

nucleotide by an RNA cap 2'O-methyltransferase (MTase-I). Several alphabaculoviruses have a MTase-I gene. The gene of AcMNPV has been found to stimulate late gene expression in transient assays [51]. Late transcripts are terminated by the polymerase at U-rich sequences present in their 3'UTR, and subsequently the enzyme adds adenosine residues independently of template [52].

The most expressed very late genes in AcMNPV are those encoding polyhedrin and P10. Their transcription depends on a TAAG initiation promoter but their high level of expression depends on the presence of an AT-rich sequence known as the "burst" sequence, located between the TAAG and the translational start codon [53-54]. The burst sequence binds very late expression factor-1 (VLF1; [55]) originally identified in a temperature sensitive AcMNPV mutant defective in occluded virus production [56]. VLF1 is a baculovirus core gene that is essential for the packaging of DNA into normal nucleocapsids.

7. Cellular responses to infection and changes in host gene expression

Early in infection, baculoviruses produce cell cycle arrest at G2/M or S phase, prior to viral DNA replication [57]. The AcMNPV early transcription coactivator IE2 is considered to be involved in regulation of cell-cycle [38]. The progression of infection is accompanied by profound changes in the expression of cellular genes. The host protein synthesis is shutoff starting at around 12-18 hpi [58]. This was found to be mostly the result of a reduction in the levels of transcripts rather than in translation of mRNAs [59], though the actual mechanism of the decrease in the steady-state level of host messages is not known.

Despite host genes are eventually down-regulated at late times, Nobiron and co-workers [60] found that the transcript of a cognate heat shock protein (hsc) 70 gene was transiently up-regulated early in AcMNPV-infected Sf9 cells. In a comprehensive study of gene expression profile of Sf21 cells using microarrays designed from an EST database of *S. frugiperda*, Salem and co-workers [61] confirmed the general shutoff of host transcription over time of AcMNPV infection, but interestingly, they found that about 25% of host genes were slightly up-regulated at 6 hpi. The expression of heat shock proteins (HSPs) of the 70-kDa family in infected cells was followed by western analysis [62]. The results of this study showed changes in the cellular pattern of HSP/HSC70s. Moreover, the infection potentiated the response to heat shock, boosting the HSP/HSC70s content of cells several-fold in comparison with uninfected cells.

The actual level of cellular proteins during infection may vary with a different kinetics of that of the steady-state level of their mRNAs. For example, in a study by Rasmussen and Rohrmann [63], the level of TBP in AcMNPV-infected Sf9 cells, revealed constant until 72 hpi. In other study, TBP was actually found to increase between 16 and 72 hpi in Sf21 and TN368 cells, and to co-localize with viral DNA replication centers within the nucleus [64]. Therefore, TBP appears not be targeted for degradation as it is in other viral systems. However, the functional significance of its increment is unclear, given that it coincides with decreasing levels of transcripts synthesized by RNA pol II.

Currently, due to the relevance of AcMNPV as vector for the expression of proteins in cultured insect cells, it is of special interest to understand the global shutoff of host protein synthesis. In this system, the expression of foreign proteins is driven by the promoter of polyhedrin gene, which is most active at very late times of infection. By this time many processes and pathways appear highly compromised, and the expression of certain classes of proteins may be severely affected, especially those involved in trafficking through the ER and Golgi.

Baculoviruses induce apoptosis of infected cells [39]. Programmed cell death functions as an antiviral defense response to prevent production of virus progeny and spreading of the infection. To counteract the apoptotic response, baculoviruses encode antiapoptotic genes. P35 is a potent antiapoptotic protein of AcMNPV that inhibits the activity of effector caspases. The results of experiments using an AcMNPV *p35* mutant that causes apoptosis upon infection in Sf21 cells showed that apoptosis is triggered by replication of the viral DNA [65]. Apoptosis induced by this mutant was inhibited when each one of the AcMNPV genes required for replication was independently silenced by RNAi. Silencing of these genes also inhibited shutoff of host proteins synthesis, suggesting that both processes are linked. These cellular responses resemble that of vertebrates which arises as consequence of cell cycle arrest or DNA damage. In a recent report Huang and co-workers [66] presented evidence indicating that infection of Sf9 cells with AcMNPV induces a DNA damage response which is required for efficient replication of the virus.

8. Baculoviral microRNAs

MicroRNAs (miRNAs) are small RNA molecules of ~20-22 nt that regulate gene expression posttranscriptionally in a sequence dependent way. miRNAs have been widely described in animals and plants and regulate expression of protein coding genes involved in numerous processes. Genes coding for miRNAs are transcribed by the RNA pol II. The primary transcript (pri-miRNA) containing a hairpin loop is processed by the RNase III-like enzyme Drosha releasing the precursor miRNA (pre-miRNA). The pre-miRNA is a ~80 nt molecule that contains an imperfect hairpin loop and is exported to the cytoplasm by Ran-GTP dependent Exportin 5. Once in the cytosol, the pre-miRNA loop is cleaved by another RNAse III enzyme, Dicer, leaving the RNA duplex consisting of the mature miRNA and its complement (miR-NA*). One of these strands (the mature miRNA) is then incorporated in the RNA-Induced Silencing Complex (RISC), which is then ready to target the specific mRNA and either represses its translation or degrades it [67].

Viruses were also found to encode miRNAs. Strikingly, nearly all the virus encoded miRNAs were reported from DNA viruses, especially those that have a nuclear cycle, with access to the microRNA processing proteins. The majority of the viral miRNAs described belong to herpesviruses. Interestingly, studies of virus-host interactions revealed a complex miRNA regulation with both viral and host microRNAs regulating both viral and host mRNA targets [68-69]. Regarding insect viruses miRNAs, little is known yet. Two viruses, belonging to *Ascoviridae* and *Baculoviridae*, were reported to code for miRNAs. The first report of a miRNA encoded by an insect virus was from the *Heliotis virescens* ascovirus (HzAV). This virus codes for a miRNA that targets viral DNA polymerase and regulates viral replication [70]. More recently, Singh and colleges [71] presented a study in which they found and validated four miRNAs encoded by *Bombyx mori* nucleopolyhedrovirus (BmNPV): *bmnpv-miR-1, -2, -3* and *-4*. This was achieved by sequencing small RNAs obtained from infected tissues of *B. mori* larvae followed by *in silico* analysis and validation using northern blot hybridization, stem-loop RT-PCR and poly(A)-tailed RT-PCR. Interestingly, closely related baculoviruses were found to contain these miRNA in their genomes in conserved positions. All four BmNPV miRNAs are present with 100% identity in AcMNPV, BomaNPV and PlxyMNPV whereas three miRNAs were conserved in RoMNPV and only one in MaviNPV. In contrast of what occurs in animals and plants (miRNAs coded in intergenic regions or introns), these micro-RNAs were found in genomic locations completely overlapping viral ORFs, either in the coding or the complementary strand. *In silico* predictions revealed putative targets, either viral or from the host. Viral predicted targets include *dna binding protein, chitinase, bro-I, bro-III, lef8, fusolin, DNA polymerase, p25* and *ORF 3* of BmNPV. Cellular predicted target genes encode proteins related to antiviral defense mechanisms, such as prophenoloxidase and hemolin, or proteins that play an important role in small RNA-mediated gene regulation like GTP binding nuclear protein Ran, DEAD box polypeptides and eukaryotic translation initiation factors [72].

A further study on *bmnpv-miR-1* revealed the sequence dependent interaction of this miRNA with cellular Ran mRNA. The GTP-binding nuclear protein Ran is an essential component of the Exportin-5-mediated nucleocytoplasmic transport machinery involved in the transport of small RNAs from the nucleus to cytoplasm. Downregulation of Ran by the expression of *bmnpv-miR-1* in viral infection triggers the reduction of the host small RNA population and increasing of the viral load in infected *B. mori* larvae. In this way BmNPV counteracts the small RNA mediated defense of its host for its effective proliferation [63].

9. Persistent infections

It is known that some viruses are capable of persisting in their hosts without causing disease. This can be accomplished by producing either a latent or a persistent infection. The main difference between both is that during latent infection the virus is not replicating and keeping a minimal gene expression while in persistent infections all the genes are expressing, at low levels, without causing any symptom. Herpesviruses can establish latent infections in specific cell types [73]. This state is characterized by a unique transcriptional program that involves the expression of latency-associated transcripts (LATs) as the only viral products synthesized in large quantities. The virus is maintained as an independent quiescent genetic material within the host cell nucleus. An alternative mechanism is observed in measles virus by which the virus remains at low levels with the production of viral proteins. This is usually referred to as persistent infection [74].

The White Spot Syndrome Virus (WSSV) is a non-occluded virus pathogenic to shrimp, phylogenetically related to baculoviruses. It was found at very low levels in asymptomatic shrimps. The virus may reside within cells in a quiescent state as in a latent infection or causing a persistent infection [75]. Similarly, a nudivirus was found infecting persistently the cell line IMC-Hz1, derived from the corn earworm *Heliothis zea*.

Baculoviruses are highly lytic, causing a lethal disease in infected larvae. Epizootics caused by these viruses can reduce dramatically their host population [76]. Persistence of baculoviruses in the environment is mainly thought to be due to the OBs that protect virions from UV light and allow horizontal transmission. But there seems to exist another way for baculoviruses to persist in the environment at low host densities. Baculoviruses can cause sublethal infections, and so be vertically transmitted from adult to offspring [77-79] or may as well become persistent or latent [80]. A laboratory colony of *Mamestra brassicae* was found to harbour an occult infection by the baculovirus MbMNPV with expression of viral genes at a low level [81]. Later, Burden *et al.* demonstrated the persistence of this virus in naturally occurring field populations of *M. brassicae*. RT-PCR analysis showed the presence of *polyhedrin* transcripts in asymptomatic larvae, indicating a covert infection [80-81]. Similar results were obtained using *ie1* as a target [81]. Moreover, these studies revealed that covert infections could be induced to produce overt infections when infecting these larvae with another baculovirus. This means that the persistently infecting virus retains its ability to produce a lethal disease in the larva.

There is accumulating evidence of persistent baculoviral infections. Kemp et al [83] detected baculoviral presence (CfMNPV, CfDEFMNPV and a GV) in laboratory and field populations of *Choristoneura fumiferana*. Also, there were baculoviruses (SeMNPV and MbNPV) identified in *Spodoptera exigua* populations that could be reactivated to full lethal forms [84]. A study in field populations of *Spodoptera exempta* showed that virtually all the insects collected in the field were positive for *S. exempta* nucleopolyhedrovirus (SpexNPV) DNA and 60% of these insects had transcriptionally active virus, suggesting that SpexNPV is transmitted vertically at extremely high levels in field populations of *S. exempta* and can maintain a persistent infection without obvious symptoms [85].

On the whole, baculoviruses seem to use different strategies to persist in nature: on one hand OBs permit their subsistence outside the host for horizontal transmission while, on the other hand, they can persist as covert infections in the host, allowing vertical transmission too. Moreover, these covert infections can be triggered to overt infections producing the typical lethal disease in the host. Nevertheless, the mechanisms of reactivation of these sublethal infections remain to be elucidated.

10. Concluding remarks

Among nuclear DNA viruses, baculoviruses have developed a unique strategy to synthesize late mRNAs which consists in having their own DNA-directed RNA polymerase. This enzyme recognizes viral late promoters that are different to promoters responsive to the cellular RNA polymerase. By this means, the infected cell produces high levels of viral proteins at times of the infectious cycle at which the cellular protein synthesis is mostly shutdown. A late viral progeny with a distinct phenotype is embedded in proteinaceous occlusion bodies (OBs) that assemble after overexpression of the major OB protein. In order to exploit their high protein synthesis capacity, baculoviruses have been developed as vectors for expression of heterologous proteins in insect cells. This system is continuously evolving to new biotechnological applications. However, there is still a lack of knowledge about the molecular mechanisms governing the complex baculovirus infectious cycle. A better understanding of these mechanisms would also benefit the development of baculovirus as biopesticides. To this regard, the array of viral factors involved in regulation of gene expression is an important component of the specific virus-host interactions that determinate the susceptibility to the virus of different cell types within a host and of different hosts within a range of insect species.

Author details

Marcelo F. Berretta[1], M. Leticia Ferrelli[2], Ricardo Salvador[1], Alicia Sciocco[1] and Víctor Romanowski[2]

*Address all correspondence to: mberretta@cnia.inta.gov.ar

1 Instituto de Microbiología y Zoología Agrícola, INTA Castelar, Argentina

2 Instituto de Biotecnología y Biología Molecular, Facultad de Ciencias Exactas, Universidad Nacional de La Plata, CONICET, Argentina

References

[1] Carstens EB, Ball LA. Ratification vote on taxonomic proposals to the International Committee on Taxonomy of Viruses (2008). Arch Virol. 2009;154(7):1181-8. Epub 2009/06/06.

[2] Herniou EA, Arif BM, Becnel JJ, Blissard GW, Bonning B, Harrison R, et al. Baculoviridae. In: King AMQ, Adams MJ, Carstens EB, Lefkowitz EJ, editors. Virus taxonomy: classification and nomenclature of viruses: Ninth Report of the International Committee on Taxonomy of Viruses. San Diego: Elsevier Academic Press; 2011. p. 163-73.

[3] Herniou EA, Jehle JA. Baculovirus phylogeny and evolution. Curr Drug Targets. 2007;8(10):1043-50. Epub 2007/11/06.

[4] Szewczyk B, Lobo de Souza M, Batista de Castro ME, Lara Moscardi M, Moscardi F. Baculovirus Biopesticides. In: Stoytcheva M, editor. Pesticides - Formulations, Effects, Fate: InTech; 2011. p. 808.

[5] Nagamine T, Kawasaki Y, Abe A, Matsumoto S. Nuclear marginalization of host cell chromatin associated with expansion of two discrete virus-induced subnuclear compartments during baculovirus infection. J Virol. 2008;82(13):6409-18. Epub 2008/04/25.

[6] Rohrmann G. The baculovirus replication cycle: Effects on cells and insects. Baculovirus Molecular Biology: Second Edition [Internet]. Bethesda (MD): National Center for Biotechnology Information 2011.

[7] Wilson ME, Miller LK. Changes in the nucleoprotein complexes of a baculovirus DNA during infection. Virology. 1986;151(2):315-28. Epub 1986/06/01.

[8] Passarelli AL, Guarino LA. Baculovirus late and very late gene regulation. Curr Drug Targets. 2007;8(10):1103-15. Epub 2007/11/06.

[9] Vaughn JL, Goodwin RH, Tompkins GJ, McCawley P. The establishment of two cell lines from the insect Spodoptera frugiperda (Lepidoptera; Noctuidae). In Vitro. 1977;13(4):213-7. Epub 1977/04/01.

[10] Hink WF. Established insect cell line from the cabbage looper, Trichoplusia ni. Nature. 1970;226(5244):466-7. Epub 1970/05/02.

[11] Ayres MD, Howard SC, Kuzio J, Lopez-Ferber M, Possee RD. The complete DNA sequence of Autographa californica nuclear polyhedrosis virus. Virology. 1994; 202: 586-605.

[12] Friesen PD. Regulation of baculovirus early gene expression. In: Miller LK, editor. The Baculoviruses. New York and London: Plenum Press; 1997. p. 141-70.

[13] Guarino LA, Summers MD. Functional mapping of a trans-activating gene required for expression of a baculovirus delayed-early gene. J Virol. 1986;57(2):563-71. Epub 1986/02/01.

[14] Miele SAB, Garavaglia MJ, Belaich MN, Ghiringhelli PD. Baculovirus: Molecular Insights on their Diversity and Conservation. Int J Evol Biol. 2011;2011:15.

[15] Carson DD, Guarino LA, MD. S. Functional mapping of an AcNPV immediately early gene which augments expression of the IE-1 trans-activated 39K gene. Virology 1988;Feb;162(2):444-51.

[16] Slack JM, Blissard GW. Identification of two independent transcriptional activation domains in the Autographa californica multicapsid nuclear polyhedrosis virus IE1 protein. J Virol. 1997;71(12):9579-87. Epub 1997/11/26.

[17] Rodems SM, Friesen PD. Transcriptional enhancer activity of hr5 requires dual-palindrome half sites that mediate binding of a dimeric form of the baculovirus transregulator IE1. J Virol. 1995;69(9):5368-75. Epub 1995/09/01.

[18] Olson VA, Wetter JA, Friesen PD. Baculovirus transregulator IE1 requires a dimeric nuclear localization element for nuclear import and promoter activation. J Virol. 2002;76(18):9505-15. Epub 2002/08/21.

[19] Habib S, Hasnain SE. A 38-kDa host factor interacts with functionally important motifs within the Autographa californica multinucleocapsid nuclear polyhedrosis virus homologous region (hr1) DNA sequence. J Biol Chem. 1996;271(45):28250-8. Epub 1996/11/08.

[20] Landais I, Vincent R, Bouton M, Devauchelle G, Duonor-Cerutti M, Ogliastro M. Functional analysis of evolutionary conserved clustering of bZIP binding sites in the baculovirus homologous regions (hrs) suggests a cooperativity between host and viral transcription factors. Virology. 2006;344(2):421-31. Epub 2005/10/04.

[21] Berretta M, Romanowski V. Baculovirus homologous regions (hrs): pleiotropic functional cis elements in viral genomes and insect and mammalian cells. Current Topics in Virology. 2008;7:47-56.

[22] Rohrmann G. The AcMNPV genome: Gene content, conservation, and function. Baculovirus Molecular Biology, 2nd ed. Bethesda (MD): National Library of Medicine (US), NCBI; 2011.

[23] Stewart TM, Huijskens I, Willis LG, Theilmann DA. The Autographa californica multiple nucleopolyhedrovirus ie0-ie1 gene complex is essential for wild-type virus replication, but either IE0 or IE1 can support virus growth. J Virol. 2005;79(8):4619-29. Epub 2005/03/30.

[24] Grula MA, Buller PL, Weaver RF. alpha-Amanitin-Resistant Viral RNA Synthesis in Nuclei Isolated from Nuclear Polyhedrosis Virus-Infected Heliothis zea Larvae and Spodoptera frugiperda Cells. J Virol. 1981;38(3):916-21. Epub 1981/06/01.

[25] Fuchs LY, Woods MS, Weaver RF. Viral Transcription During Autographa californica Nuclear Polyhedrosis Virus Infection: a Novel RNA Polymerase Induced in Infected Spodoptera frugiperda Cells. J Virol. 1983;48(3):641-6. Epub 1983/12/01.

[26] Rice WC, Miller LK. Baculovirus transcription in the presence of inhibitors and in nonpermissive Drosophila cells. Virus Res. 1986;6(2):155-72. Epub 1986/11/01.

[27] Passarelli AL, Miller LK. Three baculovirus genes involved in late and very late gene expression: ie-1, ie-n, and lef-2. J. Virol. 1993; 67: 2149-2158.

[28] Rapp JC, Wilson JA, Miller LK. Nineteen baculovirus open reading frames, including LEF-12, support late gene expression. J Virol. 1998;72(12):10197-206. Epub 1998/11/13.

[29] Kool M, Ahrens CH, Goldbach RW, Rohrmann GF, Vlak JM. Identification of genes involved in DNA replication of the Autographa californica baculovirus. Proc Natl Acad Sci U S A. 1994;91(23):11212-6. Epub 1994/11/08.

[30] Lu A, Miller LK. The roles of eighteen baculovirus late expression factor genes in transcription and DNA replication. J Virol. 1995;69(2):975-82. Epub 1995/02/01.

[31] Mikhailov VS, Rohrmann GF. Baculovirus replication factor LEF-1 is a DNA primase. J Virol. 2002;76(5):2287-97. Epub 2002/02/12.

[32] de Andrade Zanotto PM, Krakauer DC. Complete genome viral phylogenies suggests the concerted evolution of regulatory cores and accessory satellites. PLoS One. 2008;3(10):e3500. Epub 2008/10/23.

[33] Mikhailov VS, Okano K, Rohrmann GF. Structural and functional analysis of the baculovirus single-stranded DNA-binding protein LEF-3. Virology. 2006;346(2):469-78. Epub 2005/12/27.

[34] Wu Y, Carstens EB. A baculovirus single-stranded DNA binding protein, LEF-3, mediates the nuclear localization of the putative helicase P143. Virology. 1998;247(1): 32-40. Epub 1998/07/31.

[35] Pearson M, Bjornson R, Pearson G, Rohrmann G. The Autographa californica baculovirus genome: evidence for multiple replication origins. Science. 1992;257: 1382-1384

[36] Nagamine T, Kawasaki Y, Matsumoto S. Induction of a subnuclear structure by the simultaneous expression of baculovirus proteins, IE1, LEF3, and P143 in the presence of hr. Virology. 2006;352(2):400-7. Epub 2006/06/20.

[37] Taggart DJ, Mitchell JK, Friesen PD. A conserved N-terminal domain mediates required DNA replication activities and phosphorylation of the transcriptional activator IE1 of Autographa californica multicapsid nucleopolyhedrovirus. J Virol. 2012;86(12):6575-85. Epub 2012/04/13.

[38] Prikhod'ko EA, Miller LK. Role of baculovirus IE2 and its RING finger in cell cycle arrest. J Virol. 1998;72(1):684-92. Epub 1998/01/07.

[39] Clem RJ, Fechheimer M, Miller LK. Prevention of apoptosis by a baculovirus gene during infection of insect cells. Science. 1991;254(5036):1388-90. Epub 1991/11/29.

[40] Lu A, Miller LK. Differential requirements for baculovirus late expression factor genes in two cell lines. J Virol. 1995;69(10):6265-72. Epub 1995/10/01.

[41] Lin G, Blissard GW. Analysis of an Autographa californica nucleopolyhedrovirus lef-11 knockout: LEF-11 is essential for viral DNA replication. J. Virol. 2002; 76(6): 2770-9.

[42] Guarino LA, Xu B, Jin J, Dong W. A virus-encoded RNA polymerase purified from baculovirus-infected cells. J. Virol. 1998; 72:7985-7991.

[43] Guarino, LA, Dong W, Jin J. In vitro activity of the baculovirus late expression factor LEF-5. J. Virol. 2002; 76: 12663-12675.

[44] Lin G, Blissard GW. Analysis of an Autographa californica multicapsid nucleopoly-hedrovirus lef-6-null virus: LEF-6 is not essential for viral replication but appears to accelerate late gene transcription. J. Virol. 2002;76: 5503-5514.

[45] Guarino LA, Mistretta TA, Dong W. Baculovirus lef-12 is not required for viral repli-cation. J. Virol. 2002;76(23):12032-43.

[46] Morris TD, Miller LK. Mutational analysis of a baculovirus major late promoter. Gene. 1994;140(2):147-53. Epub 1994/03/25.

[47] Thiem SM, Miller LK. Identification, sequence, and transcriptional mapping of the major capsid protein gene of the baculovirus Autographa californica nuclear poly-hedrosis virus. J Virol. 1989;63(5):2008-18. Epub 1989/05/01.

[48] Flores-Jasso CF, Valdes VJ, Sampieri A, Valadez-Graham V, Recillas-Targa F, Vaca L. Silencing structural and nonstructural genes in baculovirus by RNA interference. Vi-rus Res. 2004;102(1):75-84. Epub 2004/04/08.

[49] Guarino LA, Jin J, Dong W. Guanylyltransferase activity of the LEF-4 subunit of ba-culovirus RNA polymerase. J Virol. 1998;72(12):10003-10. Epub 1998/11/13.

[50] Jin J, Dong W, Guarino LA. The LEF-4 subunit of baculovirus RNA polymerase has RNA 5'-triphosphatase and ATPase activities. J Virol. 1998;72(12):10011-9. Epub 1998/11/13.

[51] Li L, Harwood SH, Rohrmann GF. Identification of additional genes that influence baculovirus late gene expression. Virology. 1999;255(1):9-19. Epub 1999/03/02.

[52] Jin J, Guarino LA. 3'-end formation of baculovirus late RNAs. J Virol. 2000;74(19): 8930-7. Epub 2000/09/12.

[53] Ooi BG, Rankin C, Miller LK. Downstream sequences augment transcription from the essential initiation site of a baculovirus polyhedrin gene. J Mol Biol. 1989;210(4): 721-36. Epub 1989/12/20.

[54] Weyer U, Possee RD. Analysis of the promoter of the Autographa californica nuclear polyhedrosis virus p10 gene. J Gen Virol. 1989;70 (Pt 1):203-8. Epub 1989/01/01.

[55] Yang S, Miller LK. Activation of baculovirus very late promoters by interaction with very late factor 1. J Virol. 1999;73(4):3404-9. Epub 1999/03/12.

[56] McLachlin JR, Miller LK. Identification and characterization of vlf-1, a baculovirus gene involved in very late gene expression. J Virol. 1994;68(12):7746-56. Epub 1994/12/01.

[57] Ikeda M, Kobayashi M. Cell-cycle perturbation in Sf9 cells infected with Autographa californica nucleopolyhedrovirus. Virology. 1999;258(1):176-88. Epub 1999/05/18.

[58] Carstens EB, Tjia ST, Doerfler W. Infection of Spodoptera frugiperda cells with Autographa californica nuclear polyhedrosis virus I. Synthesis of intracellular proteins after virus infection. Virology. 1979;99(2):386-98. Epub 1979/12/01.

[59] Ooi BG, Miller LK. Regulation of host RNA levels during baculovirus infection. Virology. 1988;166(2):515-23. Epub 1988/10/01.

[60] Nobiron I, O'Reilly DR, Olszewski JA. Autographa californica nucleopolyhedrovirus infection of Spodoptera frugiperda cells: a global analysis of host gene regulation during infection, using a differential display approach. J Gen Virol. 2003;84(Pt 11): 3029-39. Epub 2003/10/24.

[61] Salem TZ, Zhang F, Xie Y, Thiem SM. Comprehensive analysis of host gene expression in Autographa californica nucleopolyhedrovirus-infected Spodoptera frugiperda cells. Virology. 2011. Epub 2011/02/01.

[62] Lyupina YV, Dmitrieva SB, Timokhova AV, Beljelarskaya SN, Zatsepina OG, Evgen'ev MB, et al. An important role of the heat shock response in infected cells for replication of baculoviruses. Virology. 2010;406(2):336-41. Epub 2010/08/17.

[63] Rasmussen C, Rohrmann GF. Characterization of the Spodoptera frugiperda TATA-binding protein: nucleotide sequence and response to baculovirus infection. Insect Biochem Mol Biol. 1994;24(7):699-708. Epub 1994/07/01.

[64] Quadt I, Mainz D, Mans R, Kremer A, Knebel-Morsdorf D. Baculovirus infection raises the level of TATA-binding protein that colocalizes with viral DNA replication sites. J Virol. 2002;76(21):11123-7. Epub 2002/10/09.

[65] Schultz KL, Friesen PD. Baculovirus DNA replication-specific expression factors trigger apoptosis and shutoff of host protein synthesis during infection. J Virol. 2009;83(21):11123-32. Epub 2009/08/27.

[66] Huang N, Wu W, Yang K, Passarelli AL, Rohrmann GF, Clem RJ. Baculovirus infection induces a DNA damage response that is required for efficient viral replication. J Virol. 2011;85(23):12547-56. Epub 2011/09/16.

[67] Bartel DP. MicroRNAs: genomics, biogenesis, mechanism, and function. Cell. 2004;116(2):281-97. Epub 2004/01/28.

[68] Takane K, Kanai A. Vertebrate virus-encoded microRNAs and their sequence conservation. Jpn J Infect Dis. 2011;64(5):357-66. Epub 2011/09/23.

[69] Grundhoff A, Sullivan CS. Virus-encoded microRNAs. Virology. 2011;411(2):325-43. Epub 2011/02/01.

[70] Hussain M, Taft RJ, Asgari S. An insect virus-encoded microRNA regulates viral replication. J Virol. 2008;82(18):9164-70. Epub 2008/07/11.

[71] Singh J, Singh CP, Bhavani A, Nagaraju J. Discovering microRNAs from Bombyx mori nucleopolyhedrosis virus. Virology. 2010. Epub 2010/08/31.

[72] Singh CP, Singh J, Nagaraju J. A baculovirus-encoded miRNA suppresses its host miRNAs biogenesis by regulating the Exportin-5 co-factor Ran. J Virol. 2012. Epub 2012/05/18.

[73] Cohrs RJ, Gilden DH. Human Herpesvirus Latency. Brain Pathology. 2001;11(4): 465-74.

[74] Rima BK, Paul Duprex W. Molecular mechanisms of measles virus persistence. Virus Research. 2005;111(2):132-47.

[75] Khadijah S, Neo SY, Hossain MS, Miller LD, Mathavan S, Kwang J. Identification of White Spot Syndrome Virus Latency-Related Genes in Specific-Pathogen-Free Shrimps by Use of a Microarray. Journal of Virology. 2003;77(18):10162-7.

[76] Cory JS, Myers JH. The Ecology And Evolution Of Insect Baculoviruses. Annual Review of Ecology, Evolution, and Systematics. 2003;34(1):239-72.

[77] Goulson D, Gory JS. Sublethal Effects of Baculovirus in the Cabbage Moth, Mamestra brassicae. Biological Control. 1995;5(3):361-7.

[78] Kukan B. Vertical transmission of nucleopolyhedrovirus in insects. J Invertebr Pathol. 1999;74(2):103-11. Epub 1999/09/16.

[79] Burden JP, Griffiths CM, Cory JS, Smith P, Sait SM. Vertical transmission of sublethal granulovirus infection in the Indian meal moth, Plodia interpunctella. Mol Ecol. 2002;11(3):547-55. Epub 2002/03/29.

[80] Burden JP, Nixon CP, Hodgkinson AE, Possee RD, Sait SM, King LA, et al. Covert infections as a mechanism for long-term persistence of baculoviruses. Ecology Letters. 2003;6(6):524-31.

[81] Hughes DS, Possee RD, King LA. Evidence for the presence of a low-level, persistent baculovirus infection of Mamestra brassicae insects. J Gen Virol. 1997;78 (Pt 7): 1801-5. Epub 1997/07/01.

[82] Burden JP, Possee RD, Sait SM, King LA, Hails RS. Phenotypic and genotypic characterisation of persistent baculovirus infections in populations of the cabbage moth

(Mamestra brassicae) within the British Isles. Arch Virol. 2006;151(4):635-49. Epub 2005/12/06.

[83] Kemp EM, Woodward DT, Cory JS. Detection of single and mixed covert baculovirus infections in eastern spruce budworm, Choristoneura fumiferana populations. J Invertebr Pathol. 2011;107(3):202-5. Epub 2011/05/28.

[84] Murillo R, Hussey MS, Possee RD. Evidence for covert baculovirus infections in a Spodoptera exigua laboratory culture. Journal of General Virology. 2011;92(5): 1061-70.

[85] Vilaplana L, Wilson K, Redman E, Cory J. Pathogen persistence in migratory insects: high levels of vertically-transmitted virus infection in field populations of the African armyworm. Evol Ecol. 2010;24(1):147-60.

Chapter 4

Genetic Engineering of Baculoviruses

Santiago Haase, Leticia Ferrelli,
Matías Luis Pidre and Víctor Romanowski

Additional information is available at the end of the chapter

http://dx.doi.org/10.5772/56976

1. Introduction

Baculoviruses are arthropod-specific, enveloped viruses with circular, supercoiled double-stranded DNA genomes [1]. They infect *Lepidoptera* (butterflies and moths), *Hymenoptera* (sawflies) and *Diptera* (mosquitoes) [2]. While many viruses are studied because of their damaging effects, the study of baculoviruses was stimulated by their potential utility to control insect pests [3]. Later, the utility of baculovirus as gene expression vectors was evidenced and a new research area emerged [4]. A major step forward was the development of bacmid technology [5] (the construction of bacterial artificial chromosomes containing the genome of the baculovirus) which allows the manipulation of the baculovirus genome in bacteria. With this technology, foreign genes can be introduced into the bacmid by site- directed recombination or by transposition. Baculoviruses have been used to explore fundamental questions in molecular biology such as the nature of programmed cell-death [6]. Moreover, the ability of baculoviruses to transduce mammalian cells led to the consideration of their use as gene therapy and vaccine vectors. Strategies for genetic engineering of baculoviruses have been developed to meet the requirements of new application areas, and the establishment of new genetic modification systems is still necessary when an unexplored experimental system is to be addressed. The aim of this chapter is to detail the areas of application of the baculovirus in basic molecular biology and applied biotechnology and the strategies used to generate genetically modified baculoviruses according to each area of study.

1.1. Molecular biology of baculoviruses

Baculovirus genomes consist of a circular, double-stranded DNA molecule with a size ranging from 80 to over 180 kbp and encoding 90 to 180 genes.

The name "baculovirus" is derived from the latin *"baculum"* (stick), denoting the rod-shaped nucleocapsids that are 230– 385 nm in length and 40–60 nm in diameter [1]. The virions are enveloped and present two phenotypes: occluded virions (OVs) and budded virions (BVs). These two types of virions differ in the origin and composition of their envelopes and their functions in the virus life cycle. In both, the genome is complexed with multiple copies of a small basic protein (p 6.9) which neutralizes the negative charge of the DNA and this structure is protected by other proteins forming the nucleocapsid. The OVs are enclosed in a paracrystalline matrix forming occlusion bodies (OBs), which are orally infectious. Their morphology was initially used to define two major groups or genera of the Baculoviridae: the Nucleopolyhedrovirus (NPVs) and the Granulovirus (GVs). NPV OBs, called polyhedra, are about 0.6–2 μm in size and their major occlusion protein is called polyhedrin. GV OBs, also known as granules or capsules, are oval-shaped with diameters in the range of 0.2–0.4 μm (Figure 1). OBs are highly stable and can resist most normal environmental conditions thereby allowing virions to remain infectious for very long periods of time.

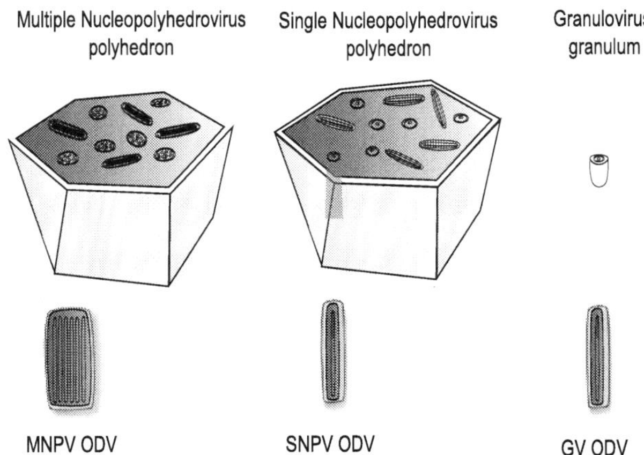

Figure 1. On the basis of the OB morphology, baculovirus were originally divided in two major groups: the Nucleopolyhedrovirus (NPVs) and the Granulovirus (GVs). NPVs occlusion bodies are called polyhedra and their major occlusion protein is called polyhedrin and GV occlusion bodies and granules or capsules for GVs.

Based on the fusion protein present in the BVs NPVs have been further classified into two groups: type I NPVs contain GP64, a low-pH-dependent membrane fusion protein required for virus entry and cell-to-cell transmission [7,8,9,10], BVs of group II NPVs and GVs lack a homolog of GP64, and membrane fusion during viral entry is triggered by F protein [11,12]. It was found that the entry of baculoviruses in mammalian cells is mediated by GP64. In contrast, baculoviruses with F protein cannot transduce these cells [13].

The natural cycle of infection by AcMNPV in insect larvae is summarized in Figure 2. Caterpillars ingest polyhedra that contaminate their food. The polyhedrin matrix is dissolved in the alkaline environment of the larvae midgut releasing ODVs (occlusion derived virions). These virions enter midgut cells after fusion with membrane epithelial cells. The virions are uncoated and enter the nucleus where viral genes are expressed.

The gene transcription of baculovirus has been divided in sequential phases: immediate early, delayed early, late and very late. Immediate early genes are recognized by host transcription factors and viral proteins are not necessary at this stage. Transcription of delayed early genes requires activation by products of immediate early genes. The delayed early phase is followed by the synthesis of DNA and the late gene products of the virus [14]. There is a close relationship between the DNA replication and the switch to late gene transcription, and it is believed that these events are physically connected. In the late phase, that occurs following the initiation of viral DNA replication, nucleocapsid structural proteins are synthesized, including glycoprotein GP64 playing a crucial role in the horizontal infection by BV [15]. During the very late phase the production of infectious BV is greatly reduced. Nucleocapsids interact with nuclear membranes and eventually become enveloped usually in groups of a few particles. Envelopment of the nucleocapsids appears to be an essential primary step in the process of occlusion of nucleocapsids by the very late protein-polyhedrin. The occlusion continues until eventually the nucleus becomes filled with occlusion bodies. As occlusion proceeds, fibrillar structures begin to accumulate in the nucleus (sometimes also in the cytoplasm). These structures are composed mostly of a single polypeptide named p10, which is a very late protein [16]

The function of fibrillar structures is not clear but they may play a role in the controlled cellular disintegration in caterpillars [17,18]. In the terminal stages of infection two viral proteins, chitinase and cathepsin, act together to facilitate host cuticle breakdown [19]. After death the caterpillar liquefies and releases polyhedra which can infect other insects. At the end of the infection, OBs may account for over 30% of the dry weight of the larvae [20].

2. Baculovirus expression vectors

2.1. Introduction

The high levels of expression of the very late genes has been exploited to design the first vectors for foreign gene expression based on baculoviruses. They are especially suitable regarding safety (not harmful to non-target organisms) and easy containment in the laboratory. Baculovirus vectors are used to infect insect cells or larvae where high levels of recombinant proteins are produced; the eukaryotic environment provides appropriate post-translational modifications in comparison with prokaryotic expression systems. Insect cells to be used in the baculovirus expression system are derived from lepidopteran insects and are relatively easy to grow. No control of oxygen atmosphere is required. Moreover, insect cells can be adapted to serum-free media and production of recombinant protein can be scaled up to pilot plant or larger bioreactors [21, 22]. The lepidopteran insect cells used are also normally free of human pathogens. Thereby, the proteins produced in the baculovirus virus expression system can be used for functional studies, vaccine preparations or diagnostics.

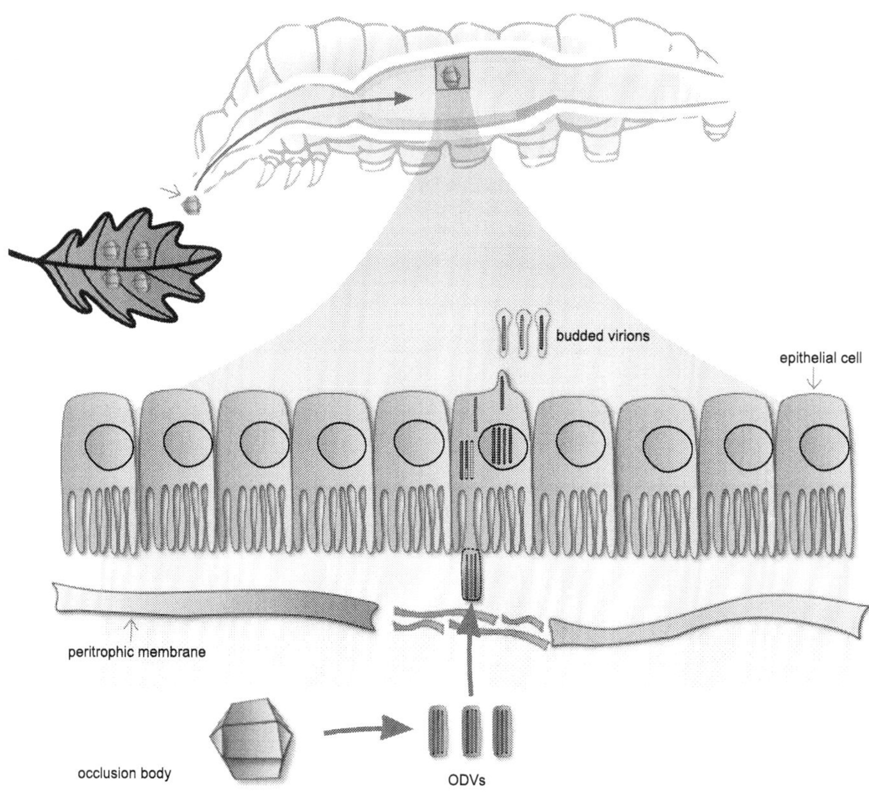

Figure 2. *Per os* infection of baculoviruses. A cross sectional representation of the anatomy of an insect larva is depicted. A baculovirus occlusion body (OB) is ingested in contaminated food. OBs pass through the foregut and enter the midgut where they dissolve in the alkaline midgut lumen and release occlusion derived viruses (ODVs).

A baculovirus expression vector (BEV) is a recombinant baculovirus that has been genetically modified to lead the expression of a foreign gene. BEVs are viable in insect cell culture and sometimes in larvae, depending on the baculovirus genes deleted in the process of the recombinant virus generation. In BEVs, the foreign gene coding sequence is usually placed under the transcriptional control of a viral promoter. For this reason usually viral factors are required for the transcription of the foreign gene.

The most commonly used process for cloning recombinant baculovirus is briefly summarized in Figure 3. Baculovirus genomic DNA and a transfer plasmid are cotransfected into an insect cell culture. Double homologous recombination between viral DNA and transfer plasmid causes the allelic replacement that incorporates the recombinant gene in the baculovirus genome. Clonal purification requires several plaque passages. After this, viral stocks can be produced and amplified for recombinant protein production [31]. Insect cells are used for purification of many proteins, including therapeutic and vaccine peptides. Larvae are used to reduce production costs, or when recombinant baculovirus are to be tested as bioinsecticides. Finally, baculovirus can be used for transduction of mammalian cells, for production of therapeutic proteins, or to transduce organisms for gene therapy or vaccination.

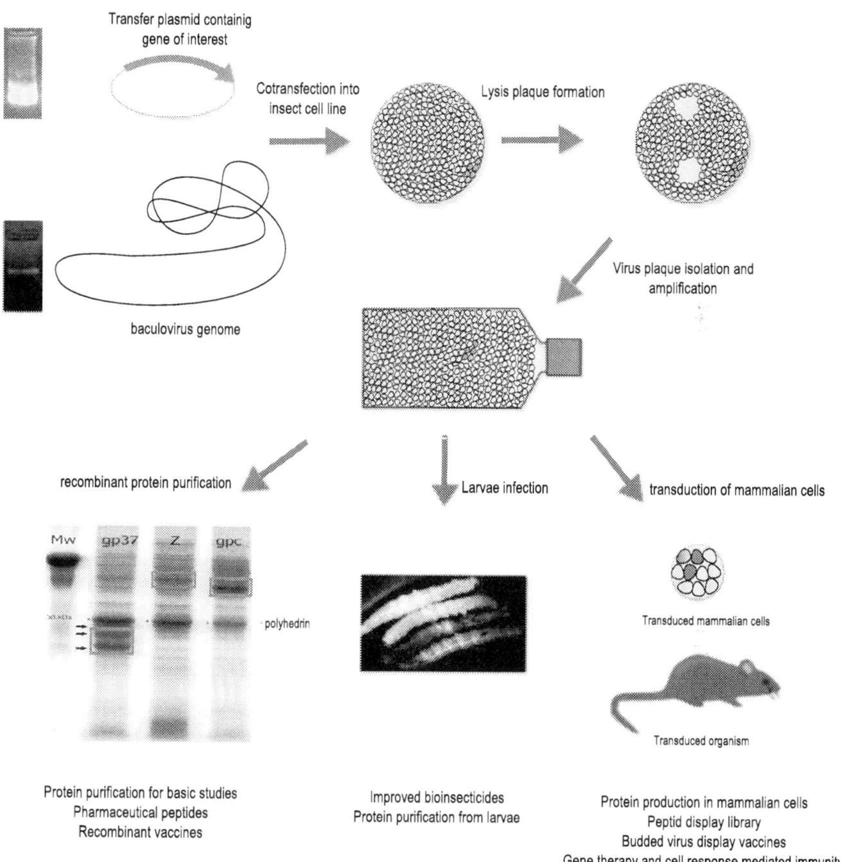

Figure 3. General process for baculovirus cloning. Genomic DNA and a transfer plasmid are cotransfected into an insect cell culture. Recombinant virus propagates causing lysis plaques. Virus is isolated from lysis plaques and amplified. Virus can be used for recombinant protein production in insect cells or larvae. Furthermore, baculovirus can be used for transduction of mammalian cells or whole animals.

2.2. Historical perspective

The major component of the polyhedrovirus OB is polyhedrin. This polypeptide comprises about 25% of total cell protein in the late phase of baculovirus infection [23]. This was the main property of baculovirus that led to their consideration as expression vectors. In the first studies of baculovirus, AcMNPV polyhedrin gene was located and cloned. Then, a plasmid containing the sequences of the polyhedrin gene (polh) and its flanking regions was constructed and subsequently the polyhedrin open reading frame (ORF) was replaced with the reporter gene β-galactosidase [24]. This is the simplest version of a transfer plasmid, which contains the strong polh promoter and upstream and downstream flanking sequences, but lacks the polyhedrin ORF which is usually replaced with the coding sequence of a foreign gene of interest. Transfer plasmid and viral DNA are cotransfected into cultured insect cells, where allelic replacement of polyhedrin can occur via homologous recombination involving the flanking viral sequences present in both DNAs.

This strategy was devised because the baculovirus genome is large [25]; and in vitro ligation of foreign DNA fragments with restriction enzyme-digested viral DNA has been successful only in few cases [26, 27]. Other strategies that have been explored include enzymatic recombination in vitro [28] and homologous recombination in yeast followed by selection [29]. Although these methods are ingenious, neither has become part of the mainstream baculovirus technology.

Allelic replacement is a consequence of double homologous recombination between viral DNA and transfer plasmid, and occurs at a frequency of only about 1% [30]. Thus, the viral progeny is a mixture of recombinant and wild type virus, which needs to be resolved in plaque assays. The recombinant progeny is occlusion negative (*occ-*) and produces polyhedrin- negative plaques (cell plaques with no polyhedral OBs), whereas wild-type progeny produces polyhedrin-positive plaques (*occ+*). Clonal purification requires several plaque passages. After this, viral stocks can be produced and amplified to infect cell cultures for recombinant protein production.

2.3. Strategies to simplify the isolation of recombinant baculoviruses

As mentioned before, following the strategy described above, the proportion of the progeny virus population derived from the cotransfection experiment is less than 1% [31]. Plaque purification of recombinant clones requires a tedious search for *occ-* plaques. Good quality microscope and experienced eye are necessary, and few recombinant viruses can be screened simultaneously because of the number of titrations required. To circumvent these problems, several modifications in the parental viral genome were carried out in order to simplify the isolation of recombinant baculoviruses, with the aim of reducing the parental virus yields in the progeny of the co-transfected insect cells.

One of the most successful strategies is the use of a linearized parental genome (in principle, no virus can be recovered) instead of the circular viral DNA (fully infectious). The addition of a unique naturally infrequent restriction site in the baculoviral genome allows the digestion with the adequate restriction enzyme and digested parental DNA is cotransfected with transfer vector [32]. As linearized parental DNA has a reduced infectivity compared with its circular counterpart, frequencies of recombinant progeny rise to about 30%. The baculoviral genome can also be modified to contain restriction sites on both sides of a cassette containing a reporter gene coding sequence (such as β-galactosidase) under the control of a baculoviral promoter, so that double digestion with this enzyme removes the reporter gene cassette. The presence of two restriction sites reduces the frequency of undigested circular DNA genome. Moreover, if undigested or single digested-repaired parental DNA produces progeny, those few parental plaques stain blue in the presence of X-gal and can be easily discarded. This strategy was exploited in the AcMNPV BEV system. In this virus, the genome was modified to contain Bsu36I sites on both sides of the β-galactosidase sequences, so that digestion with this enzyme removed the gene and also part of a virus gene (ORF 1629) that encodes a structural protein [33]. By removing part of the essential ORF 1629 gene, the virus is unable to form infectious particles efficiently even if the double digested linear DNA is repaired and recircularized in insect cells. In contrast, a process of homologous recombination repairs the deletion of ORF 1629 while simultaneously inserting the foreign gene in place of β-galactosidase. Several commercial systems make use of the repair of the deletion in ORF 1629. Among them, Bac-to-Bac® (Invitrogen) and flashBacTM (Oxford Expression Technologies) quickly gained popularity in the scientific community.

2.4. Bacmid technology

A major step forward in the technology of baculovirus genetic engineering has been the development of baculovirus genomes capable of replicating in a bacterial host as bacterial artificial chromosomes (Figure 4). These recombinant baculoviruses are called bacmids, and they have been modified to contain classical bacterial artificial chromosomes replicons and selection markers for selection in bacteria. BAC vectors contain a fragment of *E. coli* fertility factor (F- factor) replicon (*miniF*) and are maintained as circular supercoiled extrachromosomal single copy plasmid in the bacterial host [57, 58]. BACs can accept inserts up to 300 Kb in length. The principal advantage BACs have over other high insert capacity vectors like yeast artificial chromosomes (YAC) and mammalian artificial chromosomes is stability of insert propagation over multiple generations.

Once transferred into the bacterial host, the baculovirus genome can be manipulated easily through site-specific recombination, Rec-A mediated homologous recombination or transposition. Once the recombinant bacmid is generated and the presence of transgene and the absence of the parental bacmid in the bacterial colonies are verified, e.g., by PCR, the DNA from those colonies is purified and used to transfect susceptible insect cells. As was mentioned above, naked genomic DNA from baculovirus can efficiently establish infection when it reaches the cell nuclei. BV particles can be recovered from culture supernatant and used as inoculum to produce high titer stocks.

Various commercial transfer vectors are available and compatible with bacmid systems to allow expression of one or two proteins (*e.g.*, pFastBac1TM and pFastBacDualTM from InvitrogenTM). Some are designed to add tags and signal peptide sequences fused to the protein of interest in order to facilitate their purification. In addition, transfer vectors that are compatible with GatewayTM and TOPO® cloning technologies have been developed.

A problem frequently found when working with BAC systems is the presence of parental bacmid background even in the same colony where the recombinant bacmid is found, despite the antibiotic-based selection and blue-white screening. This then requires a new transformation of bacterial cells with mixed DNA and the screening of newly replated colonies. To avoid this requirement, a negative-selection system has been developed that makes use of the *sacB* marker [59]. When the transposon is not integrated in the bacmid genome, the bacterial cell will be killed in presence of sucrose due to the expression of *sacB* gene, which encodes an enzyme that metabolizes sucrose to a toxic compound.

The first bacmid developed contained the AcMNPV genome. Later, bacmid systems were developed for *Bombyx mori* NPV, *Helicoverpa armigera* single-nucleocapsid nucleopolyhedrovirus *(HearSNPV)* [61] and *Cydia pomonella* granulovirus (CpGV) [60] (the first report of a granulovirus bacmid).

Later, bacmid technology was exploited to develop a system that allows the generation of recombinant baculoviruses with negligible background. This system relies on homologous recombination in insect cells between a transfer vector containing a gene to be expressed and a replication-deficient AcMNPV bacmid. The deficiency of AcMNPV is due to a deletion in the essential gene *orf1629*, and homologous recombination between bacmid DNA and transfer vector (containing *orf1629*) repairs this deleted gene [62, 63, 64]. Therefore, only recombinant virus can replicate and no further selection is required, facilitating the rapid production of multiple recombinant viruses on automated platforms in a one-step procedure. Several commercial vectors (flashbackTM, Oxford Expression Technologies Ltd., BacMagicTM (Merck), BaculoOneTM (PAA), etc.) follows this principle. All these systems claim that no plaque purification of baculovirus is required, although it is recommended (there is a possibility that defective genomes can be replicated when a replication competent viral genome resides in the same cell). Later, bacmids using this selection system were improved for protein expression, carrying additional deletions in cathepsin (*v-cath*) and p10 gene [65].

2.5. Improving protein quality and quantity in baculovirus expression systems

2.5.1. Introduction

The proteins to be expressed using recombinant baculovirus and insect cells are commonly of mammalian origin and, as it happens in other expression systems, the expression levels and the conformation and posttranslational modifications vary among individual proteins. The principal purpose of a protein expression system is not only the production of large quantities of recombinant protein, but also the production of a recombinant protein that resembles the native protein. One of the most difficult challenges in expression systems is the expression of transmembrane proteins. The correct expression of complex transmembrane proteins that cross the membrane several times is even more difficult. In order to improve the quality in the routing, the post-translational modifications and the stability of recombinant proteins, several modifications have been carried out that address these limitations of baculovirus expression

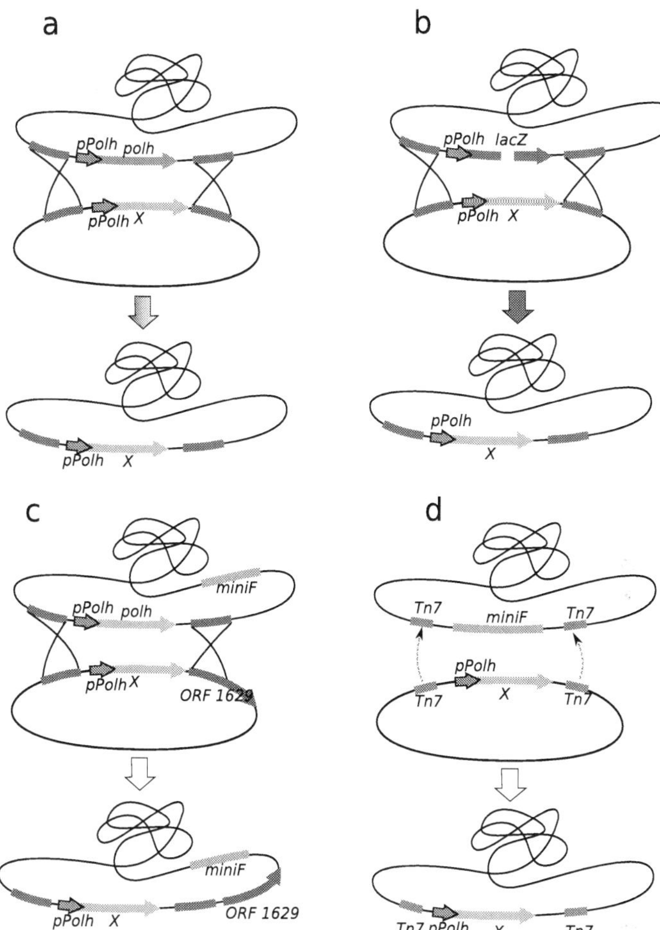

Figure 4. Baculovirus expression vectors over time. Various methods exist to generate recombinant baculoviruses expressing a foreign gene (gene X). Historically, recombinant baculoviruses were generated through homologous recombination (1). Subsequently, linearized vectors were developed to increase the percentage of recombinants (2). Bacmid technology allowed the maintenance of defective baculoviruses as bacterial artificial chromosomes. Homologous recombination with transfer vector in cells repairs the essential gene (3). Bacmid technology also allowed the generation of recombinant baculovirus by in vitro transposition (4).

system. It is important to notice, however, that the protein is produced in the context of a viral infection. Since certain protein processing pathways are compromised by baculovirus infection, the capacity of host cells to correctly route, fold and modify the recombinant protein is affected. This intrinsic limitation must be recognized and baculovirus expression system must be regarded as a transient expression system.

2.5.2. Heterologous DNA properties and codon usage

Although the promoter elements that control the transcription of the heterologous gene are derived from baculovirus, it is important to consider the effect of introducing heterologous or artificial 5′ and 3′untranslated regions (UTR). The 5′UTRs of baculovirus are short AT-rich sequences. Therefore, the introduction of GC-rich sequences upstream of the ORF may have a negative effect on the heterologous gene transcriptional levels. The choice of 3′UTR, including polyA sequences may also determine the heterologous gene expression levels. As expected, p10 polyA signals are more efficient than the widely used SV40 terminator [66]. There are no in-depth studies comparing the influence of codon usage on translation levels in baculovirus expression systems. However, no strong bias in alternative codon frequency has been observed in baculovirus coding sequences, suggesting that the codon optimization is unlikely to improve significantly the translation levels [67].

2.5.3. Deletion of baculovirus genes to prevent proteolytic cleavage

Chitinase, the product of the gene *chiA*, is an enzyme that breaks down the chitin exoskeleton of the insect host, together with cathepsin (V-CATH, encoded by *v-cath*) at the end of the infection, ensuring the dispersal of the viral occlusion bodies [68]. As those genes have a specific function in the context of the infection in the insect, they are not required for the propagation of the virus in cultured insect cells. Chitinase is produced at high levels and stored in the endoplasmic reticulum, and it may interfere in the secretory apparatus of the host cell. On the other hand, cathepsin is a protease that is made as an inactive precursor (PRO-V-CATH). PRO-V-CATH can be activated when preparing protein samples for SDS-PAGE, leading to the degradation of the recombinant protein. Bacmids were developed with deletions in the genes *chiA* and *v- cath*, resulting in higher levels of secreted protein [69, 70].

2.5.4. Secretion of proteins

Many secretory pathway proteins have N-terminal signal peptides that direct the protein correctly through the ER and the Golgi system and ultimately to the surface of the cell. If the signal peptide is not adequately recognized, the protein may be not targeted to the cell surface, and the misfolded protein is also prone to degradation as it may be recognized by quality control systems [71].

Native signal peptides of mammalian proteins may be replaced by signal peptides derived from insect proteins such as the signal peptide of honey bee melittin [72] or derived from baculovirus proteins, such as the GP64 signal peptide. Although the introduction of insect signal peptides normally targets the protein to cell surface, it does not always lead to a correct folding of the protein.

2.5.5. Glycosylation of proteins in the baculovirus-insect cell system

Glycosylation is a common covalent chemical modification that can affect many protein properties, including intracellular trafficking, biological function, immunological properties and biochemical stability. One of the most advantageous features of the baculovirus-insect cell system is that it can produce glycosylated proteins. However, the protein glycosylation pathways of lepidopteran cells differ from those of higher eukaryotes [73]. N-glycosylation begins in insect cells with the transfer of the oligosaccharide Glc3Man3GlcNAc2 (where Glc, Man and GlcNAc refer to glucose, mannose and N-acetylglucosamine, respectively) from a lipid complex to an asparagine residue in the polypeptide chain in the ER lumen. As the protein passes through the ER and Golgi system, enzymes trim and add different sugar moieties to this N-linked glycan. In this step is where insect and mammalian cells start to vary. This results in glycoproteins with simple oligomannose sugar chains in insects, while in mammals complex sugar groups with terminal sialic acids are added.

Differences in glycosylation patterns may affect the folding and targeting of recombinant glycoproteins and their immunological properties. Moreover, differences in glycosylation may even lead to protein degradation [74].

To overcome the limitations of the baculovirus insect-cell system in glycosylation, a series of transgenic cell lines derived from lepidopteran Sf9 and High Five cells expressing genes for the enzymes required to produce the complex mammalian glycosylation patterns were developed [75, 76]. Genes incorporated include bovine β-1,4-galactosyl transferase and rat α-2,6 sialyltransferase. The introduction of these enzyme resulted in the incorporation of galactosyl and sialyl residues in the produced proteins.

2.5.6. Expression of cytosolic and ER processing enzymes

Proteins to be secreted are translated in the cytosol and can translocate across the ER membrane either by a cotraslational or post-traslational mechanism. Transport in mammalian cells is primarily cotraslational, and in yeast both post-traslational and cotraslational mechanisms are used. In insect cells, the predominant mechanism is still not known. In mammalian cells, the cytosolic chaperone hsp70 is believed to contribute to the traslocation of proteins by interacting with nascent polypeptides and preventing their aggregation. Coexpression of immunoglobulin G (IgG) and human hsp70 resulted in higher levels of soluble IgG precursor. As a consequence, mature IgG secreted levels increased [77].

In the ER chaperones also assist the folding of polypeptides by preventing improper aggregations and conformations. In mammalian cells, immunoglobulin heavy chain binding protein (BiP) is an ER chaperone that interacts with several polypeptides destined for secretion and may be involved in the translocation or proteins across the ER membrane. When recombinant BiP was coexpressed in insect cells with IgG the soluble and secreted IgG levels were increased [78]. Other additional chaperones, such as calnexin and calreticulin, can also assist folding and assembly of membrane proteins in BEVs. Catalytic enzymes in the ER also collaborate by accelerating the folding. Disulfide bond formation occurs in the oxidizing ER compartment, with the catalytic action of protein disulfide isomerase (PDI). Studies demonstrated that overexpression of PDI increases the folding and secretion of IgG in insect cells [79]. It has been observed that co-expression of foldases appears to work more efficiently when the corresponding genes are provided by the baculovirus vector than those integrated in the genome of transgenic cell lines. This observation may be related to the phenomenon of host genome transcriptional shut down known to occur during baculovirus infection.

2.6. Improving baculovirus genome stability

A major drawback that limits the application of baculovirus for large-scale production is the accumulation of defective interfering (DI) particles upon serial viral cell culture passages. DI particles are not able to propagate autonomously due to deletion of large portions of their genomic DNA, but can co-propagate in the presence of viable virus [80]. Since deletion often includes the inserted foreign gene of interest, when DI particles proportion increases, recombinant protein expression levels decreases. The accumulation of DI particles can be reduced by the practice of infecting at low multiplicities of infection (MOI).

Genetic engineering strategies have been developed to prevent the accumulation of DI particles. In *Spodoptera exigua* MNPV (SeMNPV), it has been observed that DI particles are enriched in a non-*hr ori* fragment. Removal of this *non-hr ori* from the genome of the baculovirus prevented the formation of DI particles up to 20 cell culture passages [81]. Removal of an AcMNPV *non- hr ori* had the same effects on genomic stability. It was also observed that when a large foreign fragment of DNA is cloned in baculovirus DNA genome in which no selection pressure exists, the addition of an *hr* (which functions as origin of replication) may prevent the loss of the foreign DNA [82].

3. Baculoviruses as bioinsecticides

3.1. Introduction

The basis of modern baculovirology was stimulated by the potential utility of baculoviruses to control insect pests [34]. Baculoviruses are highly infectious and selective pathogens (their host range is usually limited to one species), are very safe to people and wildlife and long term crop protection can be established [35]. Despite these advantageous features, the application of baculovirus as bioinsecticides has not still matched their potential. Although the use of baculovirus bioinsecticides was hampered by their slow speed of action when compared with fast-killing chemical insecticides, they gained increasing acceptance as they were considered for long term protection of crops, in the framework of integrated pest management.

Up to date, the most successful project was implemented in Brazil where over two million hectares of soybean were controlled by baculovirus AgMNPV [36, 37]. However, it is important to notice that a series of factors contributed to the success of AgMNPV as bioinsecticide. First, AgMNPV is highly pathogenic and only one application is sufficient to control the pest over the production cycle. In second place, *Anticarsia gemmatalis* was the most important plague in soybean crops in Brazil, and other plagues did not cause significant economic damage. Finally, the application of AgMNPV was promoted by Brazilian state and the integrated pest management governmental programs facilitated the public acceptance of alternatives to chemical insecticides. Despite this favorable unique context, the success of Brazilian project revitalized the interest in baculovirus as bioinsecticides and many countries and private companies begun to develop new programs of baculovirus control and the search of novel baculoviruses.

3.2. Genetic improvement of baculovirus insecticides

In the search of increasing the commercial fitness of baculovirus as bioinsecticides, strategies to improve the baculovirus pesticide parameters by means of genetic engineering were developed.

Slow action of baculoviruses often limits its practical application and many strategies aimed to improving the timing of the pest killing or paralyzation by baculovirus. The first strategies were based on the interference of host physiology with insect hormones. When a diuretic hormone gene was introduced into *B. mori* baculovirus genome, recombinant BmNPV killed larvae about 20% faster than wild type virus [38]. The expression of this hormone by baculovirus causes the infected larvae to rapidly lose water.

Another strategy was based upon the control of juvenile hormone. In lepidoptera, this hormone controls the onset of metamorphosis at the final molt. The expression of juvenile hormone esterase decreases the concentration of the hormone [39, 40]. A reduction in the levels of juvenile hormone (JH) early in the last larval instar has been shown to initiate metamorphosis and lead to a cessation of feeding behavior. If this juvenile hormone esterase (JHE) is inhibited, the concentration of JH remains high enough to keep the larva in the feeding stage, resulting in giant insects. Another approach used consists in the deletion of the virus-encoded ecdysteroid glucosyltransferase gene [41]. The product of the *egt* gene normally prevents larval molting during infection increasing feeding activity of infected larvae. The EGT enzyme inactivates hormone ecdysone by transferring sugar molecules. The inactivation of this hormone results in an increased food consumption, allowing the virus to maximize the viral progeny. The infection with an *egt* defective recombinant AcMNPV resulted in a 30% faster killing of larvae and significant reduction in food consumption.

The degree of improvement that can be achieved by gene deletion alone appears limited. For this reason, several research lines have focused on the use of gene insertion technology in order to achieve more substantial improvement in the performance of viral insecticides.

Among the strategies that have been explored to date, the insertion of insect-specific toxins is the most promising one for development of commercially viable baculovirus insecticides [42]. In nature, insect predators and parasites use venoms to immobilize their prey. Although arthropod venoms are composed of a mixture of toxins that may have activity against organisms other than insects, it is possible to isolate genes that target insects with high specificity.

Although the first experiments using an insect-specific toxin of the scorpion *Buthus eupeus* [42] did not show an improvement in the speed of action of the recombinant baculovirus, the use of other scorpion toxin genes resulted in significant enhancement of virus insecticidal performance. One of the most promising insect-specific toxins used for the generation of recombinant baculovirus is the product of the gene AaIT of the scorpion *Androctonus australis*. The product of this gene is a small peptide (70 amino acids) that interacts with voltage- dependent sodium channels causing rapid paralysis in insects. Moreover, AaIT has no activity on vertebrate nervous tissue and is nontoxic to mice. When AaIT toxin was introduced into

AcMNPV, the speed of kill increased by about 40% and the feeding damage was reduced by about 60% [43].

Another paralytic toxin that holds promise is the TxP-I toxin, a component of the venom of the predatory straw itch mite *Pyemotes tritici* [44,45]. The mechanism of action this toxin has not been studied in depth, although it is related to voltage-dependent calcium channels (VDCC). The mean time to death of larvae infected with AcMNPV recombinant baculovirus expressing TxP-I under the control of p10 very late promoter was reduced by 50-60% compared to larvae infected with the wild-type strain, depending on virus dose and larval instar [46].

The choice of the promoter that controls the transcription of the heterologous toxic gene is very important. Although *polh* and *p10* very late promotes provides high levels of transcription, early and late viral promoters or constitutive promoters can result in an earlier accumulation of the toxin, causing more significant reductions in the speed of paralysis of the larvae. A chimeric promoter constructed by insertion of the p6.9 promoter downstream of the *polh* promoter was found to be more effective than *polh* promoter alone [47]. Another promoter tested was the constitutive *Drosophila hsp70* heat-shock protein gene promoter [48]. Despite the lower levels of toxin accumulation, the results obtained with this promoter were comparable to those obtained with the p6.9 promoter. The choice of the promoter must be considered from a biosafety perspective. Evidence indicates that recombinant baculoviruses expressing toxin genes are not pathogenic to vertebrates, and that the probability of horizontal transfer of the toxin gene to vertebrates is very low. Moreover, as it was mentioned above, specific arthropod toxins have no effect on vertebrate neural system. Despite these arguments, it is desirable to select promoters that are not functional in vertebrates.

3.3. Strategies for modifying host range of baculoviruses

A primary advantage of baculovirus bioinsecticides is their host specificity. In contrast to chemical insecticides that may harm vertebrates or kill arthropods indiscriminately, baculo- viruses target specific populations of insect pests. This feature makes them compatible with classical biological controls in integrated pest management strategies and makes particularly useful for controlling insect pests in environmentally sensitive areas. Although bioinsecticides are attractive from an ecological perspective, their limited host range is undesirable from an economical point of view. Since many different baculoviruses may be needed to control complexes of simultaneous insect pests, costs would be excessively high. For this reason, many researchers have studied the possibility of modifying the baculovirus host range while maintaining their safety for vertebrates and nontarget arthropods.

3.4. Determinants of virus host range

The host range of any virus is determined by its ability to enter the cells and tissues of a host organism, replicate and release new infectious virus particles. The virus host range is fre- quently determined by the presence of suitable receptors that facilitate virus attachment and entry into a host cell. This does not appear to be the case for baculoviruses. Baculoviruses are able to enter nonpermissive insect and even mammalian cells. This indicates that if receptors are used by baculoviruses, they are common to insect and mammalian cells [49, 50, 51, 52]. In nonpermissive insect cells, reporter gene expression was observed from early baculovirus promoters, but expression from very late baculovirus promoters was limited. Expression from late baculovirus promoters varied among nonpermissive insect cell lines. As mentioned before, it was established that the transcription from late baculovirus promoters requires the viral DNA replication [53]. These findings indicate that in nonpermissive insect cells viral DNA is delivered to the nucleus, the site of baculovirus replication, although replication is restricted in a cell specific manner. For this reason, the viral genes that determine the host range are likely to be related with the process of DNA replication.

One of the first steps forward in baculovirus host range alteration was the generation of a recombinant AcMNPV capable of replicate in nonpermissive *B. mori* cells and larvae. This was achieved by replacing the endogenous *p143* gene, which encodes an essential protein with homology to DNA helicases by a hybrid *p143* gene [54]. The hybrid *p143* gene resulted from the homologous recombination between AcMNPV and BmNPV *p143* genes, and differed from AcMNPV *p143* only in four amino acids. How these changes in *p143* affected AcMNPV host range is still not well understood. Infection of *B. mori* BmN cells by wild type AcMNPV induces protein synthesis arrest [55]. This suggests that AcMNPV *p143* or perturbations in the cell caused by AcMNPV *p143* or its activity may induce a cellular response. This example demonstrates that baculovirus host range can be manipulated through genetic engineering. However, it is important to notice that BmNPV and AcMNPV are closely related baculoviruses showing on average ORF amino acid sequence identities of about 93%. Although deletion of a gene critical for replication in one host can reduce the virus host range, in many cases, the insertion or modification of a single gene will not be sufficient to expand host range. The expanded AcMNPV host range resulting from *p143* recombination with BmNPV is probably a singular case. Functional complementation studies have conducted to the identification of other viral elements that may result in host- specific interaction. Those elements include the homologous regions *(hrs)*. *Hrs* consist of repeated units of about 70 base pairs with an imperfect 30 base pairs palindrome near their center, and have been implicated both as transcriptional enhancers and origins of DNA replication for a number of baculoviruses. It was demonstrated that *hrs* interact with host and viral factors in a species-specific way. In an interesting work baculoviruses were analyzed by bioinformatics in the search of genes subject to positive selection pressure (when the rate of nonsynonymous substitutions per potential nonsynonymous site in a gene is greater than the rate of synonymous substitutions per potential synonymous site, the gene is said to be undergoing positive selection). Since most genes appear to be subject to negative selection most of the time, this method can be used to identify viral genes involved in adapting to new or current hosts [56].

Another relevant topic to be addressed in the development of baculovirus recombinants with expanded host range is the selection of appropriate promoters for the expression of heterologous genes. If selected candidate genes for expansion of the host range are to be incorporated in the baculovirus genome under the control of their own promoters, it is necessary to evaluate the functionality of these promoters in this context.

4. Mammalian cells transduction and BacMam systems

4.1. Introduction

Initial interest in baculoviruses as gene delivery vectors for mammalian cells was driven by their good biosafety profile [84]. Compared to other human-derived viral gene delivery vectors, the safety requirements for handling baculoviruses are relatively low. Baculoviruses are so exceptionally adapted to their natural hosts that they pose no threat to vertebrate organisms. They are unable to replicate in mammalian cells, can be manipulated in laboratories at BSL1/2 levels and can be easily inactivated [85]. Moreover, insect larvae in the wild are infected via the gut by occluded baculoviruses and polyhedrin-deleted recombinant virus used to transduce mammalian cells does not efficiently infect larvae. The viruses are unstable outside of the laboratory, so they are environmentally contained as well.

Baculovirus entry into mammalian cells was suggested to depend on electrostatic interactions, heparin sulfate and phospholipids, but the exact cell surface molecules for baculovirus docking remained unknown [86]. It was also proposed that clathrin-mediated endocytosis and macropinocytosis play roles in baculovirus entry [87, 88]. Contradictorily, a recent study [89] discovered that (1) baculovirus entered cells into vesicles devoid of clathrin; (2) macropino-cytosis-related regulators imparted no significant effects on virus transduction and (3) the internalization and nuclear uptake were affected by the regulators of clathrin-independent entry. These data unveiled a baculovirus entry pathway independent of clathrin-mediated endocytosis and macropinocytosis and suggested that phagocytosis might play a role, which echoed the observations reported previously [90]. Moreover, other recent studies reported that baculovirus transduction related to direct fusion pathway induced by a short pH trigger [91]. Nevertheless, one consensus is that baculovirus envelope protein gp64 is pivotal for entry because blocking gp64 can abrogate the baculovirus ability to transduce mammalian cells and activate dendritic cells [92]. Very recently, it has been demonstrated that 6-O- and N-sulfated syndecan-1 promotes baculovirus binding and entry into mammalian cells. [93].

Numerous cell lines have been transduced [94], including primary cells in vitro and human livers ex vivo and the capability of baculovirus as a gene therapy vector has been studied. More recent studies have described the use of AcMNPV vectors in the form of BVs for in vivo targeting of different organs including brain and liver [95], and stem cells for tissue engineering [96].

The term BacMam refers to baculoviruses in which a mammalian promoter is used to drive heterologous gene expression in mammalian cells following viral transduction (Figure 5). Since the viral genome can stably accommodate an insert sequence of at least 40 kb, BacMams are particularly suitable for expression of multimeric complexes. Unless a selection force is applied, gene expression in transduced cells is transient and can usually last for up to 4 days. However, the expression can even be prolonged to 16 days. For viruses carrying a selectable marker, stable cell lines can also be established upon selection [97].

BacMams have been used as delivery vehicles to mammalian cells for many polypeptide genes, including secreted [98] and transmembrane proteins [99, 100, 101]. When high MOIs are used,

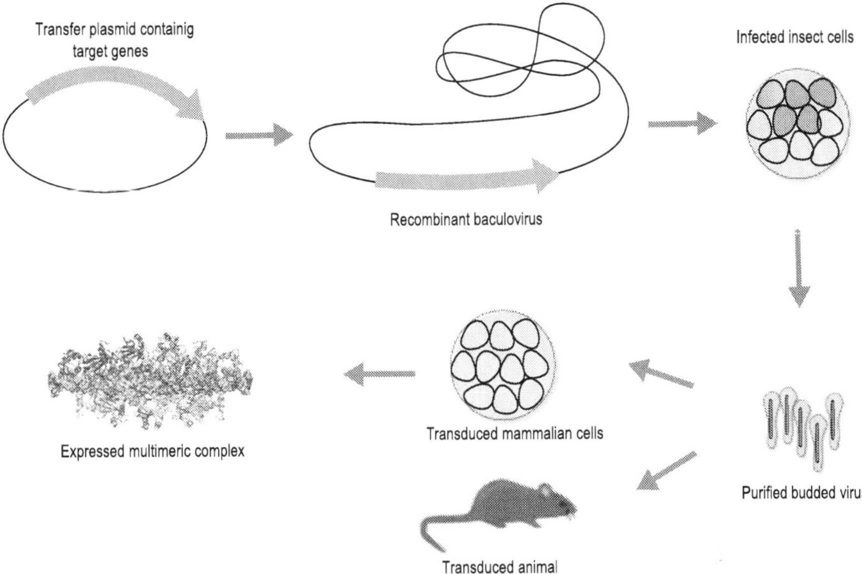

Figure 5. Production and application of BacMam viruses. Target gene sequences cloned into a transfer plasmid containing a mammalian cell-active expression cassette are transferred to baculovirus DNA via recombination. The viral DNA is transfected into insect cells where virus production occurs. Budded virus are clarified from the insect cell culture medium. The stock virus is used to transduce mammalian cells and expression of the recombinant protein(s) is usually validated 24–48 h later. The effects of the expression can be tested with this transient assay. BacMam vectors can be also used to transduce mammalian organisms.

transduction efficiencies near 100% can be reached. With this high transduction efficiency and flexibility, the technology easily enables coexpression of several genes with multiple baculoviruses and modulation of expression level by dosing and timing. This flexibility is especially relevant in studies of multimeric complex functional proteins and also in assays of processes where mix-and-match coexpression experiments with a number of cofactors and interacting partners are necessary.

4.2. Available vectors for BacMam development

The vectors used for the development of BacMams are derivatives from AcMNPV transfer vectors. The most widely used system for the generations of BacMam are based on the Bac-to-Bac system (InvitrogenTM) for baculovirus generation. With this system the recombinant baculoviral genome is constructed in *E. coli*, via a transfer vector. The gene of interest is first subcloned into a BacMam transfer vector, which is then transformed into a special *E. coli* strain DH10Bac to generate the recombinant viral DNA. The viral DNA is then used to transfect insect cells in order to generate the recombinant virus. The entire process is simple and easy to perform, allowing generation of multiple viruses simultaneously. With the procedure, recombinant BacMams can be generated in less than 2 weeks. The Bac-to-Bac system [102] uses the Tn7-mediated site-specific transposition reaction to direct integration of expression cassettes contained in the transfer vector into a baculovirus backbone vector (bacmid) preexisting in the *E. coli* DH10Bac strain. In this case, the bacmid is a mini-F replicon with the baculovirus genome and has a kanamycin resistance marker. In addition, the *E. coli* strain contains a helper plasmid that expresses the Tn7 transposase gene. The system was designed in such a way that the recombinant Tn7 transposon from the transfer vector will be integrated into a mini-attTn7 in the *lacZα* gene fragment contained within the recombinant viral genome, causing inactivation of the α-complementation of *lacZ*. The desired recombinant transformants will be resistant to tetracycline, kanamycin, and gentamicin and can be easily distinguished from nonrecombinants by blue/white selection on X-gal plates.

The BacMam transfer vectors described here are derivatives of pFastBac1 of the Bac-to-Bac system (InvitrogenTM). Originally, the AcMNPV *polh* promoter of pFastBac1 was deleted for the introduction of cassettes containing a mammalian promoter. Later, the CMV immediate early promoter was inserted to allow expression of the cloned cDNA sequences in mammalian cells. The vector pFastBacMam-1, in addition, contains a neomycin resistance gene driven by the SV40 promoter. The neomycin resistance marker allows selection of stable cell lines following BacMam transduction. Using this vector a new version was constructed (pFast-Backmam-NA) to accommodate ORFs cloned in GatewayTM vectors (Invitrogen).

4.3 Strategies to improve baculovirus transduction

4.3.1. Surface display via gp64 fusion or expression of heterologous protein

Heterologous peptides can be inserted between the signal peptide and the mature domain of the envelope fusion protein GP64, and this feature has been exploited for surface display of peptides to improve the virus transduction [103, 104], for ligand-directed targeting if an appropriate ligand is chosen [105, 106]. When a short peptide motif from gp350/220 of Epstein–Barr virus (EBV, which naturally infects B cells) was displayed as GP64 fusion peptide on the baculovirus envelope [107], the efficiency of transduction to B lymphocytes was increased. Another paradigm is the display of the immunoglobulin Fc region on the baculovirus surface [108]. Fc receptors (FcRs) are membrane proteins that bind to the Fc region of antibody and mediate the phagocytosis and antigen presentation. The Fc display allows for specific baculovirus targeting to cell lines and antigen presenting cells (APCs) expressing FcRs, hence augmenting the vaccine effect. The display system also allows for the surface presentation of functional membrane proteins to simplify subsequent isolation.

Aside from the gp64-aided display, expression of vesicular stomatitis virus G protein (VSVG) [109], influenza virus neuraminidase [110], *Spodoptera exigua* multiple nucleopolyhedrovirus F protein, single chain antibody fragments and human endogenous retrovirus envelope protein [111] in insect cells also leads to incorporation of the protein into baculovirus envelope. Among these strategies, display of VSVG or heterologous peptide/protein via the VSVG anchor is the most widely adopted and can tremendously enhance baculovirus transduction in vitro and in vivo.

Serum complement proteins (e.g. C5b-9) inactivate baculovirus, hence constituting a major hurdle in the in vivo use of baculovirus. The inactivation problem has been circumvented by the use of complement inhibitors [112] or by displaying human DAF (decay accelerating factor) via gp64 [113]. The DAF-displaying baculovirus caused lower levels of inflammatory cytokines IL-1β, IL-6, and IL-12p40 in macrophages and mitigated liver inflammation in mice when compared with the control virus. These results demonstrate that DAF display offers protection to the baculoviral vector against complement inactivation and attenuates complement-mediated inflammation injury.

4.3.2. Surface modification via capsid display, chemical coupling or electrostatic interactions

Other than the display on the envelope, heterologous protein has been displayed on the capsid by fusion with the major capsid protein VP39. The VP39 fusion with enhanced green fluorescent protein (eGFP) neither interferes with the virus assembly nor affects the virus titer, thereby enabling intracellular baculovirus trafficking and biodistribution monitoring [114]. Similarly, the ZnO binding peptide has been fused to the N-terminus of VP39 while retaining the viral infectivity and conferring the ability to bind nanosized ZnO powders [115]. Besides, by fusing the protein transduction domain (PTD) of human immunodeficiency virus (HIV) TAT protein (a protein responsible for nuclear import of HIV genome) with VP39, the engineered baculovirus results in improved transduction of various mammalian cells

Baculovirus can also be chemically conjugated with compounds such as polyethylene glycol (PEG) alone and folate [116] to improve the transduction of folate receptor-positive KB cells. Additionally, baculoviral vectors have been coated with positively charged polyethylenimine (25 kDa) through electrostatic interactions. The modification imparts baculoviral vectors resistance to human and rat serum-mediated inactivation in vitro and elevates in vivo transduction in the liver and spleen after tail vein injection into mice.

5. Baculovirus display strategies

5.1. Introduction

Recently, a novel molecular biology tool was established by the development of baculovirus surface display [117-123], using different strategies for presentation of foreign peptides and proteins on the surface of budded virions. This eukaryotic display system enables presentation of large complex proteins on the surface of baculovirus particles and has thereby become a versatile system in molecular biology.

The baculovirus system offers great potential as an eukaryotic surface display system, since the post-translational modification of the recombinant proteins is efficient and high transfection rates can be reached. These features are important for the generation of efficient surface display libraries. The principal applications of such strategies are ligand screening of surface expression libraries, for example epitope mapping, antigen display for induction of specific antibodies and presentation of proteins that increase binding to mammalian host cells.

Moreover, display strategies play an important role, as they may be used to enhance the efficiency and specificity of viral binding and entry to mammalian cells. In addition, baculovirus surface display vectors have been engineered to contain mammalian promoter elements designed for gene delivery both in vitro and in vivo. Moreover, baculovirus capsid display has recently been developed; this holds promise for intracellular targeting of the viral capsid and subsequent cytosolic delivery of desired protein moieties. Finally, the viruses can accommodate large insertions of foreign DNA and replicate only in insect cells. Together, these are attributes that are very likely to make them important tools in functional genomics and proteomics.

Display of foreign proteins or peptides on the surface of various virus particles has been valuable in a number of areas within life sciences, ranging from basic research such as protein structure–function studies to diagnostics and gene therapy. One of the most successful examples of display technology is the isolation of antibodies from large combinatorial libraries displayed on the surface of the bacteriophages [124]. The versatile principle of phage display is based on the direct physical linkage between genotype and phenotype. This linkage enables the selection of basically any protein with the desired characteristics, such as increased binding affinity or improved catalytic properties from a suitable display library [125]. Phage display comprises some severe limitations imposed by expression in the bacterial host, however, for example when large complex eukaryotic proteins that require glycosylation or particular protein folding are under study.

Over the past few years, the ability to present large complex glycoproteins on the surface of AcMNPV, has been developed into a versatile system in molecular biology. Expression of proteins or peptides on the baculoviral surface, or more recently also on the viral capsid, without compromising replication in insect cells, has shown to be useful for important applications, both *in vivo* and *in vitro*. The major envelope glycoprotein of AcMNPV is generally known as gp64. The corresponding gene encodes a type I integral membrane glycoprotein with an amino-terminal signal sequence and a carboxy-proximal transmembrane domain. The GP64 protein occurs on the viral particle as a disulphide-linked oligomer, most likely a trimer, and is responsible for viral cell entry mediated by acid-triggered membrane fusion. Structural studies on the GP64 protein have identified separate domains responsible for oligomer formation and membrane fusion. These structural characteristics of gp64 make it a good candidate as a presentation platform for the development of a eukaryotic-based viral surface display system. Modification of viral surface structures by display techniques has enabled the use of baculovirus for enhanced targeting to mammalian cells in vitro. Based on the fact that surface display may interfere with baculovirus infectivity, and that molecules which are displayed on the baculovirus envelope end up in the lysosomes of the mammalian cell and subsequent acid-induced fusion of the viral envelope in the endosomes, an approach for display of foreign protein moieties on the capsid of AcMNPV was recently developed. This system allows for presentation of desired proteins as fusions with the baculovirus major capsid protein VP39. By contrast, molecules displayed on the baculovirus capsid should escape endosomes and thereby follow the capsid through the cytoplasm into the nucleus of transduced mammalian cells. Ideally, capsid display should thus enable transfer of functional molecules into the cytoplasm and/or the nucleus of the target cells.

5.2. Baculoviral display cloning

In the first constructions used to display peptides on the surface of the budded virions the foreign open reading frames were fused to the complete GP64 coding sequence, with the parental baculovirus retaining a wild type GP64 copy [126]. The foreign genes were cloned between the gp64 signal peptide and the mature gp64 peptide. The mechanism of incorporation into the viral particle probably involves the oligomerisation of the fusion construct with wild-type GP64. Until now, only small peptides have been inserted into the protein gp64 [127, 128]. When entire proteins were inserted the virus budding efficiency decreased drastically, and titres similar to those of gp64-deletion mutants were obtained [129]. By comparison of different positions within the gp64 sequence using specific antibody epitopes, it was found that the surface probability of the inserted peptide strongly depends on the position, structural framework and the adjacent amino acids [128]. Incorporation of the fusion protein onto the viral surface usually represents only a small proportion of the total fusion protein and the levels of incorporation into the budded virus are variable and cannot be predicted. The position not only affects the viral titres obtained, but also influences the presentation of the epitope. In addition to the oligomerization domain and fusion domain, the N-terminal part of the protein also contains essential structural or sequential motifs that are more sensitive to changes than the rest of the protein.

Different promoters for the GP64 fusion protein have been evaluated to increase incorporation rates and presentation of the displayed peptide [130]. It was noticed that the use of early promoters resulted in more complete post-translational processing of glycoproteins; but the level of fusion protein detected on the surface of cells and budded virus particles was significantly enhanced when strong, very late polyhedrin promoter was used. High concentrations of the target protein are required on the cell surface in order to reach a signal-to-noise ratio that allows cell sorting to be performed by fluorescence-activated cell sorting, which, at the moment, is the only practical technique for selecting specific clones from baculovirus surface display libraries.

As an alternative to using either the entire GP64 or portions of GP64 protein as the scaffold for protein presentation, the coat protein of a different virus, vesicular stomatitis virus (VSV), or its membrane anchor domain, has also been evaluated. It was shown that by using this strategy, incorporation eGFP was extremely high [131]. The avidity of the display virus increased significantly, without putting a direct limit on the size of the target gene. In the latter cases, wild-type gp64 was still expressed in order to maintain efficient infectivity.

6. Conclusions and perspectives

The study of baculoviruses is a traditional field in virology. In particular, genetic engineering of AcMNPV emerged in the 1980s, and several systems for various purposes have been developed. However, although genetic engineering of baculoviruses seems to be a thoroughly explored area, much work is still required to fully exploit the advantages of the system.

Vectors and cells with many advantageous characteristics have been developed; yet, it would be tantalizing to assemble all these features in a single system. As mentioned before, the addition of an IRES to the transfer vector to couple the recombinant ORF to an essential BV ORF enhanced the genetic stability of the recombinant virus providing sustained recombinant protein expression. However, this feature is still not commercially available and is not compatible with many commercial systems. Other alternatives that have been explored and may be assembled in new generation systems include: selection by rescue of a lethal gene deletion, deletion of baculovirus chitinase and cathepsin, expression of chaperones and other folding proteins and expression of mammalian glycosylation pathway proteins.

The use of transgenic cell lines for expression of recombinant proteins is a convenient alternative to baculovirus infection, since the protein is not expressed in an infection context. Selection systems for the generation of transgenic insect cell lines may be optimized. Systems based on site-specific recombination would increase the rate of transgenic cell generation, thus simplifying clonal cell isolation. Negative selection systems should be explored as well. Additionally, the development of inducible expression systems would be very convenient, since they are convenient for expression of proteins that affect cell physiology.

The improvement of baculovirus bioinsecticides by means of genetic engineering is a challenging subject. Genetic stability of recombinant baculovirus is an issue, and it is addressed by strategies such as the addition of an IRES (as mentioned before) and by deleting small regions with high recombination rates. Many genes from various sources are being tested for their ability to increase baculovirus biopesticidal propierties, although small RNA-mediated silencing will probably emerge as an important alternative to foreign gene expression approach. Host range modification is even more challenging. To address this question a systematic study could start by replacing each of the baculovirus genes with related baculovirus homologs in search of functional complementation. Other various approaches may be envisaged. Also, bioinformatics studies in search of genes subjected to positive pressure are valuable to provide candidates of host-specific interaction genes. These bioinformatics studies should be updated with the recently sequenced baculovirus genomes. The use of baculovirus to transduce mammalian cell lines and mammalian organism bring baculovirus in the gene therapy and vaccine fields. One of the most challenging objectives in this area is the programming of viral particles to target specific tissues or cell types. In this direction, the replacement of the baculovirus fusion protein by other fusion proteins have shown to modify baculovirus BV tropism, and the development of targeted baculovirus is crucial for exploiting their potential as gene therapy vectors.

From this overview of the field, it is clear that there is room for many strategies and approaches to improve the various applications of genetically engineered baculoviruses.

Author details

Santiago Haase, Leticia Ferrelli, Matías Luis Pidre and Víctor Romanowski

*Address all correspondence to: shaase@biol.unlp.edu.ar

Instituto de Biotecnología y Biología Molecular, Universidad Nacional de la Plata, Conicet, Argentina

References

[1] Rohrmann GF. Baculovirus Molecular Biology: Second Edition [Internet]. Bethesda (MD): National Center for Biotechnology Information (US); 2011. Available from: http://www.ncbi.nlm.nih.gov/books/NBK49500/

[2] Herniou EA, Arif BM, Becnel JJ, Blissard GW, Bonning B, Harrison R, Jehle JA, Theilmann DA, Vlak JM: Baculoviridae. In Virus taxonomy: classification and nomenclature of viruses: Ninth Report of the International Committee on Taxonomy of Viruses. Edited by King AMQ, Adams MJ, Carstens EB, Lefkowitz EJ. San Diego: Elsevier Academic Press; (2011) pp. 163–173.

[3] Steinhaus, E.A. Insect Pathology, An Advanced Treatise (E.A. Steinhaus, ed.) Vol 2 (1963), Academic Press, New York.

[4] Smith GE, Summers MD, Frazer M J: Production of human beta interferon in insect cells infected with a baculovirus expression vector. (1983) Mol Cell Bio, 3: 2156-2165.

[5] Baculovirus Expression Vectors: A Laboratory Manual by David R. O'reilly, Lois Miller, Verne A. Luckow Spiral (1992). Oxford University Press, Usa.

[6] Miller, L. K., Kaiser, W. J., and Seshagiri, S. Baculovirus regulation of apoptosis. (1998) Sem. Virol. 8, 445–452.

[7] Monsma SA, Oomens AG, Blissard GW. The GP64 envelope fusion protein is an essential baculovirus protein required for cell-to-cell transmission of infection. (1996) J Virol; 70;4607–16.

[8] Blissard GW. Baculovirus–insect cell interactions. (1996) Cytotechnology; 20; 73-93.

[9] Hefferon KL, Oomens AG, Monsma SA, Finnerty CM, Blissard GW. Host cell receptor binding by baculovirus GP64 and kinetics of virion entry. (1999) Virology; 258; 455-68.

[10] Kingsley DH, Behbahani A, Rashtian A, Blissard GW, Zimmemberg J. A discrete stage of baculovirus GP64-mediated membrane fusion. (1999) Mol Biol Cell;10;4191–200.

[11] Lijkel WF, Westenberg M, Goldbach RW, Blissard GW, Vlak JM, Zuidema D. A novel baculovirus envelope fusion protein with a proprotein convertase cleavage site. (2000) Virology; 275; 30–41.

[12] Pearson MN, Groten C, Rohrmann GF. Identification of the Lymantria dispar nucleopolyhedrovirus envelope fusion protein provides evidence for a phylogenetic division of the Baculoviridae. (2000) J. Virol.; 74; 6126-6131.

[13] Liang C, Song J, Chen X. The GP64 protein of Autographa californica multiple nucleopolyhedrovirus rescues Helicoverpa armigera nucleopolyhedrovirus transduction in mammalian cells. (2005) J Gen Virol; 86; 1629–35.

[14] Hefferon KL, Miller LK. Reconstructing the replication complex of AcMNPV. (2002) Eur J Biochem; 269; 6233-6240.

[15] Whitford M, Stewart S, Kuzio J, Faulkner P. Identification and sequence analysis of a gene encoding gp67, an abundant envelope glycoprotein of the baculovirus Autographa californica nuclear polyhedrosis virus. (1989) J. Virol.; 63; 1393-1399.

[16] Van Der Wilk F, Van Lent JWM, Vlak JM. Immunogold detection of polyhedrin, p10 and virion antigens in Autographa californica nuclear polyhedrosis virus-infected Spodoptera frugiperda cells.(1987) J Gen Virol; 68; 2615-2624.

[17] Williams GV, Rohel DZ, Kuzio J, Faulkner P. A cytopathological investigation of Autographa californica nuclear polyhedrosis virus p10 gene function using insertion–deletion mutants. (1989) J Gen Virol; 70; 187-202.

[18] Van Oers MM, Flipsen JTM, Reusken CBEM, Vlak JM. Specificity of baculovirus p10 functions. (1994) Virology; 200; 513– 23.

[19] Hawtin RE, Zarkowska T, Arnold K, Thomas CJ, Gooday GW. Liquefaction of Autographa californica nucleopolyhedrovirus infected insects is dependent on the integrity of virus-encoded chitinase and cathepsin genes. (1997) Virology; 238; 243-253.

[20] Miller LK, Lingg AJ, Bulla LAJ. Bacterial viral and fungal insecticides. (1983) Science; 219;715-721.

[21] Inlow, D., Shauger, A., and Maiorella, B. Insect cell culture and baculovirus propagation in protein-freemedium (1989) J. Tissue Cult. Methods 12, 13-16.

[22] Chan, L. C. L, Greenfield, P. F., and Reid, S. Optimizing fed-batch production of recombinant proteins using the baculovirus expression vector system. (1998) Biotechnol. Bioeng. 59, 178–188

[23] Adang, MJ and Miller, LK. Molecular cloning of DNA complementary to the mRNA of the baculovirus Autographa californica nuclear polyhedrosis virus: Location and gene products of RNA transcripts found late in infection. (1982) J. Virol 44: 782-793.

[24] Pennock GD, Shoemaker C, Miller LK. Strong and regulated expression of Escherichia coli beta-galactosidase in insect cells with a baculovirus vector. (1984) Mol Cell Biol. Mar; 4: 399–406.

[25] Ayres, M. D., Howard, S. C., Kuzio, J., Lopez-Ferber, M., and Possee, R. D. The complete sequence of Autographa californica nuclear polyhedrosis virus. (1994) Virology 202: 586-605

[26] Ernst, W. J., Grabherr, R. M. and Katinger, H. W. Direct cloning into the Autographa californica nuclear polyhedrosis virus for generation of recombinant baculoviruses. (1994) Nucleic Acids Res. 22, 2855–2856.

[27] Lu, A. and Miller, L. K. Generation of recombinant baculoviruses by direct cloning. (1996) Biotechniques 21, 63–68.

[28] Peakman, T. C., Harris, R. A., and Gewert, D. R. Highly efficient generation of recombinant baculoviruses by enzymatically medicated site-specific in vitro recombination. (1992) Nucleic Acids Res. 20, 495–500.

[29] Patel, G., Nasmyth, K., and Jones, N. A new method for the isolation of recombinant baculovirus. (1992) Nucleic Acids Res. 20, 97–104.

[30] Smith, G.E., Fraser, M.J. and Summers, M.D. Molecular engineering of the Autographa californica nuclear polyhedrosis virus genome: Deletion mutations within the polyhedrin gene. (1983) J. Virol. 46: 584-593.

[31] Vialard, J., Lalumiere, M., Vernet, T., Briedis, D., Alkhatib, G., Henning Henning, D., Levin, D., and Richardson C.,. Synthesis of the membrane fusion and hemagglutinin proteins of measles virus, using a novel baculovirus vector containing the beta- galactosidase gene (1990) J. Virol. 64: 37-50.

[32] Monaco A.P., Larin Z. YACs, BACs, PACs and MACs: Artificial chromosomes as research tools. (1994) Trends Biotechnol; 12: 280–286.

[33] Kitts, P. A., Ayres, M. D., and Possee, R. D. Linearization of baculovirus DNA enhances the recovery of recombinant virus expression vectors. (1990) Nucleic Acids Res. 18, 5667-5672.

[34] Moscardi F. Assessment of the application of baculoviruses for control of Lepidoptera. (1999) Annu Rev Entomol; 44; 257-289.

[35] Fuxa JR, Richter AR, Ameen AO, Hammock BD. Vertical transmission of TnSNPV, TnCPV, AcMNPV, and possibly recombinant NPV in Trichoplusia ni. (2002) J Invertebr Pathol; 79; 44-50.

[36] Moscardi F, Morales L, Santos B. The successful use of AgMNPV for the control of velvet bean caterpillar, Anticarsia gemmatalis, in soybean in Brazil. (2002) Proceedings of the VIII international on invertebrate pathology and microbial control and

XXXV annual meeting of the Society for Invertebrate Pathology. Brazil, Foz do Iguassu; pp. 86– 91.

[37] Moscardi F, Santos B. Produçao comercial de nucleopoliedrosis virus de Anticarsia gemmatalis Hubner (Lep: Noctuidae) em laboratorio. (2005) Proceedings of the IX Simposio de Controle Biologico. Brazil. Recife;. p. 42.

[38] Maeda S. Increased insecticidal effect by a recombinant baculovirus carrying a synthetic diuretic hormone gene. (1989) Biochem Biophys Res Commun; 165; 1177–1183.

[39] Hammock BD, Bonning BC, Possee RD, Hanzlik TN, Maeda S. Expression and effects of the juvenile hormone esterase in a baculovirus vector. (1990) Nature (London); 344; 458-461.

[40] Ichinose R, Kamita SG, Maeda S, Hammock BD. Pharmacokinetic studies of the recombinant juvenile hormone esterase in Manduca sexta. (1992) Pestic Biochem Physiol; 42; 13-23.

[41] O'Reilly DR, Miller LK. Improvement of a baculovirus pesticide by deletion of the egt gene. (1991) Biotechnology; 9;1086– 9.

[42] Carbonell, L.F., Hodge, M.R., Tomalski, M.D. and Miller, L.K. Synthesis of a gene coding for an insecticide-specific scorpion toxin neurotoxin and attempts to express it using baculovirus vectors. (1988) Gene 73:409-418

[43] Inceoglu AB, Kamita SG, Hinton AC, Huang Q, Severson TF, Kang K-d, and B.D. Hammock. Recombinant baculoviruses for insect control. (2001) Pest Manag Sci; 57;981– 7.

[44] Tomalski, M.D., Bruce, W.A., Travis, J., and Blum, M.S. Preliminary characterization of toxins from the straw itch mite, Pyemotes tritici, which induce paralysis in the larvae of a moth. (1988) Toxicon 26: 127-137.

[45] Tomalski, M. D., Kutney, R., Bruce, W. A., Brown, M. R., Blum, M. S., & Travis, J. Purification and characterization of insect toxins derived from the mite, Pyemotes tritici. (1989).Toxicon 27:1151-1167

[46] Burden JP, Hails RS, Windass JD, Suner MM, Cory JS. Infectivity, speed of kill, and productivity of a baculovirus expressing the itch mite toxin txp-1 in second and fourth instar larvae of Trichoplusia ni. (2000) J Invertebr Pathol; 75; 226-236.

[47] Sun X, Wang H, Sun X, Chen X, Peng C, Pan D, et al. Biological activity and field efficacy of a genetically modified Helicoverpa armigera SNPV expressing an insect-selective toxin from a chimeric promoter. (2004) Biol Control; 29;124– 37.

[48] Lu, A., S. Seshagiri, and L.K. Miller. Signal sequence and promoter effects on the efficacy of toxin-expressing baculoviruses as biopesticides. (1996) Biol. Control; 7: 320-332

[49] Morris TD, Miller LK: Characterization of productive and nonproductive AcMNPV Infection in selected insect cell lines. (1993) Virology, 197: 339-348.

[50] Morris TD, Miller LK: Promoter influence on baculovirusmediated gene expression in permissive and non-permissive insect cell lines. (1992) J. Virol., 66: 7397-7405.

[51] Carbonell LF, Klowden MJ, Miller LK: Baculovirus-mediated expression of bacterial genes in dipteran and mammalian cells. (1985) J. Virol., 56:153-160.

[52] Boyce FM, Bucher NLR: Baculovirus-mediated gene transfer into mammalian cells. (1996) Proc Nat Acad Sci USA; 93: 2348–2352

[53] Beniya H, Funk CJ, Rohrmann GF, Weaver RF: Purification of a virus-induced RNA polymerase from Autographa califomica nuclear polyhedrosis virus-infected Spodoptera frugiperda cells that accurately initiates late and very late transcription in vitro. 1996 Virology, 216: 12-19.

[54] Croizier G, Croizier L, Argand 0, Poudevigne D. Extension of Autographa califomica nuclear polyhedrosis virus host range by interspecific replacement of a short DNA sequence in the p143 helicase gene. (1994) Proc Nat Acad Sci USA, 91: 48-52

[55] Kamita SG, Maeda S: Abortive infection of the baculovirus Autographa californica nuclear polyhedrosis virus in Sf-9 cells after mutation of the putative helicese gene. (1996) J Virol, 70: 6244-6250.

[56] Harrison, RL, Bonning, BC. Application of maximum-likelihood models to selection pressure analysis of group I nucleopolyhedrovirus genes. (2004) Journal of General Virology, 85, 197–210.

[57] Shizuya H., Birren B., Kim U.J., Mancino V., Slepak T., Tachiiri Y., Simon M. Cloning and stable maintenance of 300-kilobase-pair fragments of human DNA in Escherichia coli using an F-factor-based vector. (1992) Proc. Natl. Acad. Sci. U. S. A.; 89:8794–8797.

[58] O'Connor M., Peifer M., Bender W. Construction of large DNA segments in Escherichia coli. (1989) Science.; 244: 1307–1312.

[59] Airenne KJ, Peltomaa E, Hytönen VP, Laitinen OH, Ylä-Herttuala S. Improved generation of recombinant baculovirus genomes in Escherichia coli. (2003) Nucleic Acids Res; 31(17)e101.

[60] Hilton S, Kemp E, Keane G, Winstanley D. A bacmid approach to the genetic manipulation of granuloviruses. (2008) J Virol Methods.; 152: 56-62.

[61] Wang H., Deng F., Pijlman G. P., Chen X., Sun X., Vlak J. M., Hu Z. Cloning of biologically active genomes from a Helicoverpa armigera single-nucleocapsid nucleopolyhedrovirus isolate by using a bacterial artificial chromosome. (2003) Virus Res 97, 57–63.

[62] Kitts, P. A. and Possee, R. D. A method for producing recombinant baculovirus expression vectors at high frequency. (1993) Biotechniques 14, 810–817.

[63] Je, Y.H., Chang, I.H., Roh, J.Y., and Jin, B.R. Generation of baculovirus expression vector using defective Autographa californica nuclear polyhedrosis virus genome maintained in Escherichia coli for Occ+ virus production. (2001) Int. J. Indust. Entomol 2, 155-160.

[64] Je, Y.H., Jin, B.R., Park, H.W., Roh, J.Y., Chang, J.H., Seo, S.J., Olszewski, J.A., O'Reilly, D.R., and Kang, S.K. Baculovirus expression vectors that incorporate the foreign protein into viral occlusion bodies. (2003) Biotechniques 34, 81-87.

[65] Kaba, S. A., Salcedo, A. M., Wafula, P. O, Vlak, J. M., and van Oers, M. M. Development of a chitinase and v-cathepsin negative bacmid for improved integrity of secreted recombinant proteins. (2004) J. Virol. Methods; 122, 113–118

[66] van Oers, M.M., Thomas, A.A., Moormann, R.J., Vlak, J.M., Secretory pathway limits the enhanced expression of classical swine fever virus E2 glycoprotein in insect cells. (2001) J. Biotechnol. 86,31-38

[67] Jiang, Y., Deng, F., Wang, H., Hu, Z., An extensive analysis on the global codon usage pattern of baculoviruses. (2008) Arch. Virol. 153, 2273–2282.

[68] Hawtin, R.E., Zarkowska, T., Arnold, K., Thomas, C.J., Gooday, G.W., King, L.A., Kuzio, J.A., Possee, R.D., Liquefaction of Autographa californica nucleopolyhedro-virus-infected insects is dependent on the integrity of virus-encoded chitinase and cathepsin genes. (1997) Virology; 238, 243–253.

[69] Kaba, S.A., Salcedo, A.M., Wafula, P.O., Vlak, J.M., van Oers, M.M., Development of a chitinase and v-cathepsin negative bacmid for improved integrity of secreted recombinant proteins. (2004) J. Virol. Meth. 122, 113–118.

[70] Hitchman, R.B., Possee, R.D., Siaterli, E., Richards, K.S., Clayton, A.J., Bird, L.E., Owens, R.J., Carpentier, D.C.J., King, F.L., Danquah, J.O., Spink, K.G., King, L.A., Improved expression of secreted and membrane-targeted proteins in insect cells. (2010) Biotechnol. Appl. Biochem. 56, 85–93.

[71] Nene, V., Inumaru, S., McKeever, D., Morzaria, S., Shaw, M., Musoke, A., Characterization of an insect cell-derived Theileria parva sporozoite vaccine antigen and immunogenicity in cattle. (1995) Infect. Immun. 63, 503–508.

[72] Tessier, D.C., Thomas, D.Y., Khouri, H.E., Laliberte, F., Vernet, T., Enhanced secretion from insect cells of a foreign protein fused to the honeybee melittin signal peptide. (1991) Gene 98, 177–183.

[73] Shi, X., Jarvis, D.L.,. Protein N-glycosylation in the baculovirus-insect cell system (2007). Curr. Drug Targets 8, 1116–1125.

[74] Helenius, A., Aebi, M., Roles of N-linked glycans in the endoplasmic reticulum. (2004) Annu. Rev. Biochem. 73, 1019–1049.

[75] Hollister, J.R., Jarvis, D.L., Engineering lepidopteran insect cells for sialoglycoprotein production by genetic transformation with mammalian beta 1,4-galactosyltransferase and alpha 2,6-sialyltransferase genes. (2001) Glycobiology 11, 1–9.

[76] Seo, N.S., Hollister, J.R., Jarvis, D.L., Mammalian glycosyltransferase expression allows sialoglycoprotein production by baculovirus-infected insect cells. (2001) Protein Expr. Purif. 22, 234–241.

[77] Yokoyama N, Hirata M, Ohtsuka K, Nishiyama K, Fujii K, Fujita M, Kuzushima K, Kiyono T, Tsurumi T. Co-expression of human chaperone Hsp70 and Hsdj or Hsp40 co- factor increases solubility of overexpressed target proteins in insect cells, (2000) Biochimica et Biophysica Acta (BBA) - Gene Structure and Expression 1493: 119-124.

[78] Hsu TA, Eiden JJ, Bourgarel P, Meo T, Betenbaugh MJ. Effects of co-expressing chaperone BiP on functional antibody production in the baculovirus system. (1994) Protein Expr Purif.; 5: 595-603.

[79] Hsu T-A, Watson S, Eiden JJ, Betenbaugh MJ: Rescue of immunoglobulins from insolubility is facilitated by PDI in the baculovirus expression system. (1996) Protein Expr Purif, 7: 281-288.

[80] Kool, M., Voncken, J.W., van Lier, F.L., Tramper, J., Vlak, J.M., Detection and analysis of Autographa californica Nuclear Polyhedrosis Virus mutants with defective interfering properties. (1991) Virology 183, 739–746

[81] Pijlman, G.P., Dortmans, J.C., Vermeesch, A.M., Yang, K., Martens, D.E., Goldbach, R.W., Vlak, J.M., Pivotal role of the non-hr origin of DNA replication in the genesis of defective interfering baculoviruses. (2002) J Virol. 76, 5605–5611.

[82] Pijlman, G.P., de Vrij, J., van den End, F.J., Vlak, J.M., Martens, D.E., Evaluation of baculovirus expression vectors with enhanced stability in continuous cascaded insect-cell bioreactors. (2004) Biotechnol. Bioeng. 87, 743–753.

[83] Pijlman, G.P., Roode, E.C., Fan, X., Roberts, L.O., Belsham, G.J., Vlak, J.M., van Oers, M.M., Stabilized baculovirus vector expressing a heterologous gene and GP64 from a single bicistronic transcript (2006) J. Biotechnol. 123, 13–21.

[84] Volkman LE, Goldsmith PA. In vitro Survey of Autographa californica Nuclear Polyhedrosis Virus Interaction with Nontarget Vertebrate Host Cells. (1983) Appl Environ Microbiol 45: 1085–1093.

[85] Kost, T.A. and Condreay, J.P. Recombinant baculoviruses as mammalian cell gene-delivery vectors. (2002) Trends Biotechnol. 20, 173–180

[86] Duisit[†], G., S. Saleun[†], S., Douthe, S., Barsoum, J., Chadeuf, G., Moullier, P. Baculovirus vector requires electrostatic interactions including heparan sulfate for efficient gene transfer in mammalian cells. (1999) J. Gene Med. 1: 93–102.

[87] Long G, Pan X, Kormelink R, Vlak JM. Functional entry of baculovirus into insect and mammalian cells is dependent on clathrin-mediated endocytosis. (2006) J. Virol. 80: 8830-8833.

[88] Matilainen, H., J. Rinne, L. Gilbert, V. Marjomaki, H. Reunanen, and C. Oker-Blom. Baculovirus entry into human hepatoma cells. (2005) J. Virol. 79: 15452–15459.

[89] Johanna P. Laakkonen, J.P., Mäkelä, A.R., Kakkonen, E., Turkki, P., Kukkonen, S. Peränen, J., Ylä-Herttuala, S., Airenne, K.J., Oker-Blom, C., Vihinen-Ranta, M., Marjomäki, V. Clathrin-independent entry of baculovirus triggers uptake of E. coli in non-phagocytic human cells. (2009) PLoS One 4: e5093.

[90] Abe, T., Hemmi, H., Miyamoto, H., Moriishi, K., Tamura, S., Takaku, H. Akira, S., Matsuura, Y. Involvement of the Toll-like receptor 9 signaling pathway in the induction of innate immunity by baculovirus. (2005) J. Virol. 79: 2847–2858.

[91] Paul, A. and Prakash, S. Baculovirus reveals a new pH-dependent direct cell-fusion pathway for cell entry and transgene delivery. (2010) Future Virology.; 5: 533-537.

[92] Schutz A, Scheller N, Breinig T, Meyerhans A. The Autographa californica nuclear polyhedrosis virus AcNPV induces functional maturation of human monocyte-derived dendritic cells. (2006) Vaccine; 24(49/50):7190–6.

[93] Makkonen, K.-E., P. Turkki, J. P. Laakkonen, S. Yla-Herttuala, V. Marjomaki, K. J. Airenne. 6-O- and N-Sulfated Syndecan-1 Promotes Baculovirus Binding and Entry into Mammalian Cells. (2013) Journal of Virology, 87 (20): 11148.

[94] Kost, T. A. and Condreay, J. P. Recombinant baculoviruses as mammalian cell gene-delivery vectors. (2002) Trends Biotechnol. 20, 173–180.

[95] Boyce, F.M. and Bucher, N. Baculovirus-mediated gene transfer into mammalian cells. (1996) Proc. Natl. Acad. Sci. U. S. A. 93, 2348–2352.

[96] Hu, Y.C. Baculovirus as a highly efficient expression vector in insect and mammalian cells. (2005) Acta Pharmacol. Sin. 26, 406–416

[97] Chen C-Y, Lin C-Y, Chen G-Y, Hu Y-C. Baculovirus as a gene delivery vector: Recent understandings of molecular alterations in transduced cells and latest applications. (2011) Biotechnol. Adv. 29:618–631.

[98] Condreay, J.P. et al. Transient and stable expression in mammalian cells transduced with a recombinant baculovirus vector. (1999) Proc. Natl. Acad. Sci. U. S. A. 96, 127–132.

[99] Clay, W. C., Condreay, J. P., Moore, L. B., et al. Recombinant baculoviruses used to study estrogen receptor function in human osteosarcoma cells. (2003) Assay Drug Dev. Technol. 1, 801–810.

[100] Ames, R., Nuthulaganti, P., Fornwald, J., Shabon, U., van-der-Keyl, H., and Elshourbagy, N. Heterologous expression of G protein-coupled receptors in U-2 OS osteosarcoma cells. (2004) Receptors Channels. 10, 117–124.

[101] Ames, R., Fornwald, J., Nuthulaganti, P., et al. BacMam recombinant baculoviruses in G protein-coupled receptor drug discovery. (2004) Receptors Channels 10, 99–107.

[102] Luckow, V. A., Lee, S. C., Barry, G. F., and Olins, P. O. Efficient generation of infectious recombinant baculoviruses by site-specific transposon-mediated insertion of foreign genes into a baculovirus genome propagated in Escherichia coli. (1993) J. Virol. 67, 4566-4579.

[103] Grabherr R, Ernst W. Baculovirus for eukaryotic protein display. (2010) Curr Gene Ther.,10: 195-200.

[104] Grabherr R, Ernst W, Oker-Blom C, Jones IM. Developments in the use of baculoviruses for the surface display of complex eukaryotic proteins. (2001) Trends in Biotechnology 19: 231–236.

[105] Kitagawa Y, Tani H, Limn CK, Matsunaga TM, Moriishi K, Matsuura Y. Ligand-directed gene targeting to mammalian cells by pseudotype baculoviruses. (2005) J Virol 79: 3639– 3652.

[106] Mäkelä AR, Matilainen H, White DJ, Ruoslahti E, Oker-Blom C. Enhanced baculovirus- mediated transduction of human cancer cells by tumor-homing peptides. (2006). J Virol 80: 6603–6611.

[107] Martyn JC, Cardin AJ, Wines BD, Cendron A, Li S, Li S, Mackenzie J, Powell M, Gowans EJ. Surface display of IgG Fc on baculovirus vectors enhances binding to antigen- presenting cells and cell lines expressing Fc receptors. (2009) Arch Virol 154: 1129–1138.

[108] Chapple, S.D.J. and Jones, I.M. Non-polar distribution of green fluorescent protein on the surface of Autographa californica nucleopolyhedrovirus using a heterologous membrane anchor. (2002) Journal of Biotechnology 95, 269–275.

[109] Borg J, Nevsten P, Wallenberg R, Stenstrom M, Cardell S, Falkenberg C, Holm C. Amino-terminal anchored surface display in insect cells and budded baculovirus using the amino-terminal end of neuraminidase. (2004) J Biotechnol 114:21–30

[110] Lee HJ, Park N, Cho HJ, Yoon JK, Van ND, Oh YK, Kim YB. Development of a novel viral DNA vaccine against human papillomavirus: AcHERV-HP16L1. (2010) Vaccine 28: 1613–1619.

[111] Georgopoulos LJ, Elgue G, Sanchez J, Dussupt V, Magotti P, Lambris JD, Tötterman TH, Maitland NJ, Nilsson B. Preclinical evaluation of innate immunity to baculovirus gene therapy vectors in whole human blood. (2009) Mol Immunol. 46: 2911–2917.

[112] Kaname Y, et al. Acquisition of complement resistance through incorporation of CD55/decay-accelerating factor into viral particles bearing baculovirus GP64. (2010) J. Virol. 84:3210–3219.

[113] Kukkonen SP, Airenne KJ, Marjomäki V, Laitinen OH, Lehtolainen P, et al. Baculovirus capsid display: a novel tool for transduction imaging. (2003) Mol Ther 8: 853–862.

[114] Lei Song, Yingying Liu, and Jinchun Chen. Baculoviral capsid display of His-tagged ZnO inorganic binding peptide. (2010) Cytotechnology 62(2):133-41.

[115] Kim YK, Choi JY, Yoo MK, Jiang HL, Arote R, Je YH, Cho MH, Cho CS. Receptor-mediated gene delivery by folate-PEG-baculovirus in vitro. (2007). J. Biotechnol. 131 (3): 353-61.

[116] Heinis, C., Huber, A., Demartis, S, Demartis S, Bertschinger J, Melkko S, Lozzi L, Neri P, Neri D. Selection of catalytically active biotin ligase and trypsin mutants by phage display. (2001) Protein Eng., 14, 1043–1052.

[117] Boublik, Y., Di Bonito, P. and Jones, I. M., Eukaryotic virus display: Engineering the major surface glycoprotein of the Autographa californica nuclear polyhedrosis virus (AcNPV) for the presentation of foreign proteins on the virus surface. (1995) Biotechnology, 13, 1079–1084.

[118] Grabherr, R., Ernst, W., Dobblhof-Dier, O. et al. Expression of foreign proteins on the surface of Autographa californica nuclear polyhedrosis virus. (1997), Biotechniques, 22, 730–735.

[119] Mottershead, D., van der Linden, I., von Bonsdorff, C. H. et al. Baculoviral display of the green fluorescent protein and rubella virus envelope proteins. (1997) Biochem. Biophys. Res. Commun., 238, 717–722.

[120] Mottershead, D. G., Alfthan, K., Ojala, K. et al., Baculoviral display of functional scFv and synthetic IgG-binding domains. (2000) Biochem. Biophys. Res. Commun., 275, 84–90

[121] Grabherr, R., Ernst, W., Oker-Blom, C. and Jones, I., Developments in the use of baculoviruses for the surface display of complex eukaryotic proteins. (2001) Trends Biotechnol., 19: 231–236.

[122] Ojala, K., Mottershead, D. G., Suokko, A. and Oker-Blom, C. Specific binding of baculoviruses displaying gp64 fusion proteins to mammalian cells. (2001). Biochem. Biophys. Res. Commun., 284, 777–784.

[123] Clackson, T. and Wells, J. A., In vitro selection from protein and peptide libraries, (1994) Trends Biotechnol., 12, 173–184.

[124] Lowman, H. B. and Wells, J. A. Affinity maturation of human growth hormone by monovalent phage display. (1993) J. Mol. Biol., 234, 564–578.

[125] Boublik, Y., Di Bonito, P. and Jones, I. M., Eukaryotic virus display: Engineering the major surface glycoprotein of the Autographa californica nuclear polyhedrosis virus (AcNPV) for the presentation of foreign proteins on the virus surface, (1995) Biotechnology, 13, 1079–1084.

[126] Ernst, W., Spenger, A., Toellner, L. et al., Expanding baculovirus surface display: Modification of the native coat protein gp64 of Autographa californica NPV. (2000) Eur. J. Biochem., 167, 4033–4039.

[127] Spenger, A., Grabherr, R., Tollner, L. et al. Altering the surface properties of baculovirus Autographa californica NPV by insertional mutagenesis of the envelope protein gp64. (2002) Eur. J. Biochem., 269, 4458– 4467.

[128] Oomens, A. G. and Blissard, G. W. Requirement for gp64 to drive efficient budding of Autographa californica multicapsid nucleopolyhedrovirus, (1999) Virology, 254, 297–314.

[129] Grabherr, R., Ernst, W., Dobblhof-Dier, O. et al. Expression of foreign proteins on the surface of Autographa californica nuclear polyhedrosis virus, (1997) Biotechniques, 22, 730–735.

[130] Chapple, S. D. and Jones, I. M., Nonpolar distribution of green fluorescent protein on the surface of Autographa californica nucleopolyhedrovirus using a heterologous membrane anchor, (2002) J. Biotechnol., 95, 269–275.

Chapter 5

Bovine Papillomaviruses — Taxonomy and Genetic Features

Michele Lunardi, Amauri Alcindo Alfieri,
Rodrigo Alejandro Arellano Otonel and
Alice Fernandes Alfieri

Additional information is available at the end of the chapter

http://dx.doi.org/10.5772/56195

1. Introduction

Papillomaviruses (PVs) are epitheliotropic viruses that cause benign proliferative lesions in the skin (warts or papillomas) and mucous membranes (condylomas) of their natural hosts. However, certain malignant epithelial lesions have been attributed to PVs, especially in cases of cervical cancer and other human urogenital tract tumors [1].

The first DNA oncovirus recognized was also the first animal PV to be identified. Known as CRPV (cottontail rabbit papillomavirus), this virus was identified in the 1930s in warts on the skin of cottontail rabbits [2].

PVs are small, non-enveloped, icosahedral viruses that replicate in the nuclei of squamous epithelial cells. The diameter of the viral particles varies between 52 and 55 nm (Figure 1) [1].

Currently, PVs constitute a widely diverse group of DNA viruses. PVs have been found in many mammal species, as well as in certain birds and reptiles. In addition to human beings (human papillomavirus – HPV), PVs have been identified in most domestic animals, including bovines (BPV), canines (CPVs), goats (Capra hircus papillomavirus – ChPV1), equines (*Equus caballus* papillomavirus EcPVs), domestic felines (*Felis domesticus* papillomavirus – FdPV), sheep (*Ovis aries* papillomavirus – OaPV), and swine (*Sus scrofa* papillomavirus – SsPV1) [3].

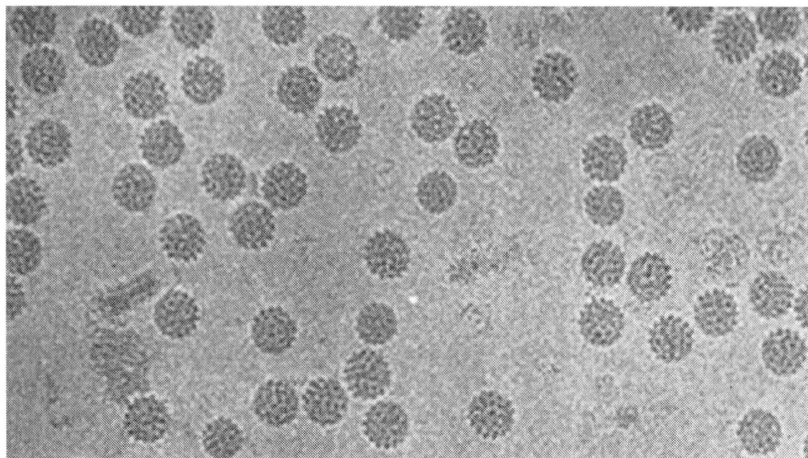

Figure 1. Electron micrograph of bovine papillomavirus type 1 virions (BPV1) (diameter of 55 nm). Source: [1].

2. Taxonomic classification of PVs

Originally, PVs were grouped with the polyomaviruses in the family *Papovaviridae*, which was justified based on such shared traits as morphologically similar non-enveloped capsids and circular double-stranded DNA genomes. Because the genomes of both groups were later found to exhibit different sizes and organizations, as well as a low similarity between their nucleotide (nt) and aminoacid (aa) sequences, PVs are currently classified in the family *Papillomaviridae* [4,5].

PVs are traditionally designated as "viral types". Each viral type represents a complete genome with the L1 gene nt sequence – which encodes the main capsid protein – exhibiting at least 10% dissimilarity compared with the same sequence from any other previously identified PV [5].

The classification of PVs into genera unites several phylogenetically related species that differ with respect to their biological properties, whereas classification based on species groups phylogenetically close viral types that also exhibit similar biological and pathological traits. In terms of nt sequence identity, these taxonomic relationships are expressed as follows: i) different genera exhibit less than 60% similarity in their L1 ORFs (open reading frames) and less than 23% to 43% when their full genomic sequences are compared; ii) different species within the same genus exhibit 60% to 70% similarity in their L1 ORFs (Table 1) [5].

Currently, the family *Papillomaviridae* contains at least 29 genera that include more than 200 PV types. The Greek alphabet is used to name the genera, which thus range from *Alphapapillomavirus* to *Dyoiotapapillomavirus*. Each species is designated according to the viral type that

Taxonomic Level	L1 ORF Identity
Genus	<60%
Species	60-70%
Viral Type	71-89%

Table 1. Relationship between diverse taxonomic levels and the identity observed on the L1 ORF nt sequences. Source: [5].

best represents it, whereas the remaining PV types classified within a single species are named as virus strains (Table 2).

Genera	Species
Alphapapillomavirus	Human Papillomavirus 2, 6, 7, 10, 16, 18, 26, 32, 34, 53, 54, 61, 90 Macaca mulata Papillomavirus 1
Betapapillomavirus	Human Papillomavirus 5, 9, 49, 92, 96 Macaca fascicularis Papillomavirus 2
Gammapapillomavirus	*Human Papillomavirus 4, 48, 50, 60, 88, 101, 109, 112, 116, 121*
Deltapapillomavirus	Alces alces Papillomavirus 1 Bos taurus Papillomavirus1 Capreolus capreolus Papillomavirus 1 Odocoileus virginianus Papillomavirus1 Ovis aries Papillomavirus 1
Epsilonpapillomavirus	*Bos taurus Papillomavirus 5*
Zetapapillomavirus	*Equus caballus Papillomavirus 1*
Etapapillomavirus	*Fringilla coelebs Papillomavirus*
Thetapapillomavirus	*Psittacus erithacus Papillomavirus 1*
Iotapapillomavirus	*Mastomys natalensis Papillomavirus 1*
Kappapapillomavirus	Oryctolagus cuniculus Papillomavirus 1 Sylvilagus floridanus Papillomavirus 1
Lambdapapillomavirus	*Canis familiaris Papillomavirus 1 and 6* Felis domesticus Papillomavirus 1 Procyon lotor Papillomavirus 1
Mupapillomavirus	*Human Papillomavirus 1 and 63*
Nupapillomavirus	*Human papillomavirus 41*
Xipapillomavirus	*Bos taurus Papillomavirus 3*
Pipapillomavirus	Mesocricetus auratus Papillomavirus 1 Micromys minutus Papillomavirus 1
Rhopapillomavirus	*Trichechus manatus latirostris Papillomavirus 1*

Genera	Species
Sigmapapillomavirus	*Erethizon dorsatum Papillomavirus 1*
Taupapillomavirus	*Canis familiaris Papillomavirus 2*
Upsilonpapillomavirus	*Tursiops truncatus Papillomavirus 1* and *2*
Phipapillomavirus	*Capra hircus Papillomavirus 1*
Chipapillomavirus	*Canis familiaris Papillomavirus 3* and *4*
Psipapillomavirus	*Rousettus aegyptiacus Papillomavirus 1*
Omegapapillomavirus	*Ursus maritimus Papillomavirus 1*
Dyodeltapapillomavirus	*Sus scrofa Papillomavirus 1*
Dyoepsilonpapillomavirus	*Francolinus leucoscepus Papillomavirus 1*
Dyozetapapillomavirus	*Caretta caretta Papillomavirus 1*
Dyoetapapillomavirus	*Erinaceus europaeus Papillomavirus 1*
Dyothetapapillomavirus	*Felis domesticus Papillomavirus 2*
Dyoiotapapillomavirus	*Equus caballus Papillomavirus 2*

Table 2. Classification of *Papillomaviridae* family. Source: [3].

PVs isolated from vertebrates are classified into 24 genera, whereas viral species that occur exclusively in birds and reptiles are grouped into three genera and one genus, respectively. The taxonomic nomenclature of animal PV types is based on the scientific name of their hosts according to the genus and species. For example, FdPV1 is the name given to PV of the domestic cat (*Felis domesticus*) type 1 [3]. An exception occurs in the case of the bovine papillomavirus, which was named *Bos taurus* papillomavirus but by consensus is usually referred to as BPV. Table 3 describes the genera and species of PVs identified in various species of domestic animals.

Because PVs are not amenable to isolation using classic cell culture techniques and do not induce a strong humoral immune response in their hosts, the taxonomic terms "strain" and "serotype" were not originally applied to this virus family. Consequently, the family classification is based on the similarities between nt sequences and a limited number of biological and medical properties [5,6].

3. Genomic organization

In the 1970s, the cloning of PVs genomes contributed substantially to the study of their biological and biochemical properties. Sequencing of the cloned genomes allowed the identification of different ORFs as probable viral genes [7].

The genomic organization of the various PVs is notably similar. A common feature of PVs is that all of the ORFs are contained on a single strand of the viral DNA. Therefore, only one of

Genera	Species	Viral Strains
Deltapapillomavirus	Bos taurus Papillomavirus 1	Bos taurus Papillomavirus 1
		Bos taurus Papillomavirus 2
	Ovis aries Papillomavirus 1	Ovis aries Papillomavirus 1
		Ovis aries Papillomavirus 2
Epsilonpapillomavirus	Bos taurus Papillomavirus 5	Bos taurus Papillomavirus 5
		Bos taurus Papillomavirus 8
Zetapapillomavirus	Equus caballus Papillomavirus 1	Equus caballus Papillomavirus 1
Lambdapapilloamvirus	Canis familiaris Papillomavirus 1	Canis familiaris Papillomavirus 1
	Canis familiaris Papillomavirus 6	Canis familiaris Papillomavirus 6
	Felis domesticus Papillomavirus 1	Felis domesticus Papillomavirus 1
Xipapillomavirus	Bos taurus Papillomavirus 3	Bos taurus Papillomavirus 3
		Bos taurus Papillomavirus 4
		Bos taurus Papillomavirus 6
		Bos taurus Papillomavirus 9
		Bos taurus Papillomavirus 10
		Bos taurus Papillomavirus 11
		Bos taurus Papillomavirus 12
Taupapillomavirus	Canis familiaris Papillomavirus 2	Canis familiaris Papillomavirus 2
		Canis familiaris Papillomavirus 7
Phipapillomavirus	Capra hircus Papillomavirus 1	Capra hircus Papillomavirus 1
Chipapillomavirus	Canis familiaris Papillomavirus 3	Canis familiaris Papillomavirus 3
		Canis familiaris Papillomavirus 5
	Canis familiaris Papillomavirus 4	Canis familiaris Papillomavirus 4
Dyodeltapapillomavirus	Sus scrofa Papillomavirus 1	Sus scrofa Papillomavirus 1
Dyothetapapillomavirus	Felis domesticus Papillomavirus 2	Felis domesticus Papillomavirus 2
Dyoiotapapillomavirus	Equus caballus Papillomavirus 2	Equus caballus Papillomavirus 2
-	-	Bos taurus Papillomavirus 7

Table 3. Papillomavirus species that infect domestic animals. Source: Adapted from [3].

the DNA strands serves as a template for transcription. The coding strand might exhibit up to 10 ORFs, which are classified according to the cell differentiation stage when they are expressed by means of the letters E (early) and L (late). The early genomic segment (E) comprises up to eight ORFs, which are expressed in epithelial cells in the early stages of maturation. The late segment (L) usually contains two ORFs that are expressed in differentiated keratinocytes. A third region without ORFs has been identified in all PV genomes and is named the LCR (long control region) or URR (upstream regulatory region).This region contains the origin of replication and elements that control transcription (Figure 2) [1].

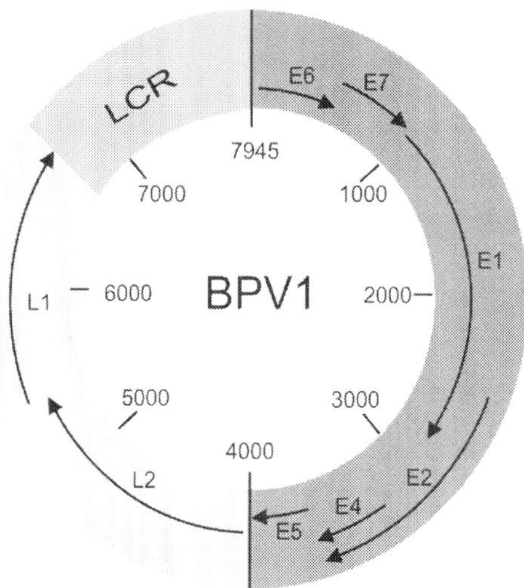

Figure 2. Schematic representation of genome of bovine papillomavirus type 1 (BPV1).

Expression of the six most common non-structural and regulatory proteins (E1, E2, E4, E5, E6, and E7), which are encoded by the early viral genome region, occurs in basal cells or during the intermediate stages of maturation. The expression of the two viral structural proteins (L1 and L2) encoded by the late genomic segment occurs in keratinocytes in the final stage of maturation [8].

4. Viral proteins

4.1. Proteins E1 and E2

Protein E1 is encoded by the largest ORF found in the early genomic segment of PVs. Significant homology among the different PVs has been found upon comparison of the amino acid sequences inferred for the proteins that are encoded by the E1 gene [9].

Together with viral protein E2, protein E1 recognizes the origin of replication and represents the central factor of PV replication. It is indispensable for the initiation of viral DNA replication. In addition to this main function, E1 participates in the recruitment of host cell replication proteins and exhibits intrinsic ATPase/helicase activity, which induces relaxation of the DNA coiling at the origin of replication and during the progression of the replication fork [9].

Transformation studies using BPV1 have shown that the presence of an intact E1 ORF is crucial for the maintenance of viral genome stability in cells through the presence of multiple genomic copies in episomal form [10,11,12].

However, the interaction between protein E1 and the origin of replication exhibits low specificity. Specific and efficient recognition of the origin of replication occurs exclusively through the cooperative binding of proteins E1 and E2 to sites adjacent to the origin of replication. Therefore, protein E2 participates in this mechanism as an aggregation factor that promotes the recruitment of helicase E1 to the origin of replication [13].

The BPV1 E2 ORF encodes a protein that comprise the central viral regulatory system and thus control genetic expression and viral replication. Protein E2 also modulates the transcription of the early viral promoters through its binding sites [14]. In addition, E2 participates in the maintenance of the viral genome in its episomal form by promoting binding between these genomes and mitotic chromosomes during cell division [15,16].

4.2. Protein E4

The non-structural protein E4 occurs abundantly in the cytoplasm of the differentiated keratinocytes of papillomas. Therefore, although the gene that encodes this protein is located in the early viral genome region, E4 is produced later in the differentiation process. The E4 protein of HPV16 has also been associated with the collapse of cytokeratin filaments, which thus suggests an auxiliary function in the process of viral exit from cells [1].

4.3. Proteins E5, E6 and E7

In humans, the oncoproteins E5, E6, and E7 encoded by the genomes of certain HPVs, represent the primary viral factors related to the onset and progression of cervical cancer. These genetic products are able to override the negative regulation of cell growth that is mediated by host cell proteins. In addition, it is believed that these viral oncoproteins promote the genomic instability observed in HPV-related cancers [17].

Binding with proteins of the retinoblastoma family is the main mechanism by which protein E7 contributes to the escape of infected cells from the negative regulatory mechanisms of cell growth. In the case of HPV, protein E7 interacts with these cellular factors and targets them for degradation [18]. The result of such binding and degradation is the release and activation of E2F transcription factors that regulate the expression of genes during S phase of the cell cycle. Efficient interaction between E7 and these factors triggers a compensatory inhibition of cell growth and apoptosis that is mediated by the p53 tumor suppressor protein-dependent pathway [17].

The targeting of protein p53 for degradation by viral protein E6 in high-risk HPVs eliminates the inhibition of cell growth in both undifferentiated and differentiated cells [17]. The actions of the viral proteins E6 and E7 to abrogate these regulatory factors of the cell cycle allow infected cells undergoing differentiation to remain in S phase. As a result, many cell cycle checkpoints are abrogated. Consequently, an accumulation of mutations and progression into cancer occurs in cells that are persistently infected by these viruses [19].

Most of the tumors of cattle affected by enzootic hematuria express the BPV2 oncoprotein E5 [20,21,22]. The onset of cellular transformation that is triggered by E5 might occur mainly through its interaction with and activation of the platelet-derived growth factor (PDGF) β receptor. Thus, a mitogenic response is induced even in the absence of PDGF [23,24].

4.4. Proteins L1 and L2

The viral capsid consists of the two structural proteins L1 (*ca.* 55 kDa) and L2 (*ca.* 70 kDa), being L1 the major capsid protein and representing approximately 80% of the total virus protein [25]. Virus-like particles (VLPs) can be produced using prokaryotic and eukaryotic systems to express combination of L1 and L2 or L1 alone [26,27]. Although L2 is not needed for viral assembly, it is incorporated in the VLPs when it is co-expressed with L1. Under cryoelectron microscopy, the morphology of the VLPs that contain only L1 appear identical to that of intact viral particles (Figure 3) [28]. The epitopes that induce the production of neutralizing antibodies are principally found on L1 but might also be present on L2 (Table 4) [29].

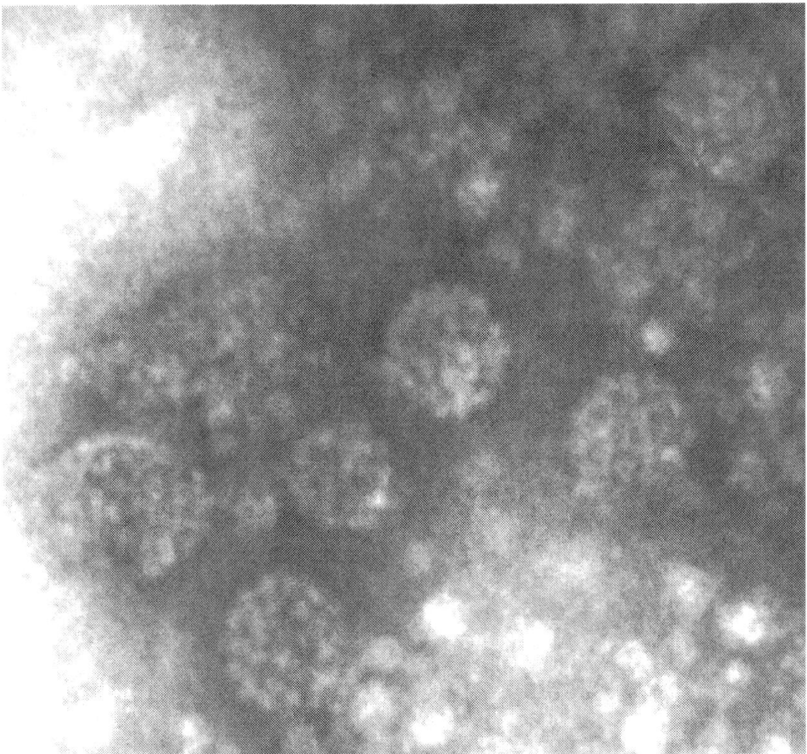

Figure 3. Electron micrograph of VLP sproduced through expression of BPV2 L1.

Viral Proteins	Approximate size (kDa)	Function / activity
E1	68.1	recognition of origin of replication / helicase activity
E2	34.3	recruitment of E1 to the origin of replication / modulation of the transcription of the early viral promoters
E4	12.5	presumed auxiliary function in the virion exit from infected cells
E5	5.2	interacts with and activates the platelet-derived growth factor (PDGF) β receptor
E6	15.8	targets p53 tumor suppressor protein
E7	13.6	binds to proteins of the retinoblastoma family
L1	55.5	component of the viral capsid
L2	50.5	component of the viral capsid

Table 4. Viral proteins and their functions.

5. Clinical conditions in cattle

Infections by different BPV types are related to several clinical conditions in cattle. The occurrence of the benign skin tumors that characterize cutaneous papillomatosis might be found in several areas of the animals' bodies. Depending on the extent of lesions, the development of the animals might be affected, they might become predisposed to secondary infections and/or infestations, and their hidescan be damaged. These possibilities are a few of the potential consequences that might result in economic losses for the beef and, even more so, dairy industries. Papillomas affecting the udders and teats of lactating cows cause difficulties with feeding calves and manual and mechanical milking, whereas secondary bacterial infections predispose the animals to clinical and/or subclinical ascending mastitis [30].

The interaction between specific BPV types and prolonged bracken (*Pteridium aquilinum*) intake has been suggested as the cause of enzootic hematuria and upper gastrointestinal tract cancers in cattle. With regard to enzootic hematuria, it is believed that latent or subclinical infections with BPV1 or BPV2 occur first in the bladder mucosa. Because the bladder represents the main target of bracken toxins, once the virus is established, infection might be reactivated, which might induce neoplasia through the immunosuppressant and carcinogenic chemical compounds present in bracken, which results in progression to malignancy [31].

Although the incidence of such tumors varies among cattle raised on bracken-infested pastures, it might be higher than 90% among adult animals [31,32].

With regard to gastrointestinal tract tumors, the immunosuppression associated with bracken intake is defining for the persistence of BPV4-induced papillomas, which might progress into malignant carcinomas under the influence of the carcinogenic elements present in bracken [33,34,35].

Therefore, although infection by these BPVs plays a central role in the pathogenesis of these cattle neoplasias, the presence of environmental and biological cofactors is essential for the development of such lesions [22,36].

6. Diversity of BPVs

Although the genomic sequences of approximately 150 HPV types have already been characterized, at the beginning of the 1980s only six BPV types (BPV1 to 6) had been identified from cases of bovine cutaneous papillomatosis and cancer [37-42].

Studies performed from the beginning of the 2000s onward to investigate the actual diversity of BPVs have indicated the existence of many BPV types, which is similar to observations made regarding the human virus. The first such study employed the generic primer pair FAP59/FAP64 on swabs of healthy skin from 19 species of vertebrates. In six of the 10 analyzed bovines that did not exhibit any clinical sign compatible with BPV infection, one or two putative new BPV types were detected. These putative new viral types were named BAA1 through BAA5 [43].

Subsequently, a study aimed at establishing the prevalence of BPV in teat papillomas and teat healthy skin used the primer pairs FAP59/FAP64 and MY09/MY11 to analyze 15 teat papillomas and 122 swabs of teat healthy skin on cattle from five Japanese prefectures [44]. That study found four previously characterized BPV types (BPV1, 3, 5, and 6), two of the previously identified putative new BPV types (BAA1 and 5), and 11 additional putative new types (named BAPV1 through 10 and BAPV11MY) among the 39 BPV-positive samples. Nevertheless, the putative new types BAA1 and BAPV7 through 10 were detected only in samples of healthy skin. In addition, during one outbreak of mammary papillomatosis that occurred in Japan and affected 560 heifers, the presence of BPV6 was confirmed in the majority of the 16 analyzed samples [45]. The previously described putative new types BAA5 and BAPV1 were also identified in these animals.

Although cutaneous papillomatosis poses a serious sanitary problem in beef and, even more so, dairy cattle, studies aimed at identifying the BPV types involved in the occurrence of skin lesions in Brazilian cattle are only sporadically performed. Recently, the detection of BPV1, 2, 6, and 8 in papillomas of cattle from the state of Parana was accomplished using generic FAP primers [46,47]. In another study, the identification of four previously undescribed putative new BPV types, named BPV/BR-UEL 2 through 5, pointed to the occurrence of considerable viral diversity among Brazilian cattle [48]. The genetic characterization of one of these new BPV types, namely, BPV/BR-UEL2, through sequencing of the full L1 gene, confirmed that it belongs to the genus *Xipapillomavirus* [49] (Figure 4).

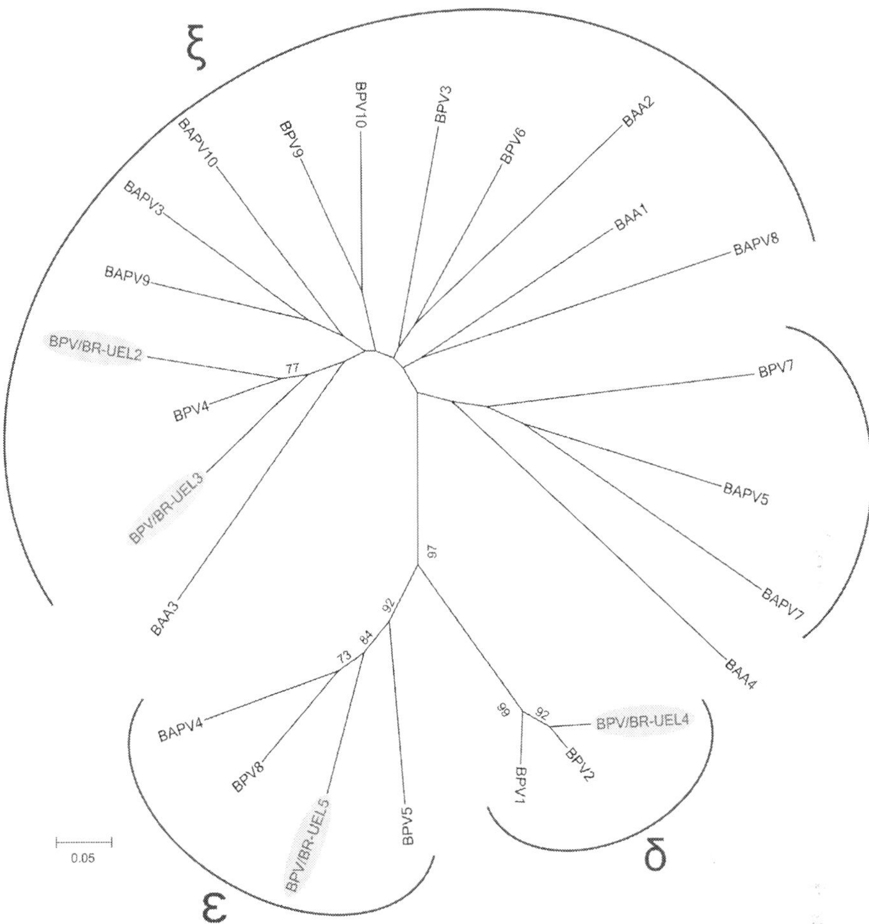

Figure 4. Phylogenetic reconstruction based on L1 ORF partial nucleotide sequences (FAP amplicons) demonstrating the classification suggested for the putative new BPV types into the genera *Deltapapillomavirus* (δ), *Epsilonpapillomavirus* (ε), *Xipapillomavirus* (ξ), and a yet unnamed genus that includes BPV7. The numbers at the internal nodes represent the bootstrap support values determined for 1000 replicates. The BPV/BR-UEL2, 3, 4, and 5 types are indicated by shading. Source: [48].

7. New BPV types

Recently, complementary analysis of several putative new BPV types through sequencing of the full viral genomes allowed the characterization of these new viral types [3]. The first such new type to be characterized was BPV7, which was initially named BAPV6. Because the nt sequence of the BPV7 L1 ORF is more closely related to PVs of the genera *Betapapillomavirus*, *Gammapapillomavirus*, and *Pipapillomavirus*, which include viruses causing skin lesions in human beings and the mucosa of hamsters, this new BPV type constitutes a new and yet unnamed genus [50].

The second recently described BPV type is BPV8, formerly known as BAPV2, which was identified in Japan. The description of this new viral type was performed together with the description of a variant named BPV8-EB, which was detected in a case of cutaneous papillomatosis in a European bison born in Italy [51]. The high degree of similarity observed between the L1 ORF sequences of BPV8 and BPV5 (75%), as well as the results of the phylogenetic analysis, were the basis for classifying this new viral type in the genus *Epsilonpapillomavirus*. In addition, the genomic structures of the early and late regions of these two different members of the genus were almost identical. The only difference exhibited between them was in the E4 ORF, which is present in BPV8 but absent in BPV5.

Recently, two BPV types, namely, BPV9 and 10, were identified from teat papillomas [52]. These new viral types were initially designated BAPV1 and BAA5 [43,44]. Phylogenetic analysis and the greater similarity of the L1 ORFs with BPV3 (74.2% and 71.2%, respectively) allowed the classification of these two new isolates in the genus *Xipapillomavirus* [52].

Hatama [53] assessed the viral genotypes present in 167 skin warts in Japanese herds through polymerase chain reaction (PCR), cloning, and sequencing. A total of 124 of the assessed lesions tested positive for BPV using PCR. Three putative new BPV types, and eight previously described BPV types (BPV1, 2, 3, 4, 5, 6, 9, and 10) were identified in the partial sequences obtained from sequencing the PCR products. The characterization of the full sequence of one of the new BPV types (BPV11) and the comparison of its L1 gene nt sequence to other members of this viral family allowed its classification in the genus *Xipapillomavirus* [53].

The complete genome sequence of an isolate identified from an epithelial tongue lesion in a Japanese bovine was recently obtained, and this isolate was named BPV12 [54]. Comparison of the BPV12 L1 gene nt sequence to other viral types isolated from cattle suggested that it should also be classified in the genus *Xipapillomavirus*.

Recently, the sequencing of the complete genome of the putative new viral type BPV-BR-UEL4, which was isolated from a skin papilloma on a cow from a herd in southern Brazil,was performed by subjecting the viral genome to rolling circle amplification (RCA), PCR, the subsequent cloning of two long amplicons, and sequencing by means of primer walking. Phylogenetic analysis based on the L1 ORF nt sequences of 45 PVs distributed among 17 genera, including the previously sequenced BPV types and PVs identified from different artiodactyl species, showed that the new viral type, named BPV13, belongs to the genus *Deltapapillomavirus*, which is generally dominated by artiodactyl PVs and also includes BPV1 and 2 (Figure 5). As previously reported for BPV1 and 2, the putative E7 protein of BPV13 does not contain a retinoblastoma tumor suppressor-binding domain. Additionally, the BPV13 E5 ORF also encodes a small transforming protein (Figure 6) [55]. The combination of these two different biological aspects has been recognized as a distinct marker for fibropapilloma development. This pathogenic mechanism appears to be unique among delta-PVs [56].

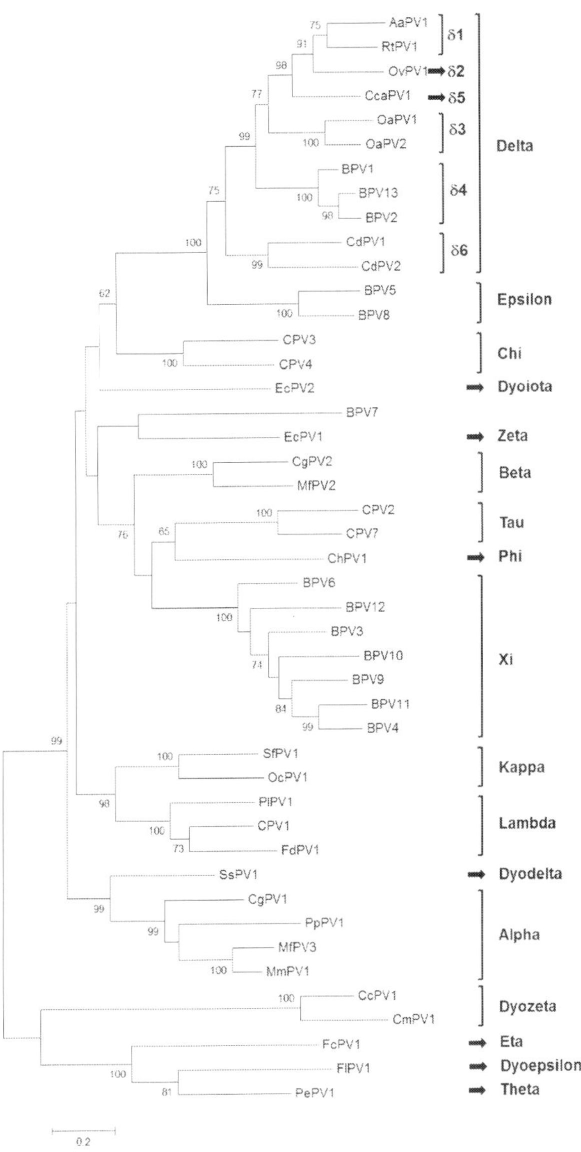

Figure 5. Phylogenetic tree based on L1 ORF nt sequences. In addition to 14 genera where animal PVs are classified, the genera *Deltapapillomavirus*, *Epsilonpapillomavirus*, and *Xipapillomavirus*, which contain BPVs, are indicated in the tree. Additionally, the six species classified within the *Deltapapillomavirus* genus are shown. The numbers at the internal nodes represent the bootstrap support values determined in 1000 replications. Source: [55]

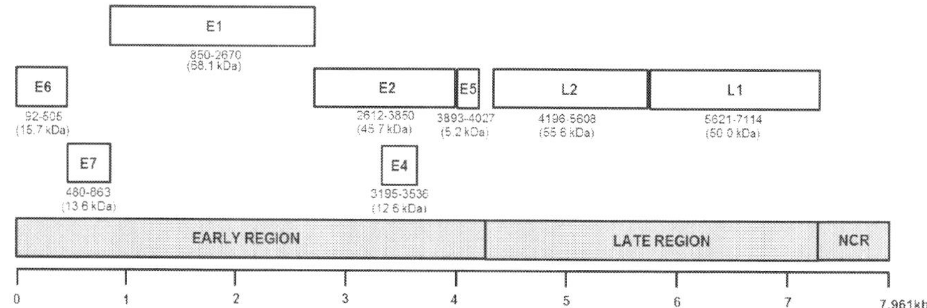

Figure 6. A diagram of the genomic organisation of BPV13. The three main regions characteristic of PV genomes are shown as grey rectangles above the ruler. The viral genome is represented as linear, and ORFs are shown as white rectangles. The numbers below each ORF indicate the nt positions of the start to stop codons and the corresponding molecular mass (in parentheses) for each putative viral protein. Source: [55].

8. Co-infections and heterologous infections

In Brazil, the association between BPV infection and the occurrence of cutaneous papillomatosis, enzootic hematuria, and upper gastrointestinal neoplasias has been confirmed in cattle [46,48,57-59]. Previous studies have found BPV in tissues other than the skin epithelium. Thus, BPV1, 2, and 4 have been identified in the embryos and female reproductive tissues of infected cows [60-63]. Similarly, BPV DNA has been detected in samples of blood, milk, urine, seminal fluid, and spermatozoa from infected cattle. These findings point to the possible participation of these fluids and cell types in BPV transmission [64].

In addition, other studies have shown the occurrence of multiple infections in cattle exhibiting several cutaneous papillomas that are caused by different BPV types and the possibility of viral co-infections in single lesions [65-68]. Additionally, the presence of several BPV types in single lesions is similar to the situation in human skin, where co-infection by more than 10 viral types is frequently detected [69].

Each papillomavirus is known to exhibit specificity for a single animal host species in which it replicates productively. However, only a few viral types are also able to infect a second animal species. In such cases, non-productive infections, that is, infections without the production of infective virions, are the result [1]. This type of infection is the case for the equine sarcoid, which can be defined as a fibroblastic locally invasive skin tumor. Sarcoid is the most frequent neoplasia affecting equine species, and it represents the best-known example of heterologous PV infection because it is caused by BPV1 and 2 [70]. In addition to horses, donkeys, and mules, skin lesions caused by these viral types have also been described in zebras and buffaloes [71,72]. Another example of heterologous PV infection is provided by the detection of DNA from FeSarPV (feline sarcoid-associated papillomavirus), a putative new PV type that was initially identified in feline sarcoids with non-productive infections, in fibropapillomas and skin samples from cattle with dermatitis [73-75]. The recent detection of FeSarPV in biological samples from cattle strengthens the hypothesis that cattle might be the natural host of this virus [73].

9. Vaccines against BPV

Immunity against BPV is considered to be type-specific, and the immune status of the infected animals is considered to be the crucial factor for clinical progression. Whereas humoral immunity prevents new infections, cellular immunity (possibly mediated by T lymphocytes) is associated with the spontaneous and immune-mediated regression of established lesions [76].

The finding that epitopes that induce the production of neutralizing antibodies are present in the structural proteins L1 and L2 explains the success of the use of these proteins in the production of vaccines [34].

The recent availability of VLP-based immunogens against HPV that are able to protect mainly against infection by HPV16 and 18 has allowed the development of the first vaccine against one of the main human neoplasias, *i.e.*, cervical cancer [77]. The data that have been collected since the implementation of the HPV vaccine are quite encouraging, and these vaccines seem to be highly efficient [78,79].

In addition, preventive vaccines have been developed for cattle that are mainly against BPV2 and 4. These viral types were selected because they represent the cutaneous and mucous BPVs, respectively, and are associated with the development of cancer in cattle [80]. A vaccine prepared with the BPV2 L1 capsid protein produced as a beta-galactosidase fusion protein in Escherichia coli induced the production of neutralizing antibodies and was able to prevent infection [81]. A similar effect was achieved using an E. coli derived BVP1 L1 protein, which protected calves against post-vaccine challenge with a homologous virus [82].

VLPs produced from the L1 or L1 and L2 genes from BPV4 have also proven to be highly immunogenic and produce powerful prophylactic vaccines. The prevention of infection during challenge with BPV4 through vaccination with L1 VLPs has shown that L1 promotes the production of neutralizing antibodies [34]. Vaccination with VLPs produced from BPV4 L1 and L2 proteins in insect cells also efficiently prevented the development of experimentally induced papillomas [83].

Because BPV does not grow in conventional cell cultures for the production of killed or attenuated live vaccines, protein expression systems, such as yeast and insect cells, have been used to produce VLP vaccines. However, the use of these systems is expensive. Recently, as has been described for other papillomaviruses (e.g., HPV16), a candidate vaccine against BPV1 consisting of L1 VLPs produced *in planta* elicited a strong and specific immune response, which demonstrated its potential as a future vaccine that could be produced at a lower cost [84].

Because the viral life cycle and the progression from benign to malignant lesion are similar in humans and animals, animal PVs and their natural hosts have represented good models for the study of HPV [30,85]. In addition, animal PVs, particularly BPV1 and 4, SfPV1, and CPV1, have also served as models for vaccines against PVs, and observation of the induction of protective immunity through the use of VLP-based vaccines in their corresponding host species has opened the way for the implementation of VLPs in HPV vaccines [86,87].

Currently, immunization through the use of L2 protein peptides has been suggested as an alternative to the use of VLP-based HPV vaccines. Curiously, residues at the N terminus of the L2 viral protein appear to represent a cross-neutralizing epitope capable of eliciting a broad-spectrum protection against many different viral types [87]. Once more, vaccination of both cattle and rabbits with L2-based vaccines has been highly protective against challenge with infectious virus [88,89], which confirms the great potential of L2 vaccines in preventing HPV infections.

10. Conclusion

BPV infection is associated with cutaneous papillomatosis, enzootic hematuria and upper gastrointestinal tract cancers in cattle. Although approximately 150 HPV types have already been characterized, only 13 BPV types had their genomes sequenced. However, in accordance with diversity observed from HPVs, the identification of numerous putative new BPV types through partial L1 gene sequences have pointed to the occurrence of a similar diversity among BPVs. Historically, together with a few animal PVs, BPVs and cattle have represented good models for the study of HPV and vaccines against this important human pathogen, fact that opened the way for the implementation of VLPs in HPV vaccines. The demonstration of protective effect of L2-based vaccines in this animal species reinforces the possibility of the future use of L2 protein peptides as an alternative vaccine to prevent HPV infections.

Acknowledgements

This work was supported by the Brazilian funding agencies CAPES, CNPq, FINEP and Fundação Araucaria (FAP/PR). A.A.A. and A.F.A. are supported by research fellowships from CNPq.

Author details

Michele Lunardi[1], Amauri Alcindo Alfieri[2]*, Rodrigo Alejandro Arellano Otonel[2] and Alice Fernandes Alfieri[2]

*Address all correspondence to: alfieri@uel.br

1 Laboratory of Veterinary Microbiology, Veterinary Teaching Hospital, Universidade de Cuiaba, Cuiaba, MT, Brazil

2 Laboratory of Animal Virology, Department of Veterinary Preventive Medicine, Universidade Estadual de Londrina, Londrina, PR, Brazil

References

[1] Howley PM, Lowy DR. Papillomaviruses. In: Knipe DM, Howley PM. (Eds.) Fields' Virology, Philadelphia: Lippincott Williams and Wilkins; 2007. p 2300-2354.

[2] Shope RE, Hurst EW. Infectious Papillomatosis of Rabbits; with a Note on the Histopathology. The Journal of Experimental Medicine 1933;58 607-624.

[3] Bernard HU, Burk RD, Chen Z, van Doorslaer K, Hausen H, de Villiers EM. Classification of Papillomaviruses (PVs) Based on 189 PV Types and Proposal of Taxonomic Amendments. Virology 2010;401 70-79

[4] van Regenmortel MHV, Fauquet CM, Bishop DHL, Calisher CH, Carsten EB, Estes MK, Lemon SM, Maniloff J, Mayo MA, Mcgeoch DJ, Pringle CR, Wickner RB. Virus Taxonomy. Seventh Report of the International Committee for the Taxonomy of Viruses. New York, San Diego: Academic Press; 2002.

[5] de Villiers EM, Fauquet C, Broker TR, Bernard HU, Zur Hausen H. Classification of Papillomaviruses. Virology 2004;324 17-27.

[6] Fauquet CM, Mayo MA, Maniloff J, Desselberger U, Ball LA. Virus Taxonomy. The Eighth Report of the International Committee on Taxonomy of Viruses. Family Papillomaviridae. Elsevier; 2005.

[7] Danos O, Katinka M, Yaniv M. Human Papillomavirus la Complete DNA Sequence: A Novel type of Genome Organization Among Papovaviridae. The EMBO Journal 1982;1 231-236.

[8] Zheng ZM, Baker CC. Papillomavirus Genome Structure, Expression, and Post-transcriptional Regulation. Frontiers in Bioscience 2006;11 2286-2302.

[9] Wilson VG, West M, Woytek K, Rangasamy D. Papillomavirus E1 Proteins: Form, Function, and Features. Virus Genes 2002;24 275-290.

[10] Sarver N, Rabson MS, Yang YC, Byrne JC, Howley PM. Localization and Analysis of Bovine Papillomavirus Type 1 Transforming Functions. Journal of Virology 1984;52 377-388.

[11] Lusky M, Botchan MR. Genetic analysis of Bovine Papillomavirus Type 1 Trans-acting Replication Factors. Journal of Virology 1985;53 955-965.

[12] Rabson MS, Yee C, Yang YC, Howley PM. Bovine Papillomavirus Type 1 3' Early Region Transformation and Plasmid Maintenance Functions. Journal of Virology 1986;60 626-634.

[13] Sanders CM, Stenlun, A. Mechanism and Requirements for Bovine Papillomavirus, Type 1, E1 Initiator Complex Assembly Promoted by the E2 Transcription Factor Bound to Distal Sites. The Journal of Biological Chemistry 2001;276 23689–23699.

[14] Spalholz B, Yang Y, Howley P. Transactivation of a Bovine Papilloma Virus Transcriptional Regulatory Element by the E2 Gene Product. Cell 1985;42 183–191.

[15] Ilves I, Kivi S, Ustav M. Long-term Episomal Maintenance of Bovine Papillomavirus Type 1 Plasmids is Determined by Attachment to Host Chromosomes, which is Mediated by the Viral E2 Protein and its Binding Sites. Journal of Virology 1999;73 4404–4412.

[16] Lehman CW, Botchan MR. Segregation of Viral Plasmids Depends on Tethering to Chromosomes and is Regulated by Phosphorylation. Proceedings of the National Academy of Sciences of the United States of America 1998;95 4338-4343.

[17] Moody CA, Laimins LA. Human Papillomavirus Oncoproteins: Pathways to Transformation. Nature Reviews. Cancer 2010;10 550-560.

[18] Dyson N, Howley PM, Munger K, Harlow E. The Human Papilloma Virus16 E7 Oncoprotein is Able to Bind to the Retinoblastoma Gene Product. Science 1989;243 934–937.

[19] Duensing S, Munger K. Mechanisms of Genomic Instability in Human Cancer: Insights from Studies with Human Papillomavirus Oncoproteins. International Journal of Cancer 2004;109 157–162.

[20] Campo MS, Jarrett WFH, Barron RJ, O'neil BW, Smith KT. Association of Bovine Papillomavirus Type 2 and Bracken Fern with Bladder Cancer in Cattle. Cancer Research 1992;53 6898-6904.

[21] Borzacchiello G, Iovane G, Marcante ML, Poggiali F, Roperto F, Roperto S, Venuti A. Presence of Bovine Papillomavirus Type 2 DNA and Expression of the Viral Oncoprotein E5 in Naturally Occurring Urinary Bladder Tumors in Cows. Journal of General Virology 2003;84 2921-2926.

[22] Roperto S, Brun R, Paolini F, Urraro C, Russo V, Borzacchiello G, Pagnini U, Raso C, Rizzo C, Roperto F, Venuti A. Detection of Bovine Papillomavirus Type 2 (BPV2) in the Peripheral Blood of Cattle with Urinary Bladder Tumors: Possible Biological Role. Journal of General Virology 2008;89 3027-3033.

[23] DiMaio D, Mattoon D. Mechanisms of Cell Transformation by Papillomavirus E5 Proteins. Oncogene 2001;20 7866-7873.

[24] Borzacchiello G, Russo V, Gentile F, Roperto F, Venuti A, Nitsch L, Campo MS, Roperto S. Bovine Papillomavirus E5 Oncoprotein Binds to the Activated Form of the Platelet-derived Growth Factor Beta Receptor in Naturally Occurring Bovine Urinary Bladder Tumours. Oncogene 2006;25 1251-60.

[25] Favre M. Structural Polypeptides of Rabbit, Bovine, and Human Papillomaviruses. Journal of Virology 1975;15 1239-1247.

[26] Rose RC, Bonnez W, Reichman RC, Garcea RL. Expression of Human Papillomavirus Type 11 L1 Protein in Insect Cells: in Vivo and in Vitro Assembly of Viruslike Particles. Journal of Virology 1993;67 1936-1944.

[27] Zhou J, Stenzel DJ, Sun, XY, Frazer IH. Synthesis and Assembly of Infectious Bovine Papillomavirus Particles in Vitro. Journal of General Virology 1993;74 763-768.

[28] Hagensee ME, Olson NH, Baker TS, Galloway DA. Three-dimensional Structure of Vaccinia Virus-produced Human Papillomavirus Type 1 Capsids. Journal of Virology 1994;68 4503-4505.

[29] Roden RB, Weissinger EM, Henderson DW, Booy F, Kirnbauer R, Mushinski JF, Lowy DR, Schiller Jt. Neutralization of Bovine Papillomavirus by Antibodies to L1 and L2 Capsid Proteins. Journal of Virology 1994;68 7570-7574.

[30] Campo MS. Animal Models of Papillomavirus Pathogenesis. Virus Research 2002;89 249-261.

[31] Roperto S, Borzacchiello G, Brun R, Leonardi L, Maiolino P, Martano M, Paciello O, Papparella S, Restucci B, Russo V, Salvatore G, Urraro C, Roperto F. A Review of Bovine Urothelial Tumors and Tumor-Like Lesions of the Urinary Bladder. Journal of Comparative Pathology 2010;142 95-108.

[32] Pamukcu AM, Price JM, Bryan GT. Naturally Occurring and Bracken Fern Induced Bovine Urinary Bladder Tumors. Veterinary Pathology 1976;13 110-122.

[33] Campo MS, O'neil BW, Barron RJ, Jarrett WF. Experimental Reproduction of the Papilloma-carcinoma Complex of the Alimentary Canal in Cattle. Carcinogenesis 1994;15 1597-1601.

[34] Campo MS. Bovine Papillomavirus and Cancer. Veterinary Journal 1997;154 175-188.

[35] Borzacchiello G, Ambrosio V, Roperto S, Poggiali F, Tsirimonakis E, Venuti A, et al. Bovine Papillomavirus Type 4 in Oesophageal Papillomas of Cattle from the South of Italy. Journal of Comparative Pathology 2003;128 203-206.

[36] Roperto S, Russo V, Borzacchiello G, Sandulli S, De Martino L, Roperto F. Infezione Batteriche Della Vescica in Bovini Affetti da Tumori Uroteliale. Proceedings of the Annual Meeting of Italian Society of Veterinary Sciences 2006;60 99-100.

[37] Pfister H, Linz U, Gissmann L, Huchthausen B, Hoffman D, Zur Hausen H. Partial Characterization of a New Type of Bovine Papillomavirus. Virology 1979;96 1–8.

[38] Campo MS, Moar MH, Jarrett WFH, Laird HM. A New Papillomavirus Associated with Alimentary Cancer in Cattle. Nature 1980;286 180-182.

[39] Campo MS, Moar MH, Laird HM, Jarrett WFH. Molecular Heterogeneity and Lesion Site Specificity of Cutaneous Bovine Papillomaviruses. Virology 1981;113 323-335.

[40] Campo MS, Coggins LW. Molecular Cloning of Bovine Papillomavirus Genomes and Comparison of their Sequence Homologies by Heteroduplex Mapping. Journal of General Virology 1982;63 255–264.

[41] Chen EY, Howley PM, Levinson AD, Seeburg PH. The Primary Structure and Genetic Organization of the Bovine Papillomavirus Type 1 Genome. Nature 1982;299 529-534.

[42] Jarrett WFH, Campo MS, O'neil BW, Laird HM, Coggins LW. A Novel Bovine Papillomavirus (BPV6) Causing True Epithelial Papillomas of the Mammary Gland Skin: a Member of a Proposed New BPV Subgroup. Virology 1984;136 255-264.

[43] Antonsson A, Hansson BG. Healthy Skin of Many Animal Species Harbors Papillomaviruses which are Closely Related to their Human Counterparts. Journal of Virology 2002;76 12537-42.

[44] Ogawa T, Tomita Y, Okada M, Shinozaki K, Kubonoya H, Kaiho I, Shirasawa H. Broad-spectrum Detection of Papillomaviruses in Bovine Teat Papillomas and Healthy Teat Skin. Journal of General Virology 2004;85 2191-2197.

[45] Maeda Y, Shibahara T, Wada Y, Kadota K, Kanno T, Uchida I, Hatama S. An Outbreak of Teat Papillomatosis in Cattle Caused by Bovine Papilloma Virus (BPV) Type 6 and Unclassified BPVs. Veterinary Microbiology 2007;121 242-248.

[46] Claus MP, Vivian D, Lunardi M, Alfieri AF, Alfieri AA. Phylogenetic Analysis of Bovine Papillomavirus Associated with Skin Warts in Cattle Herds From the State of Paraná. Pesquisa Veterinária Brasileira 2007;27 314-318.

[47] Claus MP, Lunardi M, Alfieri AF, Sartori D, Fungaro MHP, Alfieri AA. Identification of the Recently Described new Type of Bovine Papillomavirus (BPV8) in Brazilian Beef Cattle Herd. Pesquisa Veterinária Brasileira 2009;29 25-28.

[48] Claus MP, Lunardi M, Alfieri AF, Ferracin LM, Fungaro MHP, Alfieri AA. Identification of Unreported Putative new Bovine Papillomavirus Types in Brazilian Cattle Herds. Veterinary Microbiology 2008;132 396-401.

[49] Lunardi M, Claus MP, Alfieri AA, Fungaro MHP, Alfieri AF. Phylogenetic Position of an Uncharacterized Brazilian Strain of Bovine Papillomavirus in the Genus Xipapillomavirus Based on Sequencing of the L1 Open Reading Frame. Genetics and Molecular Biology 2010;33 745–749.

[50] Ogawa T, Tomita Y, Okada M, Shirasawa H. Complete Genome and Phylogenetic Position of Bovine Papillomavirus Type 7. Journal of General Virology 2007;88 1934-1938.

[51] Tomita Y, Literak I, Ogawa T, Jin Z, Shirasawa H. Complete Genomes and Phylogenetic Positions of Bovine Papillomavirus Type 8 and a Variant Type from a European Bison. Virus Genes 2007;35 243-249.

[52] Hatama S, Nobumoto K, Kanno T. Genomic and Phylogenetic Analysis of Two Novel Bovine Papillomaviruses, BPV9 and BPV10. Journal of General Virology 2008;89 158-163.

[53] Hatama S, Ishihara R, Ueda Y, Kanno T, Uchida I. Detection of a Novel Bovine Papillomavirus type 11 (BPV11) Using Xipapillomavirus Consensus Polymerase Chain Reaction Primers. Archives of Virology 2011;156 1281- 1285.

[54] Zhu W, Dong J, Shimizu E, Hatama S, Kadota K, Goto Y, Haga T. Characterization of Novel Bovine Papillomavirus Type 12 (BPV-12) Causing Epithelial Papilloma. Archives of Virology 2012;157 85-91.

[55] Lunardi M, Alfieri AA, Otonel RAA, de Alcântara BK, Rodrigues WB, de Miranda AB, Alfieri AF. Genetic Characterisation of a Novel Bovine Papillomavirus Member of the Deltapapillomavirus Genus. Veterinary Microbiology 2012, http://dx.doi.org/10.1016/j.vetmic.2012.08.030.

[56] Narechania A, Terai M, Chen Z, DeSalle R, Burk RD. Lack of the Canonical pRB-binding Domain in the E7 ORF of Artiodactyl Papillomaviruses is Associated with the Development of Fibropapillomas. The Journal of General Virology 2004;85 1243-1250.

[57] dos Santos RCS, Lindsey CJ, Ferraz OP, Pinto JR, Mirandola RS, Benesi FJ, Birgel EH, Pereira CAB, Beçak W. Bovine Papillomavirus Transmission and Chromosomal Aberrations: an Experimental Model. Journal of General Virology 1998;79 2127-2135.

[58] Wosiacki SR, Barreiros MAB, Alfieri AF, Alfieri AA. Semi-Nested-PCR for Detection and Typing of Bovine Papillomavirus Type 2 in Urinary Bladder and Whole Blood from Cattle with Enzootic Haematuria. Journal of Virological Methods 2005;126 215-219.

[59] Wosiacki SR, Claus MP, Alfieri AF, Alfieri AA. Bovine Papillomavirus Type 2 Detection in the Urinary Bladder of Cattle with Chronic Enzootic Haematuria. Memórias do Instituto Oswaldo Cruz 2006;101 635-638.

[60] de Carvalho C, de Freitas AC, Brunner O, Góes LGB, Cavalcante AY, Beçak W, dos Santos RCS. Bovine Papillomavirus Type 2 in Reproductive Tract and Gametes of Slaughtered Bovine Females. Brazilian Journal of Microbiology 2003;34 82-84.

[61] de Freitas AC, de Carvalho C, Brunner O, Birgel-Junior EH, Dellalibera AMMP, Benesi FJ, Gregory L, Beçak W, dos Santos RCS. Viral DNA Sequences in Peripheral Blood and Vertical Transmission of the Virus: a Discussion About BPV1. Brazilian Journal of Microbiology 2003;34 76-78.

[62] Yaguiu A, de Carvalho C, de Freitas AC, Góes LGB, Dagli MLZ, Jr Birgel EH, Beçak W, dos Santos, RCS. PapillomatosIs in Cattle: in Situ Detection of Bovine Papillomavirus DNA Sequences in Reproductive Tissues. Brazilian Journal of Morphological Sciences 2006;23 525-529.

[63] Yaguiu A, Dagli ML, Birgel EH, Jr., Alves Reis BC, Ferraz OP, Pituco EM, et al. Simultaneous Presence of Bovine Papillomavirus and Bovine Leukemia Virus in Different Bovine Tissues: in Situ Hybridization and Cytogenetic Analysis. Genetics and Molecular Research 2008;7 487-97.

[64] Lindsey CL, Almeida ME, Vicari CF, Carvalho C, Yaguiu A, Freitas AC, Beçak W, Stocco RC. Bovine Papillomavirus DNA in Milk, Blood, Urine, Semen, and Spermatozoa of Bovine Papillomavirus-infected Animals. Genetics and Molecular Research 2009;8 310-318.

[65] Carvalho CC, Batista MV, Silva MA, Balbino VQ, Freitas AC. Detection of Bovine Papillomavirus Types, Co-Infection and a Putative New BPV11 Subtype in Cattle. Transboundary and Emerging Diseases 2012;59 441-447.

[66] Claus MP, Lunardi M, Alfieri AA, Otonel RAA, Ferracin LM, Pelegrinelli Fungaro MH, Alfieri AF. A Bovine Teat Papilloma Specimen Harboring Deltapapillomavirus (BPV-1) and Xipapillomavirus (BPV-6) Representatives. Brazilian Archives of Biology and Technology 2009;52 87-91.

[67] Claus MP, Lunardi M, Alfieri AA, Otonel RAA, Sartori D, Fungaro MHP, Alfieri AF. Multiple Bovine Papillomavirus Infections Associated with Cutaneous Papillomatosis in Brazilian Cattle Herds. Brazilian Archives of Biology and Technology 2009;52 93-98.

[68] Schmitt M, Fiedler V, Muller M. Prevalence of BPV Genotypes in a German Cowshed Determined by a Novel Multiplex BPV Genotyping Assay. Journal of Virological Methods 2010;170 67-72.

[69] Antonsson A, Forslund O, Ekberg H, Sterner G, Hansson BG. The Ubiquity and Impressive Genomic Diversity of Human Skin Papillomaviruses Suggest a Commensalic Nature of These Viruses. Journal of Virology 2000;74 11636-11641.

[70] Chambers G, Ellsmore VA, O'Brien PM, Reid SWJ, Love S, Campo MS, Nasir L. Association of Bovine Papillomavirus with the Equine Sarcoid. Journal of General Virology 2003;84 1055-1062.

[71] van Dyk E, Bosman AM, van Wilpe E, Williams JH, Bengis RG, van Heerden J, et al. Detection and Characterisation of Papillomavirus in Skin Lesions of Giraffe and Sable Antelope in South Africa. Journal of the South African Veterinary Association 2011;82 80-85.

[72] Pangty K, Singh S, Goswami R, Saikumar G, Somvanshi R. Detection of BPV-1 and-2 and Quantification of BPV-1 by Real-Time PCR in Cutaneous Warts in Cattle and Buffaloes. Transboundary and Emerging Diseases 2010;57 185-96.

[73] Munday JS, Knight CG. Amplification of Feline Sarcoid-associated Papillomavirus DNA Sequences from Bovine Skin. Veterinary Dermatology 2010;21 341-344.

[74] Teifke JP, Kidney BA, Lohr CV, Yager JA. Detection of Papillomavirus-DNA in Mesenchymal Tumour Cells and not in the Hyperplastic Epithelium of Feline Sarcoids. Veterinay Dermatology 2003;14 47-56.

[75] da Silva MA, Carvalho CC, Coutinho LC, Reis MC, de Aragão Batista MV, de Castro RS, Dos Anjos FB, de Freitas AC. Co-infection of Bovine Papillomavirus and Feline-Associated Papillomavirus in Bovine Cutaneous Warts. Transboundary and Emerging Diseases. 2012. doi: 10.1111/j.1865-1682.2012.01307.x.

[76] Nicholls PK, Stanley MA. The Immunology of Animal Papillomaviruses. Veterinary Immunology and Immunopathology 2000;73 101-27.

[77] zur Hausen H. Papillomaviruses-to Vaccination and Beyond. Biochemistry-Moscow 2008;73 498-503.

[78] Harper DM, Franco EL, Wheeler C, Ferris DG, Jenkins D, Schuind A, Zahaf T, Innis B, Naud P, De Carvalho NS, Roteli-Martins CM, Teixeira J, Blatter MM, Korn AP, Quint W, Dubin G. Efficacy of a Bivalent L1 Virus-like Particle Vaccine in Prevention of Infection with Human Papillomavirus Types 16 and 18 in Young Women: a Randomised Controlled Trial. Lancet 2004;364 1757-1765.

[79] Villa LL, Costa RLR, Petta CA, Andrade RP, Ault KA, Giuliano AR, Wheeler CM, Koutsky LA, Malm C, Lehtinen M, Skjeldestad FE, Olsson SE, Steinwall M, Brown DR, Kurman RJ, Ronnett BM, Stoler MH, Ferenczy A, Harper DM, Tamms GM, Yu J, Lupinacci L, Railkar R, Taddeo FJ, Jansen KU, Esser MT, Sings HL, Saah AJ, Barr E. Prophylactic Quadrivalent Human Papillomavirus (types 6, 11, 16, and 18) L1 Virus-like Particle Vaccine in Young Women: a Randomised Double-blind Placebo-controlled Multicentre Phase II Efficacy Trial. Lancet Oncology 2005;6 271-278.

[80] Campo MS, Grindlay GJ, Oneil BW, Chandrachud LM, Mcgarvie GM, Jarrett WFH. Prophylactic and Therapeutic Vaccination against a Mucosal Papillomavirus. Journal of General Virology 1993;74 945-953.

[81] Jarrett WFH, Smith KT, Oneil BW, Gaukroger JM, Chandrachud LM, Grindlay GJ, McGarvie GM, Campo MS. Studies on Vaccination against Papillomaviruses - Prophylactic and Therapeutic Vaccination with Recombinant Structural Proteins. Virology 1991;184 33-42.

[82] Pilacinski WP, Glassman DL, Glassman KF, Reed DE, Lum MA, Marshall RF, Muscoplat CC, Faras AJ. Immunization against Bovine Papillomavirus Infection. Ciba Foundation Symposium 1986;120 136-156.

[83] Kirnbauer R, Chandrachud LM, ONeil BW, Wagner ER, Grindlay GJ, Armstrong A, McGarvie GM, Schiller JT, Lowy DR, Campo MS. Virus-like Particles of Bovine Papillomavirus Type 4 in Prophylactic and Therapeutic Immunization. Virology 1996;219 37-44.

[84] Love AJ, Chapman SN, Matic S, Noris E, Lomonossoff GP, Taliansky M. In Planta Production of a Candidate Vaccine against Bovine Papillomavirus Type 1. Planta. 2012. doi: 10.1007/s00425-012-1692-0.

[85] Borzacchiello G, Roperto F, Nasir L, Campo MS. Human Papillomavirus Research: Do we Still Need Animal Models? International Journal of Cancer 2009;125 739-740.

[86] Campo MS. Vaccination against Papillomavirus in Cattle. Clinics in Dermatology 1997;15 275–283.

[87] Campo MS, Roden RB. Papillomavirus Prophylactic Vaccines: Established Successes, New Approaches. Journal of Virology 2010;84 1214-1220.

[88] Gaukroger JM, Chandrachud LM, O'Neil BW, Grindlay GJ, Knowles G, Campo MS. Vaccination of Cattle with Bovine Papillomavirus Type 4 L2 Elicits the Production of Virus-neutralizing Antibodies. The Journal of General Virology 1996;77 1577-1583.

[89] Embers ME, Budgeon LR, Pickel M, Christensen ND. Protective Immunity to Rabbit Oral and Cutaneous Papillomaviruses by Immunization with Short Peptides of L2, the Minor Capsid Protein. Journal of Virology 2002;76 9798–9805.

Chapter 6

Baculovirus Display: A Novel Tool for Vaccination

Matías L. Pidre, M. Leticia Ferrelli,
Santiago Haase and Víctor Romanowski

Additional information is available at the end of the chapter

http://dx.doi.org/10.5772/55572

1. Introduction

Baculoviruses are enveloped viruses that infect insect larvae mainly from the order Lepidoptera. Their genomes are circular double-stranded DNA molecules of about 80 to 180 kbp and are packed in rod-shaped nucleocapsids with a typical size of 40-50 nm in diameter and 200-400 nm in length.

Among the numerous baculoviruses, *Autographa californica* multiplenucleopolyhedrovirus (AcMNPV) is the most widely studied and used in biotechnology.

During its infection cycle it produces two phenotypes. Occlusion derived viruses (ODV) initiate the infection at the larvae midgut. After this primary infection, the viral progeny consists of budded viruses (BV) that carry on the systemic infection in larvae. These types of virions differ in their efficiencies of infection for different cell types; ODV infect midgut epithelial cells up to 10,000 fold more efficiently than BV. In contrast, BV are up to 1,000-fold more efficient at infecting cultured cells than ODV. As the viral propagation in cell culture is mediated by BV phenotype (Rohrmann, 2011), most of the knowledge regarding baculovirus infection cycle is based on studies performed in insect cells infected by BV (Figure 1).

Cell entry is mediated by a class III viral glycoprotein located at the virion surface, Gp64, which interacts with an unknown cell receptor (Backovic & Jardetzky, 2009). This interaction triggers clathrin-dependent endosomal internalization. This internalized vesicle becomes subsequently acidified. This causes a conformational change in Gp64 that result in the fusion of the viral envelope with the endosome membrane. Thus the nucleocapsid is released in the cytoplasm and migrates to the nucleus. Once in the nucleus, DNA is uncoated and the transcriptional cascade begins (Figure 2).

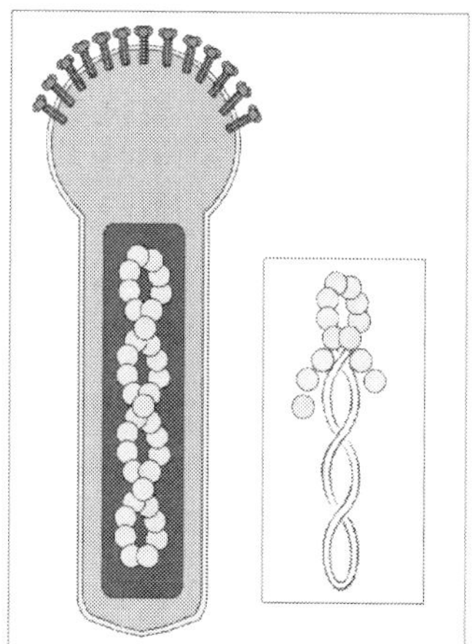

Figure 1. Structure of the budded virus.

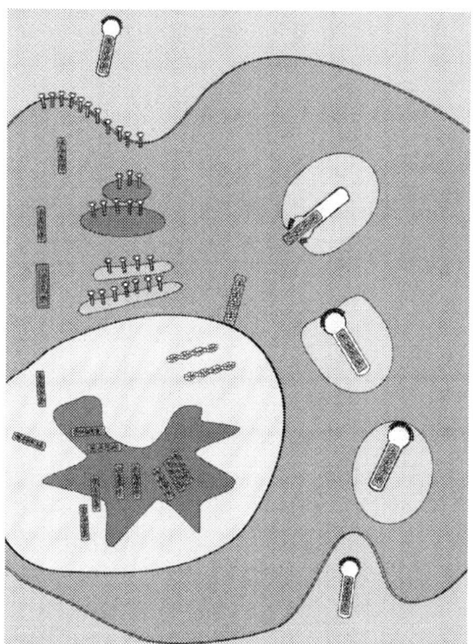

Figure 2. Baculovirus replication cycle. Infection cycle initiates when a budded virus (BV) interacts with the cell membrane and is endocytosed. When the endocytic vesicle is acidified, GP64 fusion protein, located at the BV membrane, trigger the fusion of the plasma membrane and the BV envelope releasing the nucleocapsid in the cytoplasm. The nucleocapsid is then transported to the nuclus where it transcribes its genes, replicates its DNA in the virogenic stroma where new nucleocapsids are assembled. Nucleocapsids then egress from the nucleus, travel to the cytoplasmic membrane and bud through aquiring an envelope containing the surface protein GP64.

AcMNPV genome encodes about 150 genes which are transcribed in a temporal fashion. Firstly, immediate early genes are transcribed by the host RNA-polymerase II. These genes generally encode for transcription factors, like Ie1, that aid the subsequent transcription of genes. After this early phase DNA replication occurs. Immediately after DNA replication there may be a transient period when proteins are not bound to the DNA and this might expose late promoters and facilitate their activation (Rohrmann, 2011). Baculoviruses also encode a novel RNA polymerase that transcribes late and very late genes and that recognizes the unique baculoviral promoter consensus sequence DTAAG. During the systemic infection nucleocapsids are assembled in the virogenic stroma. The envelope proteins are synthesized, translated in association with the endoplasmic reticulum, glycosylated and transported to and incorporated into the cytoplasmic membrane via the Golgi apparatus. Nucleocapsids destined to become BV exit the nucleus. They move to the cytoplasmic membrane at the site where envelope proteins (Gp64 and F protein) concentrate, and bud through obtaining their envelopes. Early in the systemic infection more BV are produced which spread the infection throughout the insect. Finally, late in infection, occluded virions are produced, and the cell dies releasing the occlusion bodies.

There are many biotechnological uses for baculoviruses. One of the most widespread is their use as insecticide agents. There have been much work on the development of baculoviruses to control insects but the acceptance and use of viruses for insect control has been limited. This can be attributed to their slow speed of kill and their limited host range. At present many research groups are working with the aim of overcoming these limitations developing novel strategies such as baculovirus-mediated expression of toxic proteins for insects. Moreover, recombinant baculoviruses have been extensively used as expression vectors in insect cell cultures. A variety of technological improvements have eliminated the tedious procedures to isolate the recombinant viruses turning the baculovirus-based expression system in a safe, easy to use and scale up system. (Kost et al., 2005).

Another application of baculovirus is their use as expression vectors for eukariotic proteins. Their ability to include quite large DNA extra fragments in their genomes and the possibility to use their very strong polihedryn promoter, which activates upon infection, make baculoviruses a very useful tool in biotechnology for the production of recombinant proteins in insect cells.

In addition, protein expression in larvae or cell culture is not the only application of baculoviruses. In fact, baculoviruses are widely used in the development of strategies for displaying foreign peptides and proteins on the virus surface as well as mammalian cell transduction using different mammalian expression cassettes. Baculovirus display consists of the expression of proteins or peptides in the surface of a baculovirus. This is achieved by fusing the protein of interest with the major baculoviral envelope glycoprotein Gp64, resulting in the localization of the chimeric protein on the viral envelope and the plasmatic membraneof infected cells. The surface displaying of antigenic epitopes make baculoviruses efficient vaccine vehicles capable of mounting a strong specific immune response.

The aim of this chapter will be to describe the biotechnological utilities of baculovirus display. Particularly, it will describe this technique for vaccination and gene delivery. It will discuss the adjuvant effects of baculoviruses and the immunity response of recombinant viruses. Moreover, other applications of baculovirus display such as gene therapy and high throughput screening of antibodies and antigenic epitopes libraries will also be addressed.

2. Baculoviral fusion proteins

Entry of enveloped viruses into host cells requires fusion of the viral envelope with the cytoplasmic membrane by the action of viral envelope fusion proteins. If the fusion occurs at the cell surface, viral fusion proteins typically act at neutral pH. On the other hand, in receptor-mediated endocytosis the major fusion protein activity is most often observed at the acidic endosomal pH (Monsma & Blissard, 1995).

In general, baculovirus fusion proteins mediate the membrane fusion at the late endosomal phase. For this reason, the major fusogenic activity was observed at low pH. Although it has been possible to identify which are the proteins that build fusogenic function, which is the cell receptor that recognizes these proteins remains a mystery.

Baculoviruses can be divided into two different groups according to the surface glycoprotein they use to mediate the fusion between the endosomal membrane and the viral envelope. One group is composed by viruses that use Gp64 as its fusogenic protein whereas the other group uses the F protein to mediate membrane fusion. This division is coincident with a phylogenetic separation of lepidopteran NPVs into the two major Groups I and II. These two groups differ significantly in gene content, most notably Group I NPVs use GP64 as their BV fusion protein, whereas Group II NPVs lack gp64 and utilize F protein (Zanotto et al., 1993).

AcMNPV is one of the most widely described baculovirus and belongs to Group I. It presents on its surface the major glycoprotein Gp64 and the residual F protein. While the F protein does not develop any specific function, Gp64 has been identified as the glycoprotein responsible for membrane fusion.

In this section it will be described the structure and function of glycoprotein Gp64 as responsible for the fusion of membranes and its biotechnological applications for the presentation of foreign antigens.

2.1. Gp64: Structure and function

Three classes of viral membrane fusion proteins havebeen identified. Class I which contain N-terminalhydrophobic fusion peptides, Class II, which fusion peptides are located in internal loops, and Class III that exhibit distinctstructural features in their architectures as well as in theirmembrane interacting fusion loops. Gp64 belongs to this latter group.

The major envelope protein of the budded virions, GP64, has been shown to mediate acid-triggered membrane fusion both in virions and when expressed alone in transfected cells. The native GP64 is a phosphoglycoprotein fatty acid acylated near the transmembrane domain (Monsma & Blissard, 1995). The Gp64 open reading frame (ORF) of AcMNPV encodes a 512 aminoacids polypeptide with 15 cysteine residues. The resulting disulfide bonds participate in the formation of the native structure.

As a member of the Class III fusion proteins, Gp64 is composed of five domains that result in a macromolecular structure very distinct from any reported class I or class II fusion protein. However, Gp64 conserves the typical characteristics of viral fusion proteins. It includes a fusion domain which mediates the fusion between the cell membrane and viral envelope; a transmembrane domain which anchors the protein in the lipidic bilayer and a multimerization domain that allows the protein to form trimmers. The detailed structure of AcMNPV Gp64 is shown in Figure 3 (Backovic & Jardetzky, 2009) Baculovirus gp64 also contains a seven residue C-terminal tail domain (CTD). Deletion of this domain does not significantly affect the ability to mediate fusion, but reduces the baculovirus titers to 50%. These data indicate that CTD is involved in virus budding (Figure 3).

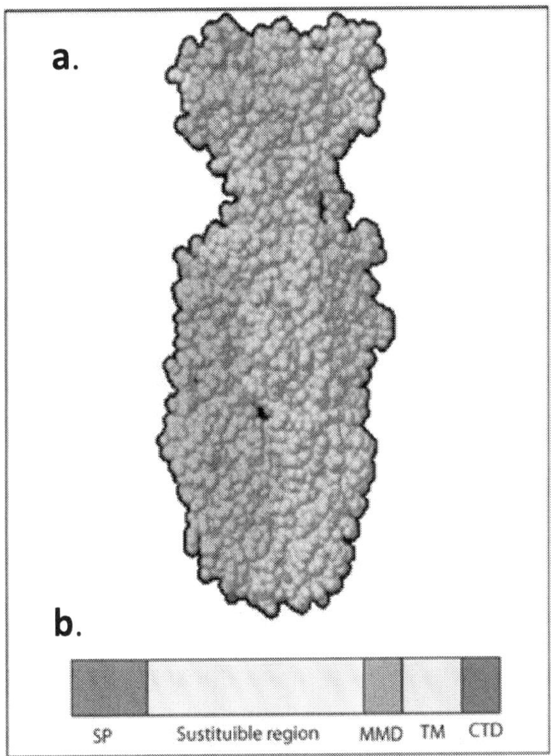

Figure 3. GP64 structure. a. Trimmeric structure of baculovirus major surface glycoprotein Gp64 obtained using the Expasy tool *Make multimer.py* in www.expasy.org. **b.** Gp64 polypeptide scheme showing different functional domains useful for antigen surface display.

Budded virions of baculoviruses enter cells by endocytosis. Gp64 is the major component of the viral envelope, and the unique protein with fusogenic activity in AcMNPV. Gp64 is triggered to induce the fusion at the low pH of endosomes. In addition Gp64 is distinguished from any other fusion protein in its ability of going through a reversible conformational change, unlike class I and class II fusion proteins, for which the postfusion conformation is thermodynamically more stable and the conformational rearrangement is irreversible.

2.2. Gp64 for protein display

Gp64 is expressed early and late in the infection of an insect cell. It is a 64 kDa protein which forms trimmers and locates in the BV envelope with a polarized distribution. As Gp64 is a transmembrane protein that exposes an outer domain, it can be used to display a selected protein on the BV surface. A chimeric Gp64 can be constructed to contain the protein of interest allowing it to be incorporated in the BV structure upon infection of insect cells (Grabherr & Ernst, 2010).

In order to facilitate the construction of a chimeric protein it was shown that is not necessary to conserve the complete structure of Gp64. The signal peptide (SP), the multimerization domain, the transmembrane (TM) and the cytoplasmic tail domain (CTD)were shown to be enough for the surface display, whereas the rest of the protein can be eliminated. This strategy avoids the need of dealing with large transfer vectors as well as permitting to increase the number of displayed proteins.

3. Baculovirus as immunogens

The innate immune system provides the first line of host defense against infection. It is extremely important to mount a strong specific immune response by expressing co-stimulating factors necessary for the activation of adaptative immunity cells.

It was shown in previous articles that inoculation of a murine macrophage cell line with budded baculovirus induces the secretion of inflammatory cytokines, such as tumor necrosis factor alpha (TNF-α), interleukin-6 (IL-6), and IL-12 (Abe et al., 2005; Chimeno Zoth et al., 2012; Han et al., 2010; Hervas-Stubbs et al., 2007).

AcMNPV induces pro-inflammatory cytokines secretion through a MyD88/TLR9-dependent signaling pathway, while other signaling molecules may participate in IFN-α production in response to AcMNPV.

Toll-like receptors (TLRs) are a family of transmembrane proteins that recognize and bind endogenous and exogenous ligands. Signaling through TLR generally culminates in the production of pro-inflammatory cytokines resulting in modulation of several aspects of the innate immune response (Han et al., 2010). In the case of baculovirus, it has been reported that BVs could induce cytokine production through the TLR9 signaling pathway in mammals.

TLR9 was shown to be responsible in vivo for immune system stimulation by oligodeoxynucleotides containing unmethylated CpG motifs. Like bacteria, AcMNPV contains a significant number of potentially bioactive CpG motifs. Indeed, a number of studies demonstrate that AcMNPV can stimulate professional Antigen Presenting Cells (APCs) by this pathway. Furthermore, Abe et al. demonstrated that internalization and endosomal maturation are required for TLR9 activation by CpG-rich DNA. They showed that the inhibition of endosomal maturation abolishes the immune system activation of AcMNPV in a dose-dependent manner. These results imply that immune system activation by AcMNPV through TLR9 requires membrane fusion via Gp64 as well as the liberation of the viral genome into cytoplasmic TLR9-containing vesicles (Figure 4.a)

On the other hand, despite BVs cannot replicate in mammalian or other vertebrate animal cells (Via et al., 1983), recent studies showed that BVs have strong adjuvant properties in mice, promoting potent humoral and CD8+ T cell adaptive responses (Abe et al., 2003; Gronowski et al., 1999). In addition, BVs induce the production of inflammatory cytokines by the *in vivo* maturation of dendritic cells (Figura 4.c).

Zoth et al. evaluated the effect of baculovirus administration on the innate immune response of chickens. They found an upregulation of IFN-γ and IL-6 in the baculovirustreated chicken spleens and a decrease of the TGF-β gene expression. These facts indicated a strong pro-inflammatory immune response. Moreover, they demonstrated that BV induced modifications in the mononuclear cells pattern of different organs using flow cytometry.

The duration of the BV-induced response is very limited. This fact represents one of the many interesting benefits of the use of baculovirus for stimulating innate immunity, because the potential damage for a strong inflammatory immune response on an extended time period could be avoided (Chimeno Zoth et al., 2012)

On the other hand, it could be presumed that baculovirus inoculation produced an indirect effect on monocytes/macrophages. Zoth et al. also showed an increase of both the mRNA and the protein levels of IFN-γ, and a priming effect of Nitric Oxyde (NO) response in splenocytes of chickens treated with baculoviruses. NO acts as a multi-functional mediator with diverse physiological and pathological roles in host defense, (MacMicking et al., 1997). The production of NO by activated monocytes/macrophages is an important innate immune response sign of cellular antiviral and bactericidal activity.

Moreover, Kitajima et al. demonstrated that AcMNPV inoculation of mice induced NK cells activation. They observed that in AcMNPV inoculated animals there was up to fourfold increase in the number of NK cells in spleen, liver, bone marrow and thymus. Furthermore, it was analyzedt he antitumor ability of AcMNPV-induced NK cells and they concluded that AcMNPV injection induces a NKT cell and IFN-γ independent NK cell cytotoxicity against tumor cells in mice (Kitajima et al., 2007) These findings will be approached in section 7.

In conclusion, the strong immune response induced by AcMNPV makes it a promising candidate for a novel, adjuvant- containing vaccine vehicle against infectious diseases (Abe et al., 2005).

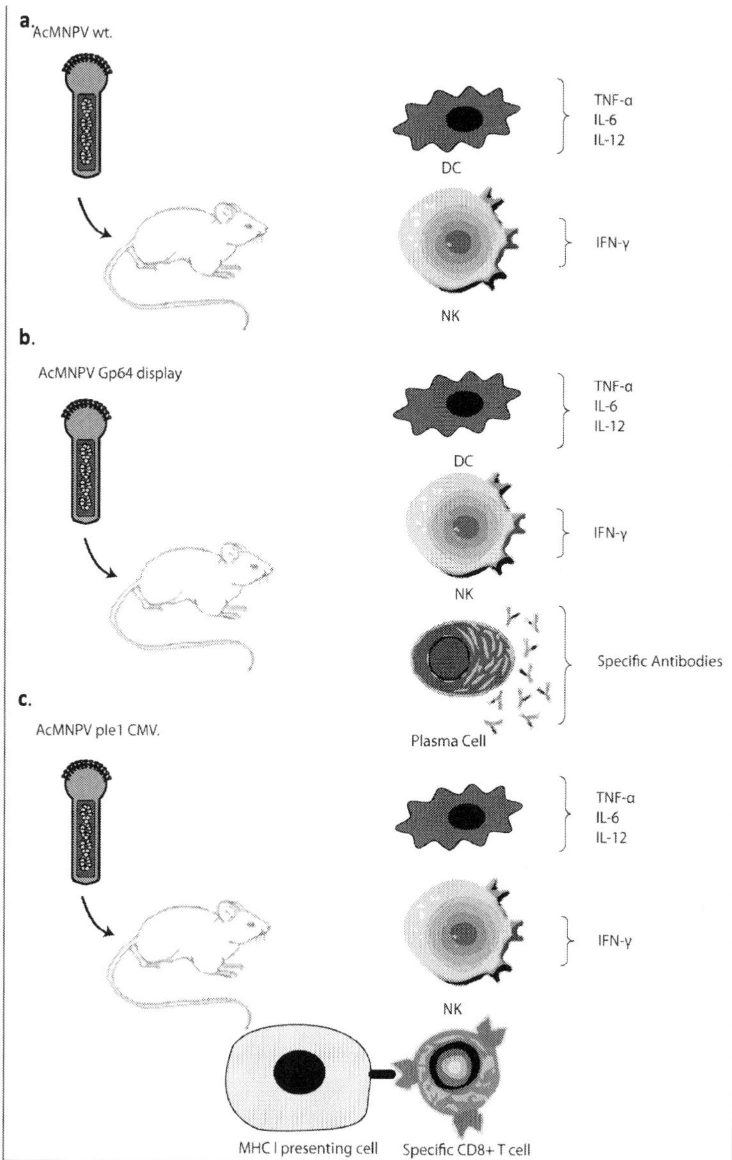

Figure 4. Immune response induced by baculovirus summarized. a. Activation of immune cells by inoculation with AcMNPV wild type. **b.** Immune response triggered by AcMNPV displaying a Gp64 fused antigen. **c.** Immune response generated by antigen coding AcMNPV under the control of CMV Ie1 promoter.

4. Baculovirus display

Eukaryotic systems represent a highly interesting model for the study of higher eukaryotic structures and interaction mechanisms because they provide posttranslation‐ al modifications and complex protein folding, in contrast to prokaryotic systems. Moreover, displaying a protein on the surface of a cell or a virus is a very successful strategy, for recreating and maturing binding properties such as antigenic recognition (Grabherr & Ernst, 2010).

Several strategies have been developed for displaying heterologous peptides or pro‐ teins on the baculovirus envelope by fusing the peptide or protein to gp64. In most instances the vector is designed with the aim of obtaining baculovirus particles that contain both wild-type gp64 and chimeric gp64 molecules. Furthermore, baculoviruses displaying proteins fused to Gp64 have proven to be very effective immunogens and they have been used successfully to generate antibody responses to a variety of displayed proteins (Kost et al., 2005).

Given that baculoviruses are able to mount a robust innate immune response by activating professional APCs, it is expected that baculovirus expressing an heterologous antigen on its surface could generate a specific response against this antigen. In fact, several works showed that baculoviruses expressing chimeric Gp64 on its surface were able to mount a very strong humoral response against the antigen displayed (Figure 4.b).

Xu et al. demonstrated in several works that baculovirus surface display of different proteins of Japanese Encephalitis Virus and swine fever virus generated high titers of specific antibodies useful for the protection against the disease. More specifically, they found that inoculation with recombinant baculoviruses produced a specific IgG response comparable with the response mounted by the preexistent attenuated vaccine and high neutralizing antibody titers against the virus (Xu et al., 2008; Xu & Liu, 2008; Xu et al., 2009; Xu et al., 2011).

Furthermore, numerous studies used baculovirus display for the development of new generation vaccines and obtained similar results to those showed by Xu et al. In this context, baculovirus surface display conferred protection and induced a strong humoral response against avian reovirus (Lin et al., 2008), human enterovirus (Meng et al., 2011), influenza (Jin et al., 2008; Prabakaran et al., 2010), malaria (Yoshida et al., 2009), etc.

In the next sub-sections the different strategies for efficient baculovirus display will be discussed. These include baculovirus display using the entire Gp64 for the generation of the chimeric proteins, baculovirus display based on single peptide insertion in Gp64 and a truncated Gp64 system with several cloning advantages will be considered (Figure 5).

Baculovirus display strategies have also been used for modification of the viral surface to command baculovirus mediated transduction of mammalian cells. In addition, capsid modifications may allow novel approaches for enhancing baculovirus mediated gene delivery. These studies will be discussedlater.

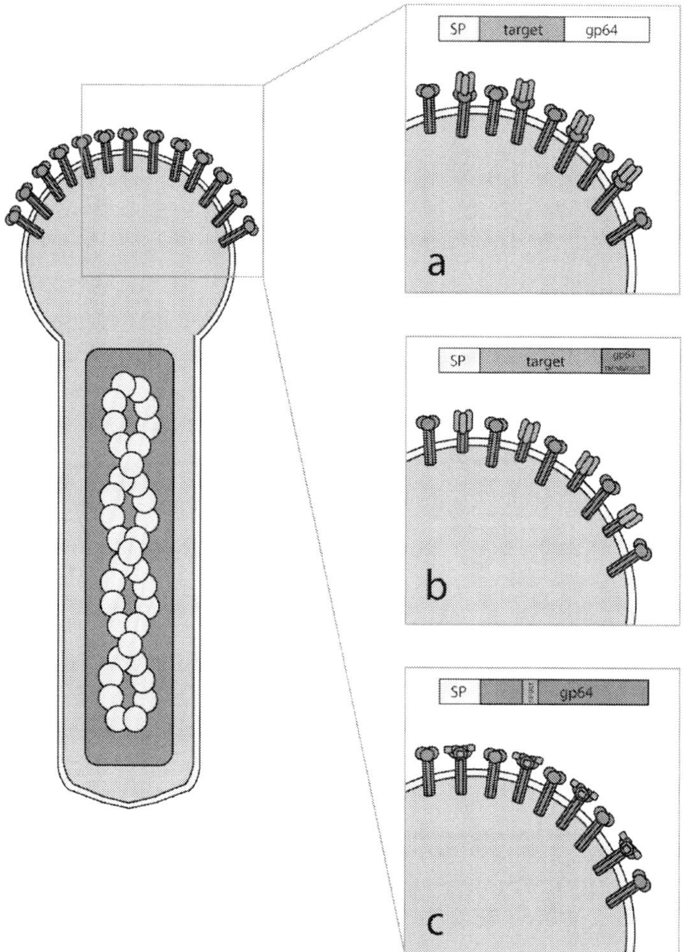

Figure 5. Different kinds of baculovirus display. a. Baculovirus surface display using the entire Gp64. **b.** Baculovirus surface display using only TM, MMD and CTD as fusion partner of the antigenic target. c. Baculovirus display using recombinant Gp64 expressing a small peptide.

4.1. Chimeric proteins using the entire Gp64

Gp64 can serve as a fusion partner that together with a chosen target protein gets incorporated into the cell membrane and into budded virions. In the first reports of baculovirus display proteins were fused to the complete gp64. In these works the target proteins were cloned into a vector providing N-terminal fusion with the gp64 signal peptide and C-terminal fusion with the full length gp64 coding region. (Boublik et al., 1995; Grabherr & Ernst, 2010). The conservation of the biological function of several proteins when they were expressed by the baculovirus display system, e.g HIV gp120, indicated that large, complex proteins could be displayed on the surface of baculovirus particles in a functional form.

The mechanism of incorporation into the viral particle was proposed to be due to oligomerization of the chimeric Gp64 with wild-type Gp64. In addition the CTD of the chimeric Gp64 may play an important role in the nucleocapsid recognition for budding process (Figure 5.a)

For the purpose of antigen display various epitopes were presented and shown to induce immune response in mice.

The advantages of this method reside in that all needed sequences for glycoprotein transport and maturation are present in the entire sequence of Gp64. Complete Gp64 fused antigens will be synthetized through the glycoprotein synthesis pathway and will be directed to plasmatic membrane and also budded virus envelope.

However, the utilization of entire Gp64 may cause some problems in the cloning process due to the length of the subsequent transfer vector.

4.2. Peptide insertion on Gp64

Another strategy consists in peptides directly engineered into the native Gp64 of AcMNPV in order to increase the avidity of the displayed target. In this case a short peptide is inserted into the sequence of the wild type Gp64, being this protein the only variant expressed in the virion, in contrast to the previous approach where both wt and the modified versions coexisted in the BV surface. It has been reported that this method resulted very efficient to mount a robust specific antibody response against the inserted peptide with a significantly increased avidity.

However, manipulating the native gp64 envelope protein may cause some problems. Given that no wild-type gp64 exists in order to guarantee functional cell fusion and virus budding, it is possible that the overall incorporation of the recombinant protein into cell membrane or viral envelope as well as viral titers decrease considerably. For this reason, insertion sites for foreign fragments must be chosen carefully. Moreover, the size of the peptides for insertion results in a limiting condition. Indeed, only small peptides have been inserted into the native gp64 with a maximum size of 23 amino acids (Figure 5.c).

Alternatively, expressing a second copy of gp64 displaying the target peptide in addition to the wild type Gp64 represents an effective solution (Grabherr & Ernst, 2010; Spenger et al., 2002).

4.3. SP, TM and CTD display systems

More recently, several reports demonstrated that using only the signal peptide region (SP), transmembrane region (TM) and the cytoplasmic tail domain (CTD) was enough for surface display on the insect cell surface as well as on the budded virions. The resulting smaller transfer vectors represented a significant improvement for the increased cloning efficiencies and the number of displayed chimeric proteins (Grabherr & Ernst, 2010; Spenger et al., 2002; Xu et al., 2009).

This method conserves the advantages of the baculovirus surface display using the entire Gp64, reducing significantly possible cloning troubles (Figure 5.b).

5. Baculovirus and cellular immunity

Apart from infecting insect cells, baculoviruses are able to transduce different types of animal cells such as human, rodent, rabbit, porcine, bovine, fish and avian cells (Hu, 2005; Hu, 2006) In addition, baculovirus can transduce embryonic stem cells, adult stem cells and induced pluripotent stem cells (Chen et al., 2011).

Baculovirus are safer than other transduction vectors because they don't integrate its DNA into host genome, nor replicates it inside the transduced cells (Chen et al., 2011; Merrihew et al., 2001). It has been demonstrated that humans do not possess pre-existing antibodies and specific T-cells against baculoviruses (Strauss et al., 2007). For this reason, baculoviruses may avoid the pre-existing immunity problem caused by other viral vectors.

Thus, the coding sequence of a protein of interest can be cloned into the viral genome under the control of a suitable promoter. Then, the inoculation of an animal with the recombinant virus results in the expression of the heterologous protein inside different cell types. The expression of a foreign protein in the cytoplasm trigger the MHC class I antigen presentation of proteasome processed peptides of the recombinant protein. In this way, joined to the adjuvancy showed by baculoviruses, transduction of animal cells may induce a strong cellular immune response (Figure 4.c).

Yoshida et al. have developed a baculovirus based dual expression system, with the aim to develop multifunctional vaccines capable of inducing strong humoral and cellular immune responses. In this study a chimeric protein was constructed with the display necessary sequences of Gp64 and the entire open reading frame of the*Plasmodium berghei* circumsporozoite protein (PbCSP) under the control of polyhedron and CMV Ie1 promoters. ELISPOT assays with splenocytes from immunized mice with the recombinant baculovirus showed significative IFN-γ secretion compared with the results for immunization with a recombinant AcMNPV without the CMV Ie1 promoter when the splenocytes was stimulated with a PbCSP synthetic peptide. In addition, this baculovirus based dual system showed to be more protective than the simple baculovirus display system (Yoshida et al., 2009)

On the other hand, Hervas-Stubbs et al. demonstrated that baculoviruses induced strong humoral and cellular immune responses by co-administration of AcMNPV wt. and a purified antigen.They showed that budded baculoviruses had strong adjuvant properties, promoting humoral and CTL responses against coadministered antigen. They observed also that baculovirus could induced DC maturation, and the production of inflammatory mediators through mechanisms primarily mediated by IFN-α and IFN-β.It has been shown previously that type

I IFNs act directly on naive B cells and CD4+ and CD8+ T cells, promoting clonal expansion and differentiation (Curtinsger 2005; (Bon & Lucchetti, 2006).

5.1. Baculovirus and mammalian cell transduction

Baculovirus entry into mammalian cells represents an important goal for immune response induction and most recently for different genic therapies. It was initially suggested that baculovirus entry depended on electrostatic interactions, heparin sulfate and phospholipids (Duisit et al., 1999; Tani et al., 2001), but the exact cell surface molecules and the involved mechanism remained unknown. Based on the mechanism of Gp64 mediated membrane fusion and the entry pathway of baculoviruses in insect cells, it was also proposed that clathrin-mediated endocytosis and macropinocytosis play roles in baculovirus entry (Long et al., 2006; Matilainen et al., 2005) In contrast, Laakkonen et al.(2008) discovered that baculovirus could enter some types of mammal cells, such as hepatic cells, by a pathway independent of clathrin-mediated endocytosis and macropinocytosis suggesting that phagocytosis might play a role (Chen et al., 2011).

These data suggest that baculovirus entry pathway varies with cell types and will be necessary more studies to elucidate the complete mechanisms. Nevertheless, all studies determined that baculovirus envelope protein gp64 is pivotal for entry and for the activation of dendritic cells (DCs) (Abe et al., 2005; Niu et al., 2008; Schutz et al., 2006).

Once inside the cells, baculovirus is transported to the endosome. Then, virions are released by the acid-triggered gp64 fusion (Kukkonen et al., 2003) and subsequently transported into the nucleus (Laakkonen et al., 2008; van Loo et al., 2001) reorganizing the actin cytoskeleton (Matilainen et al., 2005); Salminen et al., 2005). A major component of type III intermediate filaments, vimentin, also participates in intracellular trafficking (Mahonen et al., 2010). Inside the nucleus, baculoviral DNA could be recognized by the cellular transcription machinery and recombinant proteins could be expressed.

5.2. Baculovirus capsid display

As was described before in this section, several authors reported strategies in which the coding sequence of an antigen was cloned driven by the cytomegalovirus (CMV) promoter to obtain antigen specific T cell immune responses, resulting in high levels of protection against parasitic diseases.

In addition to the baculovirus surface display on the envelope, heterologous protein has been displayed on the capsid by fusion with the major capsid protein VP39 without any interference in the virus assembly (Molinari et al., 2011). Kukkonen et al. fused the enhanced green fluorescent protein (EGFP) with VP39 with the aim to improve the nuclear traffic of BV in mammalian cells, and shown no interference with virus titer (Kukkonen et al., 2003). This finding suggested the possibility of performing insertions into the inner capsid of the BV particle. VP39 is the most abundant protein of the nucleocapsid and consist in a 39 KDa polypeptide with monomers arranged in stacked rings around the nucleoprotein core (Molinari et al., 2011).

In the section 5.1 it were described the possible mechanisms of entry of baculovirus in mammalian cells. Besides the complete mechanism diverge in different cell types, endosome trafficking and Gp64 mediated fusion are always involved. Under these circumstances, it seems unlikely that the antigen displayed on the BV envelope would be able to efficiently reach the cytoplasm and consequently would be preferentially presented by MHC class II pathway. For this reason, antigen displayed on the envelope of baculovirus failed to produce a robust CD8+ T cell response, but was very effective to induce a CD4+ T and B cell responses.

However, antigen capsid display should be able to reach the cytosol and preferentially trigger MHC class I presentation pathway and mount a strong CD8+ T cell response (Molinari et al., 2011).

In this context, Molinari et al. developed a capsid display system and probed it fusing OVA with VP39 (BV-OVA) and showed that OVA could enter into the MHC class I pathway. Consequently, it was observed that inoculation of an animal model with the recombinant baculovirus triggered the activation of naive CD8+ T cells inducing an OVA-specific cytotoxic response. Though the mechanism involved in OVA MHC class I presentation was not elucidated, all these data suggest that capsid display is more convenient over envelope surface display for CTL activation. One of the proposed hypothesis consists of the possibility of the entire baculovirus capsid digestion by proteasome generating MHC class I binding peptides.

In summary, baculovirus are internalized by DCs and induce their maturation and the production of the pro-inflammatory cytokines IL-6 and IL-12 and are able to mount a type I IFN response (Section 3). Finally, Molinari et al. also examined the efficacy of the strong CTL and innate immune response elicited by baculovirus by the capacity of BV-OVA to confer protection against the classical MO5 melanoma tumor model. It was observed that inoculation with the BV-OVA protect against this tumor model.

Other researchers used capsid display as an alternative for mammalian cells transduction. In the work presented by Song et al. the ZnO binding peptide has been fused to the N-terminus of VP39 while retaining the viral infectivity and conferring the ability to bind nanosizedZnO powders (Chen et al., 2011; Song et al., 2010).

In conclusion, capsid display results in a very attractive alternative for cells transduction and for triggering MHC class I presentation of antigenic peptides. In this way, capsid display showed to be strongly effective to mount a robust cellular response against heterologous proteins promoting both IFN secretion and cytotoxicCD8+ T cells activation.

6. Baculovirus and complement

Complement is an important component of the innate immune system and plays an important role in the recognition and elimination of pathogens. Complement can be activated by three separate pathways: the classical, alternative, and lectin pathways (Ricklin et al., 2010). The classical activation pathway begins with the binding of the complement protein C1q to the pathogen surface or to antibody-antigen complex. The alternative complement activation pathway is initiated by spontaneous hydrolysis of the C3 protein into C3a and C3b and the subsequently attaching of C3b to amine and carbohydrate groups on the target surface. Finally, the lectin pathway is activated by the recognition of specific carbohydrate patterns on the pathogen surface by mannose-binding proteins. Once complement was activated, a cascade of proteolysis events of complement proteins leads to the recruitment of the membrane attack complex (MAC) and the subsequently target membrane perforation (Kaikkonen et al., 2011) (Figure 6.a.b).

On the other hand, complement must be regulated. There are two different types of complement regulators: Surface-bound regulators, and soluble regulators. Surface-bound regulators consists in a group of molecules integratedby factors that accelerate decay of the convertases (complement receptor 1, CR1; decay accelerating factor, DAF), act as a cofactor for the factor I-mediated degradation of C3b and C4b (CR1; membrane cofactor protein, MCP), or prevent the formation of the membrane attack complex (CD59) (Hourcade et al., 2000; Ricklin et al., 2010). Soluble regulators also mediate the first two functions of surface-bound regulators. C4bbinding protein (C4BP), factor H (FH) and FH like protein-1 (FHL-1) are examples of the members of this group (Kaikkonen et al., 2011) (Figure 6.c).

In this context, baculovirus engineering with the aim to confer it resistance to complement inactivation results very attractive to improve the efficiency of baculoviruses for gene delivery.

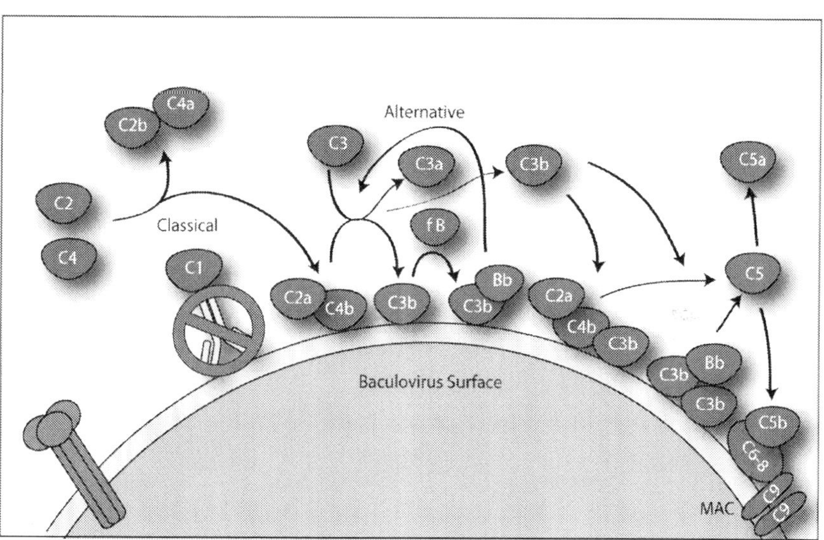

Figure 6. The two major complement activation pathways in baculoviruses: The classical pathway is triggered by the binding of C1 to antigen-bound antibody molecules. The classical pathway utilize C2 and C4 to generate the C3-convertase C4b2a. The alternative pathway is initiated by the spontaneous hydrolysis of C3. Then, the complete complement cascade (from C3 proteolysis to formation of membran attack complex (MAC)) proceeds. Adapted from Kaikkonnen et al. 2011.

6.1. Complement activation by baculoviruses

As particulated antigens, baculoviruses are vulnerable to the action of the complement. This fact was observed in several studies which demonstrate that baculovirus-mediated gene transfer into hepatocytes is strongly reduced in the presence of untreated human serum.

The complement cascade is usually activated to protect the host from foreign elements. The complement activating properties of various gene transfer vectors was demonstrated. The mechanism of complement activation by liposomes and synthetic DNA complexes depends mainly on the formulation, charge and size. Murine retroviruses are effectively lysed by primate complement triggered by the classical pathway, involving direct binding of C1q and C1s to the envelope and/or to antibody-antigen complexes. Comparatively, Hoffmann et al. found baculovirus survival in C1q-depleted human serum indicating baculovirus-mediated activation of the complement cascade through the classical pathway.

Given that there is no evidence of pre-existing anti baculovirus antibodies in human sera, this data suggests that baculoviruses activate the complement cascade by an antibody-independent activation of the classical pathway (Hofmann & Strauss, 1998).

6.2. Strategies for complement inactivation

At present there are different strategies which help to avoid complement attack during baculovirus treatment (Huser et al., 2001; Kaikkonen et al., 2010). As noted previously, the surface of baculovirus particles can easily be engineered. As an example,desired peptides or proteins can be displayed as fusion proteins (Boublik et al., 1995; Makela & Oker-Blom, 2008; Oker-Blom et al., (2003). The most widely used technique for surface engineering makes use of the trimeric major baculoviral envelope glycoprotein GP64 as a fusion partner (Kadlec et al., 2008). In this section, diverse strategies for complement inactivation mediated by baculoviruses will be discussed. In particular, the discussion will be focused in the use of polymers for baculovirus surface coating, pseudotyping of baculoviruses by the expression on VSV-G protein and surface display of eukaryotic complement inhibitors.

6.2.1. Polymer coating

With the aim to protect baculoviral vectors against complement inactivation, using polymers should be appropriated. The coating is based on the electrostatic interaction between the virus particle and the polymer. In the case of baculovirus, the negative charge of its surface allows coating with positively charged polymers such aspolyethylenimine (PEI) (Yang et al., 2009). It was observed that the 25 kDa PEI protected the virions against complement destruction resulting in a 10% to nearly 100% of vector survival in samples treated with human and rat serum, respectively. In addition,Kim et al. observed thatafter intraportal delivery the PEI-treated viruses exhibited improved transduction of liver and spleen compared to non-coated virions(Kim et al., 2009).

Additionally, another polymer, PEG (Mw 5000), has also been reported to increasebaculovirus transduction efficiency in vitro and in mouse brain and lung (Kim et al., 2007;

Kim et al., 2010; Kim et al., 2006). Although serum stability of the PEG-coatedbaculoviruses was not directly studied, these results support the notion that PEG coating can be used to protect baculovirus vectors against the immune system and prolong its survival time in circulation (Jevsevar et al., 2010).

Nevertheless, it is necessary adjust the ratiosof PEI or PEG and virus particles, and the polymer sizeto preserve virus infectivity and minimize cytotoxicity.

6.2.2. Pseudotyping

Pseudotyping consists in a process in which the natural envelope proteins of the virus are replaced with surface proteins from another virus. This strategy has been shown to mitigate the problem of complement attack (Tani et al., 2003). Unlikechemical engineering which is limited and requires extensive optimization to retain virus infectivity, pseudotyping conserves virus infectivity and allows virusevasion ofcomplement-mediated destruction.The most widely used method of pseudotyping of baculoviruses relies on the employ of the VSV-Gprotein. Several researches have shown that VSV-G is capable to improve transduction efficiency of baculovirus in vertebrate cells (Barsoum et al., 1997; Pieroni et al., 2001; Tani et al., 2003; Tani et al., 2001). VSV-G can alsoreplace GP64 and allow productive infection, replication, and propagation of thevirus in Sf9 insect cells (Kitagawa et al., 2005; Mangor et al., 2001). However, pseudotyping is typically performed by co-expressing both the desired molecule and Gp64.

Other reportsalso demonstrated increased gene delivery into mouse after direct intramuscular injection of VSV-Gpseudotypedbaculovirus (Pieroni & La Monica, 2001; Pieroni et al., 2001). Additionall, Tani et al. found that the VSV-Gmodified baculovirusexhibited greater resistance to human, rabbit, guinea pig, rat, hamster and mouse serum inactivation compared to the unmodified control baculovirus (Tani et al., 2003). Furthermore, co-display of a short transmembrane fragment of VSV-G was found to give similar complement protection as intact VSV-G (Kaikkonen et al., 2010). These results suggest that envelope modification of the baculovirus can change its immunogenic properties and protect them for complement inactivation.

6.2.3. Display of complement inhibitors

The last strategy for complement inactivation that will be discussed in this section consists in the baculovirus surface display of eukaryotic complement inhibitors. Several reports showed that genetic modificatedviruses expressing complement regulators presented an improved survival rate, unlike wild type controls. In this context, the most promising results to increase the serum stability of baculovirus vectors by genetic means have been attained by displaying complement regulating proteins fused to Gp64 on the virion surface (Huser et al., 2001; Kaikkonen et al., 2010). The first described research generated a recombinant baculovirus which expressed on its surface the DAF complement regulator. Kaikkonen et al. have recently verified the protective nature of DAF-display and studied the efficacy of other complement regulatory proteins (FHL-1, C4BP and MCP) and their combinations for complement inactivation and consequently baculovirus survival rates. (Kaikkonen et al., 2010). Their resultsconcluded that serum stability was dependent on the displayed complement regulatory protein and the source of serum.

In general, the complement regulators DAF and MCP gave the best results. Conversely, simultaneous co-display of soluble complement regulatory proteins did not provide further benefit.Best protection was gained in mouse serum (70%), while the worst protection rate was obtained with rat serum (13%). In the case of human serum, about 30% of the viral particles were still competent to transduce mammalian cells after 1 h preincubation with serum (Kaikkonen et al., 2010).

All these data suggest that engineering of baculoviral vectors for complement inactivation result very convenient not only to reduce the number of necessary inoculations for an efficient transduction, but also to avoid the undesired mortality induced by high doses of non-modified vectors.

7. Other applications

The use of baculoviruses as vectors for the generation of immunity is not the only possible application for these viruses. Their ability to transduce mammalian cells and their capacity to allow the introduction of large amounts of heterologous DNA in their genomes represent remarkable advantages. In addition to the biosafety benefits of baculovirus in comparison with other viral vectors, these features make baculoviruses as adequate vectors for in vivo animal transduction. The absence of preliminary immune cells against baculoviruses makes them a promising tool for human treatment.

In this section, two different novel applications for baculoviruses will be discussed. In first place, the use of baculoviruses for gene therapy and the goals and limitations of this practice will be analyzed. Then, the construction of displaying libraries using baculovirus display system will be exemplified.

7.1. Gene therapy

There are two different categories in which gene therapy vectors can be classified: nonviral and viral vectors. Non-viral vectors consist in polycation conjugated polymers that allow delivery of the DNA. Positively charged liposomes are one example of this type of vectors. Although these vectors are advantageous in biosafety, its application is restricted by the low efficiency in the delivery and expression of transgenes. (Verma & Somia, 1997). On the other hand, viral vectors, such as retroviral, lentiviral, adenoviral, and adeno-associated viral (AAV) vectors, have a higher efficiency in cell entry and transduction by expressing different transgenes. Advantages and disadvantages depend on each particular viral vector. The mechanism used for replication and protein expression, and the biological hazard inherent in its use are some of the features to be analyzed at the moment in which a viral vector is chosen for gene therapy. In comparison with these common viral vectors, baculoviruses possess a number of advantages.

In first place, baculovirus-mediated transduction does not present any toxic effect against mammalian cells and does not disturb cell growth even at high MOI (Gao et al., 2002; Hofmann et al., 1995). In contrast, cell proliferation may be retarded by transgene products because they could be toxic and even induce apoptosis in some cells (Detrait et al., 2002; Liu & Carstens, 1999). Furthermore, baculoviruses do not replicate in transduced mammalian cells (Kost & Condreay, 2002). These features of baculoviruses are particularly important because other viral vectors are human pathogens, and consequently represent a biological risk.

Another advantage of baculoviruses as gene therapy vectors consists in its large cloning capacity. The baculovirus (AcMNPV) genome is a large circularized DNA molecule with 130 kb of length and a maximum cloning capacity of at least 38 kb. This flexibility results particularly advantageous in contrast to retroviral and AAV vectors whose cloning capacities are limited (Hu, 2008).

In comparison with other viral vectors, baculoviruses are easy to produce. Retroviral, lentiviral, and AAV vectors require transfection of plasmids encoding essential genes into packaging cells for its production. In contrast, baculovirus can be easily propagated by infecting insect cells in suspension culture or monolayer and harvesting the supernatant 3–4 days postinfection. In addition, the construction, propagation, and handling of baculoviruses can be performed in Biosafety Level 1 laboratories without the need for specialized equipment.

Finally, one of the most important advantages is that baculoviruses do not present preexisting immunity in mammalian. One of the problems associated with other viral vectors is that most people are exposed to these viruses and develop specific humoral response. Circulating antibodies can significantly reduce the efficiency of transduction with the viral vector. The use of baculovirus vectors in gene therapy, therefore, may avoid the problem of preexisting immunity (Hu, 2008).

However, baculoviruses have a number of disadvantages as gene therapy vectors. One of these is that baculovirus induce a transient expression in mammalian cells. In vivo, transgene expression typically declines by day 7 and disappears by day 14 (Airenne et al., 2000; Lehtolainen et al., 2002). The duration of in vitro transgene expression using baculoviruses is significantly shorter than expression mediated by retroviral, lentiviral, and AAV vectors.

Baculoviral vectors differ mainly than other viral vectors in the time that the carried genes can persist in the host nucleus. In the case of retroviral, lentiviral and adenoviral vectors, viral DNA can remain into the nucleus either in an integrated or episomal form, for a longer period. In fact, Tjia et al. demonstrated that baculoviral DNA persists in the nuclei of transduced mammalian cells for only 24–48 h (Tjia et al., 1983).

Another disadvantage of using baculovirus as gene therapy vector is the inactivation by complement. As described in previous sections, contact between baculoviruses and serum complement results in rapid inactivation of budded virions. There are need several modifications for reduce the negative effect of complement in baculovirus-mediated transduction. However, the complement system is not a problem only for baculovirus. It is also a potent barrier to in vivo administration of other gene delivery systems such as liposomes, murine retrovirus, and various synthetic DNA complexes (Hu, 2008).

Additionally, baculoviruses as enveloped virus are very fragile. The envelope structure is essential for virus infectivity because of the anchored Gp64, responsible of viral and cellular membrane fusion. (Blissard & Wenz, 1992). For this reason it renders virus vulnerable to mechanical force and results in relatively low virus stability, a common problem also observed for other enveloped viruses such as retrovirus. Ultracentrifugation is often necessary for budded virions purification, but also leads to significant loss of infectivity probably because of the viral envelopes damage. Labile thermal stability, in conjunction with the tendency to be inactivated by serum complement, may further restrict the in vivo application of baculovirus gene delivery vectors.

In vivo gene therapy

Due to their ability to transduce various cell types, baculoviruses have captured increasing interest as vectors for in vivo gene delivery. Baculovirus-mediated gene delivery was tested in different tissues that including rabbit carotid artery, rat liver, rat brain, mouse brain, mouse skeletal muscle, mouse cerebral cortex and testis, and mouse liver (Hu, 2006). However, for baculovirus-mediated in vivo gene therapy in all of these tissues the complement system appears to be a significant barrier.

Baculovirus vectors have also been injected into the rodent brain where complement proteins may be absent because of the blood–brain barrier (Hu, 2008; Lehtolainen et al., 2002). After injection into the brain, baculoviruses specifically transduced the epithelium of the choroids plexus in ventricles and the obtained transduction efficiency was very high.

As discussed in previous sections, baculoviruses can be alternatively pseudotyped by displaying VSVG on the envelope. This modified virus enhanced gene transfer efficiencies into mouse skeletal muscle and the transgene expression in mice. The VSVG-modified baculovirus also exhibited greater resistance to inactivation by the complement system present in animal sera.

Moreover, it has been shown that transduction of different cell lines with a baculovirus expressing shRNAs (short-hairpin RNAs) effectively knocked down expression of the target mRNA and protein (Nicholson et al., 2005). Additionally, baculoviruses have been used to mediate RNA interference (RNAi). The recombinant baculovirus encoding RNAi sequence was efficient in suppressing expression of the target gene by 95% in cultured cells and by 82% in vivo in rat brain. These data suggest that baculoviruses may be also used as delivery vectors for RNA interference therapies (Hu, 2008; Ong et al., (2005).

7.2. Libraries

Surface display libraries represent a very useful methodology for selecting binding proteins out of defined pools of protein variants. Although prokaryotic expression systems such as phage display technology or protein targeting to the cellular surface of Escherichia coli are widely used, they fail allowing the functional display of complex proteins such as eukaryotic glycoproteins which require a high degree of modification and processing. (Ernst 1998)

Eukaryotic expression libraries, in contrast, are a powerful tool for finding new ligands, identification of cellular interaction partners and affinity maturation of antibody and antibody fragments (Grabherr & Ernst, 2010).

As discussed before, the expression of foreign proteins on the surface of insect cells, in occlusion bodies and on the baculovirus surface make baculoviruses an important resource in biotechnology. Moreover, fusion proteins with the baculoviral envelope protein Gp64 as well as different foreign membrane proteins such as the influenza virus hemagglutinin or VSV-G protein have shown to be targeted to the surface of infected insect cells in several researches about baculovirus display. Then, it is possible take advantage of baculovirus display systems with the aim to generate a surface display library for high trhoughput screening.

Ernst et al. expressed a specific antibody epitope in the context of the influenza virus hemagglutinin, randomizing the adjacent amino acid. This procedure results in the construction of a baculovirus surface display library capable to allow the selection of the displayed peptide with optimal antigenicity. Furthermore, baculovirus surface display libraries served to identify MHC class I and II mimotopes (Grabherr & Ernst, 2010; Wang, 2005).

In comparison with bacterial phage display in which cross infection does not occur and every infected cell just propagates one individual phage, in baculovirus surface display cross infection is very probably. The situation may result advantageous or disadvantageous depending the aim of the library. For the assembly of a multisubunit protein, this fact is highly advantageous. However, when the library is performed to screening different proteins, these cross infections have to be considered (Grabherr & Ernst, 2010). Adjusting the multiplicity of infection (moi) usually result convenient for avoid the cross infection problem.

In conclusion, baculovirus insect cell system consists in a highly useful tool for constructing and screening of surface display libraries, specially for the expression of eukaryotic complex proteins (Ernst et al., 1998).

8. Perspectives and conclusions

There are many biotechnological uses for baculoviruses. One of the most widespread is the use of baculoviruses as insecticide agents. Moreover, recombinant baculoviruses have been extensively used as expression vectors in insect cell cultures. A variety of technological improvements have eliminated the tedious procedures to isolate the recombinant viruses turning the baculovirus-based expression system in a safe, easy to use and scale up system (Kost et al., 2005).

In addition, protein expression in larvae or cell culture is not the only application of baculoviruses. In fact, baculoviruses are widely used in the development of strategies for displaying foreign peptides and proteins on the virus surface as well as mammalian cell transduction using different mammalian expression cassettes.

As described in this chapter, baculovirus surface display based on the generation of Gp64 chimeric proteins result in a very efficient technology capable to induce a strong immune response against specific antigens (Xu et al., 2009). The ability of baculoviruses to activate innate immune system cells guarantees the mount of a robust immune response and the generation of immunological memory. More specifically, AcMNPV induces pro-inflammatory cytokines secretion through a MyD88/TLR9-dependent signaling pathway (Abe et al., 2005; Chimeno Zoth et al., 2012).

It was showed by several authors that baculovirus surface display induced high specific antibody titers against various virus families and parasitic pathogens (Jordan et al., 2009; Meng et al., 2011; Prabakaran et al., 2010; Yoshida et al., 2009). Furthermore, it was demonstrated that many of these titers had neutralizing properties.

On the other hand, it was discussed before that baculoviruses could also transduce mammalian cells (Kost et al., 2005). This feature results very interesting because it allows intracellular expression of heterologous proteins and its subsequent presentation through the MHC class I pathway. In this context, several authors demonstrate that baculoviruses can also induce a specific cellular immune response either by cloning the desired antigen under the control of a suitable promoter, or through the capsid display technique. CTL activation and IFN-γ secretion was detected in all of these researches (Yoshida et al., 2009).

Finally, baculovirus were shown to be useful as gene therapy vectors so as to create libraries of binding proteins.

For all these reasons, we conclude that baculoviruses represent a very useful tool in biotechnology as vaccination vectors. Its adjuvant capacity makes baculoviruses in a promising alternative for the generation of immunological memory.

Author details

Matías L. Pidre, M. Leticia Ferrelli, Santiago Haase and Víctor Romanowski

*Address all correspondence to: vromanowski@gmail.com

Instituto de Biotecnología y Biología Molecular, Facultad de Ciencias Exactas, Universidad Nacional de La Plata, Conicet, Argentina

References

[1] Abe, T, Hemmi, H, Miyamoto, H, Moriishi, K, Tamura, S, Takaku, H, Akira, S, & Matsuura, Y. (2005). Involvement of the Toll-Like Receptor 9 Signaling Pathway in the Induction of Innate Immunity by Baculovirus. *Journal of Virology*, 79, 2847-2858.

[2] Abe, T, Takahashi, H, Hamazaki, H, Miyano-kurosaki, N, Matsuura, Y, & Takaku, H. (2003). Baculovirus induces an innate immune response and confers protection from lethal influenza virus infection in mice. *J Immunol* , 171, 1133-1139.

[3] Airenne, K. J, Hiltunen, M. O, Turunen, M. P, Turunen, A. M, Laitinen, O. H, Kulomaa, M. S, & Yla-herttuala, S. (2000). Baculovirus-mediated periadventitial gene transfer to rabbit carotid artery. *Gene Ther* , 7, 1499-1504.

[4] Backovic, M, & Jardetzky, T. S. (2009). Class III viral membrane fusion proteins. *Current Opinion in Structural Biology* , 19, 189-196.

[5] Barsoum, J, Brown, R, Mckee, M, & Boyce, F. M. (1997). Efficient transduction of mammalian cells by a recombinant baculovirus having the vesicular stomatitis virus G glycoprotein. *Hum Gene Ther* , 8, 2011-2018.

[6] Blissard, G. W, & Wenz, J. R. (1992). Baculovirus gp64 envelope glycoprotein is sufficient to mediate pH-dependent membrane fusion. *J Virol* , 66, 6829-6835.

[7] Bon, L, & Lucchetti, C. (2006). Auditory environmental cells and visual fixation effect in area 8B of macaque monkey. *Exp Brain Res* , 168, 441-449.

[8] Boublik, Y. Di Bonito, P. & Jones, I. M. ((1995). Eukaryotic virus display: engineering the major surface glycoprotein of the Autographa californica nuclear polyhedrosis virus (AcNPV) for the presentation of foreign proteins on the virus surface. *Biotechnology (N Y)* , 13, 1079-1084.

[9] Chen, C, Lin, Y, Chen, C. -Y, & Hu, G. -Y. Y.-C. ((2011). Baculovirus as a gene delivery vector: Recent understandings of molecular alterations in transduced cells and latest applications. *Biotechnology Advances* , 29, 618-631.

[10] Chimeno Zoth S., Carballeda, J. M., Gómez, E., Gravisaco, M. J., Carrillo, E. & Berinstein, A. ((2012). Modulation of innate immunity in chickens induced by in vivo administration of baculovirus. *Veterinary Immunology and Immunopathology* , 145, 241-247.

[11] Detrait, E. R, Bowers, W. J, Halterman, M. W, Giuliano, R. E, Bennice, L, Federoff, H. J, & Richfield, E. K. (2002). Reporter gene transfer induces apoptosis in primary cortical neurons. *Mol Ther* , 5, 723-730.

[12] Duisit, G, Saleun, S, Douthe, S, Barsoum, J, Chadeuf, G, & Moullier, P. (1999). Baculovirus vector requires electrostatic interactions including heparan sulfate for efficient gene transfer in mammalian cells. *J Gene Med* , 1, 93-102.

[13] Ernst, W, Grabherr, R, Wegner, D, Borth, N, Grassauer, A, & Katinger, H. (1998). Baculovirus surface display: construction and screening of a eukaryotic epitope library. *Nucleic Acids Res* , 26, 1718-1723.

[14] Gao, R, Mccormick, C. J, Arthur, M. J, Ruddell, R, Oakley, F, Smart, D. E, Murphy, F. R, Harris, M. P, & Mann, D. A. (2002). High efficiency gene transfer into cultured primary rat and human hepatic stellate cells using baculovirus vectors. *Liver* , 22, 15-22.

[15] Grabherr, R, & Ernst, W. (2010). Baculovirus for eukaryotic protein display. *Curr Gene Ther* , 10, 195-200.

[16] Gronowski, A. M, Hilbert, D. M, Sheehan, K. C, Garotta, G, & Schreiber, R. D. (1999). Baculovirus stimulates antiviral effects in mammalian cells. *J Virol* , 73, 9944-9951.

[17] Han, Y, Niu, M, An, L, & Li, W. (2010). Involvement of TLR21 in baculovirus-induced interleukin-12 gene expression in avian macrophage-like cell line HD11. *Veterinary Microbiology* , 144, 75-81.

[18] Hervas-stubbs, S, Rueda, P, Lopez, L, & Leclerc, C. (2007). Insect baculoviruses strongly potentiate adaptive immune responses by inducing type I IFN. *J Immunol* , 178, 2361-2369.

[19] Hofmann, C, Sandig, V, Jennings, G, Rudolph, M, Schlag, P, & Strauss, M. (1995). Efficient gene transfer into human hepatocytes by baculovirus vectors. *Proc Natl Acad Sci U S A* , 92, 10099-10103.

[20] Hofmann, C, & Strauss, M. (1998). Baculovirus-mediated gene transfer in the presence of human serum or blood facilitated by inhibition of the complement system. *Gene Ther* , 5, 531-536.

[21] Hourcade, D, Liszewski, M. K, Krych-goldberg, M, & Atkinson, J. P. (2000). Functional domains, structural variations and pathogen interactions of MCP, DAF and CR1. *Immunopharmacology* , 49, 103-116.

[22] Hu, Y. c. ((2005). Baculovirus as a highly efficient expression vector in insect and mammalian cells. *Acta Pharmacologica Sinica* , 26, 405-416.

[23] Hu, Y. C. (2006). Baculovirus Vectors for Gene Therapy. , 68, 287-320.

[24] Hu, Y. C. (2008). Baculoviral vectors for gene delivery: a review. *Curr Gene Ther* , 8, 54-65.

[25] Huser, A, Rudolph, M, & Hofmann, C. (2001). Incorporation of decay-accelerating factor into the baculovirus envelope generates complement-resistant gene transfer vectors. *Nat Biotechnol* , 19, 451-455.

[26] Jevsevar, S, Kunstelj, M, & Porekar, V. G. (2010). PEGylation of therapeutic proteins. *Biotechnol J* , 5, 113-128.

[27] Jin, R, Lv, Z, Chen, Q, Quan, Y, Zhang, H, Li, S, Chen, G, Zheng, Q, Jin, L, Wu, X, Chen, J, & Zhang, Y. (2008). Safety and immunogenicity of H5N1 influenza vaccine based on baculovirus surface display system of Bombyx mori. *PLoS One* 3, e3933.

[28] Jordan, K. R, Crawford, F, Kappler, J. W, & Slansky, J. E. (2009). Vaccination of Mice with Baculovirus-Infected Insect Cells Expressing Antigenic Proteins.

[29] Kadlec, J, Loureiro, S, Abrescia, N. G. A, Stuart, D. I, & Jones, I. M. (2008). The postfusion structure of baculovirus gp64 supports a unified view of viral fusion machines. *Nature Structural & Molecular Biology* , 15, 1024-1030.

[30] Kaikkonen, M. U, Maatta, A. I, Ylä-herttuala, S, & Airenne, K. J. (2010). Screening of Complement Inhibitors: Shielded Baculoviruses Increase the Safety and Efficacy of Gene Delivery. *Molecular Therapy* , 18, 987-992.

[31] Kaikkonen, M. U, Ylä-herttuala, S, & Airenne, K. J. (2011). How to avoid complement attack in baculovirus-mediated gene delivery. *Journal of Invertebrate Pathology* 107, SS79., 71.

[32] Kim, Y. K, Choi, J. Y, Jiang, H. L, Arote, R, Jere, D, Cho, M. H, Je, Y. H, & Cho, C. S. (2009). Hybrid of baculovirus and galactosylated PEI for efficient gene carrier. *Virology* , 387, 89-97.

[33] Kim, Y. K, Jiang, H. L, Je, Y. H, Cho, M. H, & Cho, C. S. (2007). Modification of Baculovirus for Gene Therapy. In *Communicating Current Research and Educational Topics and Trends in Applied Microbiology*. Edited by A. M. Vila: Formatex.

[34] Kim, Y. K, Kwon, J. T, Choi, J. Y, Jiang, H. L, Arote, R, Jere, D, Je, Y. H, Cho, M. H, & Cho, C. S. (2010). Suppression of tumor growth in xenograft model mice by programmed cell death 4 gene delivery using folate-PEG-baculovirus. *Cancer Gene Ther* , 17, 751-760.

[35] Kim, Y. K, Park, I. K, Jiang, H. L, Choi, J. Y, Je, Y. H, Jin, H, Kim, H. W, Cho, M. H, & Cho, C. S. (2006). Regulation of transduction efficiency by pegylation of baculovirus vector in vitro and in vivo. *J Biotechnol* , 125, 104-109.

[36] Kitajima, M, Abe, T, Miyano-kurosaki, N, Taniguchi, M, Nakayama, T, & Takaku, H. (2007). Induction of Natural Killer Cell-dependent Antitumor Immunity by the Autographa californica Multiple Nuclear Polyhedrosis Virus. *Molecular Therapy* , 16, 261-268.

[37] Kost, T. A, & Condreay, J. P. (2002). Recombinant baculoviruses as mammalian cell gene-delivery vectors. *Trends in Biotechnology* , 20, 173-180.

[38] Kost, T. A, Condreay, J. P, & Jarvis, D. L. (2005). Baculovirus as versatile vectors for protein expression in insect and mammalian cells. *Nature Biotechnology* , 23, 567-575.

[39] Kukkonen, S. P, Airenne, K. J, Marjomaki, V, Laitinen, O. H, Lehtolainen, P, Kankaanpaa, P, Mahonen, A. J, Raty, J. K, Nordlund, H. R, Oker-blom, C, Kulomaa, M. S, & Yla-herttuala, S. (2003). Baculovirus capsid display: a novel tool for transduction imaging. *Mol Ther* , 8, 853-862.

[40] Laakkonen, J. P, Kaikkonen, M. U, Ronkainen, P. H, Ihalainen, T. O, Niskanen, E. A, Hakkinen, M, Salminen, M, Kulomaa, M. S, Yla-herttuala, S, Airenne, K. J, & Vihi-

nen-ranta, M. (2008). Baculovirus-mediated immediate-early gene expression and nuclear reorganization in human cells. *Cell Microbiol* , 10, 667-681.

[41] Lehtolainen, P, Tyynela, K, Kannasto, J, Airenne, K. J, & Yla-herttuala, S. (2002). Baculoviruses exhibit restricted cell type specificity in rat brain: a comparison of baculovirus- and adenovirus-mediated intracerebral gene transfer in vivo. *Gene Ther* , 9, 1693-1699.

[42] Lin, Y. H, Lee, L. H, Shih, W. L, Hu, Y. C, & Liu, H. J. (2008). Baculovirus surface display of σC and σB proteins of avian reovirus and immunogenicity of the displayed proteins in a mouse model. *Vaccine* , 26, 6361-6367.

[43] Liu, G, & Carstens, E. B. (1999). Site-directed mutagenesis of the AcMNPV gene: effects on baculovirus DNA replication. *Virology* 253, 125-136., 143.

[44] Long, G, Pan, X, Kormelink, R, & Vlak, J. M. (2006). Functional entry of baculovirus into insect and mammalian cells is dependent on clathrin-mediated endocytosis. *J Virol* , 80, 8830-8833.

[45] MacMicking J., Xie, Q. W. & Nathan, C. ((1997). Nitric oxide and macrophage function. *Annu Rev Immunol* , 15, 323-350.

[46] Mahonen, A. J, Makkonen, K. E, Laakkonen, J. P, Ihalainen, T. O, Kukkonen, S. P, Kaikkonen, M. U, Vihinen-ranta, M, Yla-herttuala, S, & Airenne, K. J. (2010). Culture medium induced vimentin reorganization associates with enhanced baculovirus-mediated gene delivery. *J Biotechnol* , 145, 111-119.

[47] Makela, A. R, & Oker-blom, C. (2008). The baculovirus display technology--an evolving instrument for molecular screening and drug delivery. *Comb Chem High Throughput Screen* , 11, 86-98.

[48] Matilainen, H, Rinne, J, Gilbert, L, Marjomaki, V, Reunanen, H, & Oker-blom, C. (2005). Baculovirus Entry into Human Hepatoma Cells. *Journal of Virology* , 79, 15452-15459.

[49] Meng, T, Kolpe, A. B, Kiener, T. K, Chow, V. T, & Kwang, J. (2011). Display of VP1 on the surface of baculovirus and its immunogenicity against heterologous human enterovirus 71 strains in mice. *PLoS One* 6, e21757.

[50] Merrihew, R. V, Clay, W. C, Condreay, J. P, Witherspoon, S. M, Dallas, W. S, & Kost, T. A. (2001). Chromosomal integration of transduced recombinant baculovirus DNA in mammalian cells. *J Virol* , 75, 903-909.

[51] Molinari, P, Crespo, M. I, Gravisaco, M. J, Taboga, O, & Moron, G. (2011). Baculovirus capsid display potentiates OVA cytotoxic and innate immune responses. *PLoS One* 6, e24108.

[52] Monsma, S. A, & Blissard, G. W. (1995). Identification of a membrane fusion domain and an oligomerization domain in the baculovirus GP64 envelope fusion protein. *J Virol* , 69, 2583-2595.

[53] Niu, M, Han, Y, & Li, W. (2008). Baculovirus up-regulates antiviral systems and induces protection against infectious bronchitis virus challenge in neonatal chicken. *Int Immunopharmacol* , 8, 1609-1615.

[54] Oker-blom, C, Airenne, K. J, & Grabherr, R. (2003). Baculovirus display strategies: Emerging tools for eukaryotic libraries and gene delivery. *Brief Funct Genomic Proteomic* , 2, 244-253.

[55] Ong, S. T, Li, F, Du, J, Tan, Y. W, & Wang, S. (2005). Hybrid cytomegalovirus enhancer-h1 promoter-based plasmid and baculovirus vectors mediate effective RNA interference. *Hum Gene Ther* , 16, 1404-1412.

[56] Pieroni, L. La Monica, N. ((2001). Towards the use of baculovirus as a gene therapy vector. *Curr Opin Mol Ther* , 3, 464-467.

[57] Pieroni, L, & Maione, D. La Monica, N. ((2001). In vivo gene transfer in mouse skeletal muscle mediated by baculovirus vectors. *Hum Gene Ther* , 12, 871-881.

[58] Prabakaran, M, Madhan, S, Prabhu, N, Qiang, J, & Kwang, J. (2010). Gastrointestinal Delivery of Baculovirus Displaying Influenza Virus Hemagglutinin Protects Mice against Heterologous H5N1 Infection. *Journal of Virology* , 84, 3201-3209.

[59] Ricklin, D, Hajishengallis, G, Yang, K, & Lambris, J. D. (2010). Complement: a key system for immune surveillance and homeostasis. *Nature Immunology* , 11, 785-797.

[60] Rohrmann, G. F. (2011). Baculovirus Molecular Biology, 2nd edn. Edited by G. F. rohrmann. Bethesda (MD): National Library of Medicine (US), NCBI.

[61] Schutz, A, Scheller, N, Breinig, T, & Meyerhans, A. (2006). The Autographa californica nuclear polyhedrosis virus AcNPV induces functional maturation of human monocyte-derived dendritic cells. *Vaccine* , 24, 7190-7196.

[62] Song, L, Liu, Y, & Chen, J. (2010). Inorganic binding peptide-mediated immobilization based on baculovirus surface display system. *J Basic Microbiol* , 50, 457-464.

[63] Spenger, A, Grabherr, R, Tollner, L, Katinger, H, & Ernst, W. (2002). Altering the surface properties of baculovirus Autographa californica NPV by insertional mutagenesis of the envelope protein gp64. *Eur J Biochem* , 269, 4458-4467.

[64] Strauss, R, Huser, A, Ni, S, Tuve, S, Kiviat, N, Sow, P. S, Hofmann, C, & Lieber, A. (2007). Baculovirus-based vaccination vectors allow for efficient induction of immune responses against plasmodium falciparum circumsporozoite protein. *Mol Ther* , 15, 193-202.

[65] Tani, H, Abe, T, Limn, C. K, Mochizuki, R, Yamagishi, J, Kitagawa, Y, Watanabe, R, Moriishi, K, & Matsuura, Y. (2003). Baculovirus vector--gene transfer into mammalian cells]. *Uirusu* , 53, 185-193.

[66] Tani, H, Nishijima, M, Ushijima, H, Miyamura, T, & Matsuura, Y. (2001). Characterization of cell-surface determinants important for baculovirus infection. *Virology* , 279, 343-353.

[67] Tjia, S. T. zu Altenschildesche, G. M. & Doerfler, W. ((1983). Autographa californica nuclear polyhedrosis virus (AcNPV) DNA does not persist in mass cultures of mammalian cells. *Virology* , 125, 107-117.

[68] Van Loo, N. D, Fortunati, E, Ehlert, E, Rabelink, M, Grosveld, F, & Scholte, B. J. (2001). Baculovirus infection of nondividing mammalian cells: mechanisms of entry and nuclear transport of capsids. *J Virol* , 75, 961-970.

[69] Verma, I. M, & Somia, N. (1997). Gene therapy-- promises, problems and prospects. *Nature* , 389, 239-242.

[70] Wang, Y. (2005). Using a baculovirus display library to identify MHC class I mimotopes. *Proceedings of the National Academy of Sciences* , 102, 2476-2481.

[71] Xu, X, Chiou, G, Zhang, M. -T, Tong, Y. -M, Hu, D. -W, Zhang, J. -H, & Liu, M. -T. H.-J. ((2008). Baculovirus surface display of Erns envelope glycoprotein of classical swine fever virus. *Journal of Virological Methods* , 153, 149-155.

[72] Xu, X, & Liu, G. H.-J. ((2008). Baculovirus surface display of E2 envelope glycoprotein of classical swine fever virus and immunogenicity of the displayed proteins in a mouse model. *Vaccine* , 26, 5455-5460.

[73] Xu, X, Tong, G, Chiou, D. -W, Hsieh, M. -T, Shih, Y. -C, Chang, W. -L, Liao, C. -D, Zhang, M. -H, & Liu, Y. -M. H.-J. ((2009). Baculovirus surface display of NS3 nonstructural protein of classical swine fever virus. *Journal of Virological Methods* , 159, 259-264.

[74] Xu, X, Wang, G, Zhang, Z. -S, Li, Q, Zhao, Z. -C, Li, H. -N, Tong, W, & Liu, D. -W. H.-J. ((2011). Baculovirus surface display of E envelope glycoprotein of Japanese encephalitis virus and its immunogenicity of the displayed proteins in mouse and swine models. *Vaccine* , 29, 636-643.

[75] Yoshida, S, Kawasaki, M, Hariguchi, N, Hirota, K, & Matsumoto, M. (2009). A Baculovirus Dual Expression System-Based Malaria Vaccine Induces Strong Protection against Plasmodium berghei Sporozoite Challenge in Mice. *Infection and Immunity* , 77, 1782-1789.

[76] Zanotto, P. M, Kessing, B. D, & Maruniak, J. E. (1993). Phylogenetic interrelationships among baculoviruses: evolutionary rates and host associations. *J Invertebr Pathol* , 62, 147-164.

Chapter 7

Gene Delivery Systems

Andrea Pereyra and Claudia Hereñu

Additional information is available at the end of the chapter

http://dx.doi.org/10.5772/56869

1. Introduction

The present chapter will be focused on different gene delivery systems used in gene therapy approaches with the purpose of inserting into individual cells and/or tissues to treat diseases. By correcting genetic defects via genome manipulation, gene therapy can truly revolutionize medical intervention for treating monogenetic inherited/acquired diseases or polygenetic conditions.

Gene therapy has undergone a remarkable development in the last 20 years. Particularly important advances have been made in the improvement of gene transfer and expression technology, with current efforts focusing on the design of safer and long-term gene expression vectors as well as systems possessing cell-type specificity for transgene delivery and regulatability of its expression by small molecules. The foreign genetic material can be administered *in vivo*, *ex vivo* or *in vitro* depending on the nature of a disease. A successful gene therapy system must perform several functions. In all cases, the therapeutic gene must first be delivered across the cell membrane, which is a significant barrier. Once delivered inside the cell, the therapeutic gene may exist episomally or be integrated into the host genome depending on the nature of the gene transfer vector. Moreover, an important issue is the replication and segregation of the therapeutic gene during cell division in order to maintain long-lasting gene expression. These specifications will be discussed for each gene delivery system along the chapter.

- Current viral vector systems will be detailed in this chapter: Adenovirus, Retrovirus, Lentivirus, Adeno-associated Virus, Sendai Virus and Herpes Simplex Virus.

- Non-viral vector systems will be also discussed: naked DNA (including plasmids, DNA transposons), DNA-lipid complexes and nanoparticles (including Qdots and the new technology of magnetic nano-particles) which have been actively developed as additional tools for crossing the cell membrane.

• Novel systems were the viral and non-viral methods merge will be also addressed.

2. Non viral gene delivery systems

2.1. Naked DNA

Naked DNA such as plasmids remain popular as vectors for gene therapy today for their low immunogenicity and low risk of causing insertional mutagenesis. However, their episomal feature resulting in transient gene expression makes them unsuitable as gene therapy vectors when long-term gene expression is needed for treatment. DNA transposons have the properties of naked DNA and plasmids as well as the ability to insert transgenes into host chromosomes for long-term expression. DNA transposons are natural genetic elements residing in the genome as repetitive sequences that move through a direct cut-and-paste mechanism.

A simple transposon is characterized by terminal inverted repeats flanking a gene encoding transposase, an enzyme required for its translocation (Meir et al., 2011). The cut-and-paste process, called transposition, makes DNA transposons particularly attractive as gene delivery tools. To turn DNA transposons into a gene delivery tool, a two-plasmid system, consisting of a helper plasmid expressing the transposase and a donor plasmid with the terminal repeat sequences flanking genes of interest, has been developed. Using this system, transposons have been utilized extensively as genetic tools in invertebrates and in plants for transgenesis and insertional mutagenesis (Spradling et al., 1982; Hayes et al., 2003).

Plasmid DNA is an attractive alternative due to its inherent simplicity and because it can easily be produced in bacteria and manipulated using standard recombinant DNA techniques. It shows very little dissemination and transfection at distant sites following delivery and can be re-administered multiple times into mammals without inducing an antibody response against itself (Jiao et al., 1992). Also, considerable long term foreign gene expression from naked plasmid DNA (pDNA) is possible even without chromosome integration if the target cell is post-mitotic or has low mitotic rate and if an immune reaction against the foreign protein is not generated (Herweijer *et al.*, 2001; Wolff *et al.*, 1992). The poor expression levels represent a major constraint in the use of these vectors for gene transfer/therapy. However, the low efficient expression by direct injection of naked plasmids was improved by ballistic technology, cationic lipids and neutral polymers (Prud'homme et al., 2001; Gao et al., 1995; Lemieux et al., 2000) and most efficiently by electroporation (Sandri et al, 2003), an established technology that transiently permeabilizes cell membrane by short voltage pulse, allowing the uptake of a wide spectrum of biological molecules.

The stability of DNA vectors with high-molecular-weight is a central point for the improvement of gene delivery. All high-molecular-weight DNA vectors are susceptible to damage. The self-compacting option (self-entangling) can be defined as the folding of single DNA molecules into a configuration with mutual restriction by the individual segments of bent DNA. A negatively charged phosphate backbone makes DNA self-repulsive, so it is reasonable to assume that a certain number of 'sticky points' dispersed within DNA could facilitate the entangling. Tolmachov, proposes that the spontaneous entanglement of vector DNA can be enhanced by the interlacing of the DNA with sites capable of mutual transient attachment through the formation of non-B-DNA forms, such as interacting cruciform structures, inter-segment triplexes, slipped-strand DNA, left-handed duplexes (Z-forms) or G-quadruplexes. (Tolmachov, 2012).

2.2. Liposomes

The liposomes, lipids arranged in lamellar structures, are concentric bilayered vesicles surrounded by a phospholipid membrane. They are related to micelles which are generally composed of a monolayer of lipids. The amphiphilic nature of liposomes, their ease of surface modification, and a good biocompatibility profile make them an appealing solution for increasing the circulating half-life of peptides, proteins, cDNAs and siRNAs (Bhavsar et al., 2012). They may contain hydrophilic compounds, which remain encapsulated in the aqueous interior, or hydrophobic compounds, which may escape encapsulation through diffusion out of the phospholipid membrane (Figure 1).

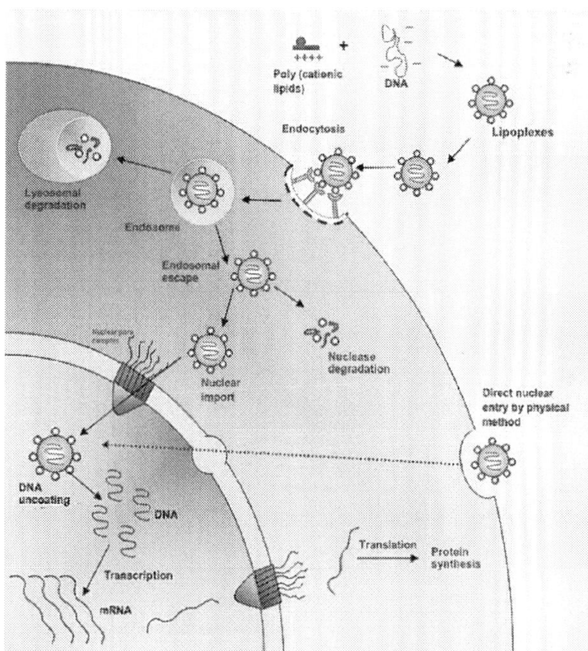

Figure 1. Non-viral gene delivery using lipoplexes. DNA is complexed with cationic liposomes and is internalized through receptor mediated endocytosis. After their internalization large amounts of complexes are degraded in the endolysosomal compartments. Only a small fraction enters into the nucleus and elicits desired gene expression. (From: Pankajakshan Divya and Devendra K. Agrawal, 2013).

Liposomes can be designed to adhere to cellular membranes to deliver drugs or cDNAs after endocytosis (Bangham et al., 1995). The first formulation was prepared in 1986 by the Christian Dior laboratories in collaboration with the Pasteur Institute (Bangham et al., 1995). Liposomes have been used as delivery systems due to their versatility, charge and surface functionalities that improve their effectiveness *in vivo*. Presumably, the lack of widespread medical impact is due to their limited biological stability. Hydrophobic nanoparticles such as unmodified liposomes are rapidly cleared via the reticuloendothelial system. Liposome formulations with prolonged circulation time have been obtained upon functionalization with PEG (polyethyleneglycol, a biologically inert polymer). The longer circulation half-life of these PEG-coated liposomes may allow a better control of therapeutic drug delivery (Gabizon A et al, 1988).

There are additional unique mesophasic structures of lipids formed as a result of lipid structure polymorphisms, which include cubic-, hexagonal- or sponge-phases. These structures have been utilized for gene, vaccine and drug delivery and provide the advantages of stability and production feasibility compared with liposomes (Shanmugam et al., 2011).

2.3. Nanoparticles and nanotechnology

Nanotechnology is a rapidly expanding field, encompassing the development of man-made materials in the 5–200 nanometer size range. This dimension vastly exceeds that of standard organic molecules, but its lower range approaches that of many proteins and biological macromolecules. The origin of nanotechnology can be traced back to 1959 when physicist Richard Feynman (1960) recognized the potential of manipulating individual atoms and molecules at the nanometer scale and suggested that materials at this scale possess unique physical properties. The nanotechnology field encompasses concepts and approaches deeply rooted in physics, polymer and colloidal chemistry, pharmaceutics, biomaterials, as well as cell and molecular biology and biophysics. The main theme of nanotechnology is the development and use of nanometer-scale materials that display unique functional properties not shown by bulk materials. Nanomaterials can interact with biological systems at a molecular and supra-molecular level, they can be coated to respond to specific cell environments and even to induce desired physiological responses in cells, while minimizing unwanted side effects. The first practical applications of nanotechnology can be traced to advances in communications, engineering, physics, chemistry, biology, robotics, and medicine. Nanotechnology has been utilized in medicine for therapeutic drug delivery and the development of treatments for a variety of diseases and disorders. An intracellular nanoparticle, consequently, may act as a drug depot within the cell and provide an intracellular sanctuary to protect therapeutic compounds from efflux or degradation.

Several synthetic strategies exist to prepare ferromagnetic iron oxide nanoparticles (Gobe et al., 1983). For maghemite (c-Fe$_2$O$_3$) and magnetite (Fe$_3$O$_4$) nanoparticles, this precipitation technique requires alkalization of a solution of metal salt with subsequent hydrolysis in microemulsions. Additionally, biosynthetic routes exist utilizing "magnetic bacteria"; the resulting nanoparticles typically range from 50 to 100 nm in diameter (Matsunaga et al., 1998). The synthesis of iron oxide nanoparticles has also been achieved by sonochemical decomposition of iron pentacarbonyl (Faraji et al., 2009), thermal decomposition of other iron complexes (Rockenberger et al., 1999), and by thermal decomposition of iron pentacarbonyl followed by oxidation. When optimized, these methods may afford monodisperse nanoparticles with sizes ranging from 3 to 20 nm for magnetite and 4 to 16 nm for maghemite (Sun et al., 2002). The circulation times of these particles can be greatly increased simply by hydrophilic surface modification with PEG. Furthermore, iron oxide nanoparticles also display fairly easy surface modification capabilities that presents an attractive prospect for direct drug or biomolecule payload attachment (Longmuir et al., 2006).

Nanoparticle-based delivery (NBD) has emerged as a promising approach to improve the efficacy and the development of new therapies (Probst et al., 2012).

2.3.1. Quantum dots (Qdots)

Quantum dots are luminescent nanoparticles typically used for imaging in biological systems. Their primary components —core (cadmium with selenium or tellurium), shell, and coating— give the photochemical properties. Qdots have small size and versatile surface chemistry and offer superb optical properties for real-time monitoring as transport vehicles at both cellular and systemic levels. Qdots offer great potential providing mechanisms for monitoring intracellular and systemic nanocarrier distribution, degradation, drug release, and clearance. They can be manufactured with diameters from a few nanometers to micrometers and a narrow size distribution using techniques requiring high annealing temperatures (Yum et al., 2009). Capping of quantum dots with ZnS has been shown to augment stability and enhance luminescence (Park et al., 2009). However, ZnS capping alone is not sufficient to fully stabilize the core, especially in biological systems. PEGylation plays a dual role in increasing biocompatibility and improving the core stability in biological systems (He et al., 2010).

Though the direct use of QDots for drug delivery remains questionable due to their potential long-term toxicity, the QDot core can be easily replaced with other organic drug carriers or more biocompatible inorganic contrast agents (such as gold and magnetic nanoparticles) based on their similar size and surface properties, facilitating translation of well characterized NBD vehicles to the clinic, maintaining NBD imaging capabilities, and potentially providing additional therapeutic functionalities such as photothermal therapy and magneto-transfection.

2.3.2. Magnetic nanoparticles (MNPs)

Magnetofection is a methodology developed in the early 2000's (Scherer et al., 2002). It is based on the association of MNPs with non-viral or viral vectors in order to optimize gene delivery in the presence of a magnetic field, and to concentrate therapeutic complexes in target areas. The association of viral vector-based gene delivery with nanotechnology now offers the possibility to develop more efficient and less invasive gene therapy strategies for a number of major pathologies and diseases *(See details in Section 3.5 of this chapter)*.

3. Viral gene delivery systems

3.1. General overview of viruses as vectors for gene delivery

For centuries the health sciences have invested sweat and tears in order to fight against viral infections affecting humans, animals and plants. It is reasonable to imagine that the idea of viruses as therapeutic agents was quite a shock when first presented. The enormous advances in Molecular Biology, Biochemistry, Genomics and Human Medicine, among others, have provided to the Virology field the necessary tools for manipulating them on their behalf.

These recombinant vectors are viruses where the genome has been altered in a controlled way by experimental manipulation. For any procedure to generate a recombinant virus the starting point is to clone and manipulate its genome. Thus DNA virus genomes may be cloned directly while RNA virus genomes may be cloned as cDNA. These molecules can then be modified by site-specific alteration, or more drastically, segments may be removed and replaced with foreign DNA sequences. Then the process must be completed by recreating infectious virus particles. This requires specific techniques and is not yet possible for all virus types (Dimmock et al., 2007).

Regarding the biological value of these viral vectors, there are constantly novel potential applications such as vaccines, carriers of nucleic acid sequences for regulating gene expression and agents for gene therapy. In order to become a therapeutic agent the DNA has to be carried into the cell and ultimately reach the nucleus; therefore it is mandatory to be provided of an strategy for membrane cross and lysosomal scape. This is something that naked DNA is very poorly equipped to achieve. By contrast, the nucleic acid that is inside an infectious virus particle can avoid these issues. First, viruses have evolved specific interactions with cell surface molecules that lead to their efficient entry and, second, if that entry involves arrival in the cytoplasm within an endocytic vesicle, then viruses have mechanisms to allow efficient escape. This process of virus-mediated gene delivery into a cell is known as transduction (Dimmock et al., 2007).

It is clear then that, to be potentially useful as a gene delivery vector, a virus should have a number of specific features (Figure 2).

However, there is no virus that can meet all the criteria for an ideal gene delivery vector and there are some significant drawbacks that will be addressed in the following sections. Thus, each application is likely to need its own vector, chosen and then tailored to fulfill the precise requirements.

3.2. Current viral vectors systems

3.2.1. Adenoviral vectors (Ad)

The discovery and initial description of Adenoviruses (AdV) took place in the early 1950s. They were first isolated from human adenoid tissue cultures (Rowe et al., 1953). Since then several different serotypes of human, avian, reptilian, amphibian and other mammalian

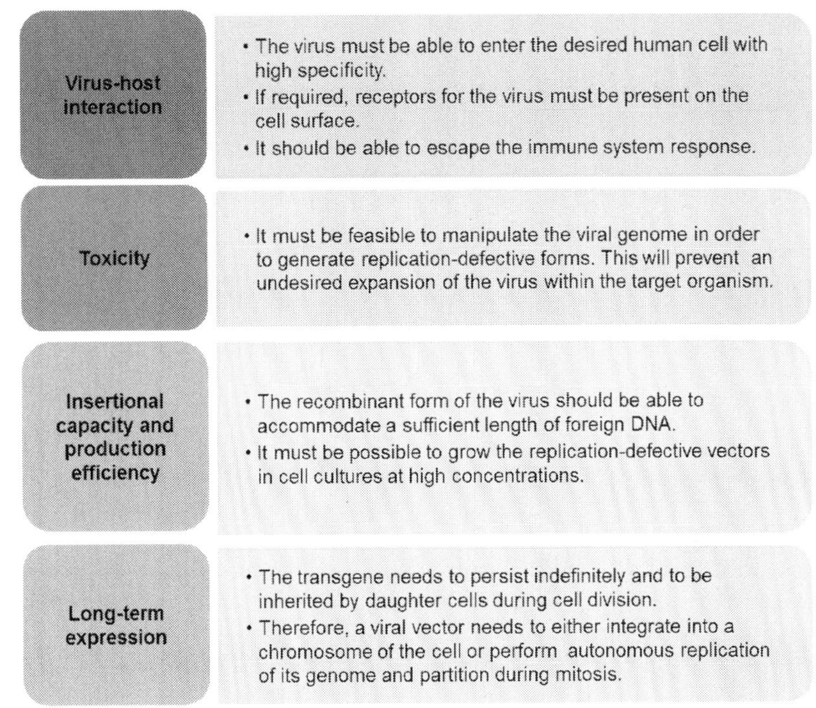

Figure 2. What does it take to be a useful viral vector? (Adapted from Dimmock N, Easton A and Leppard K; 2007)

adenoviruses have been isolated and characterized giving birth to the *Adenoviridae* family of over 50 members.

The focus on AdV for gene transfer was based on basic research. The establishing of the biology of AdV and their capacity to efficiently deliver the viral genome to the target cells became relevant then. More importantly, since AdV was not oncogenic in humans and the genomes of common AdV were completely defined and easy to modify, the production of recombinant AdV (RAdV) was achieved. In the context of gene delivery, serotypes 5 and 2 of the subgroup C have been used the most because their structure and biology is well described and there are convenient biologic reagents available to produce recombinant subgroup C gene transfer vectors in large quantities. Regarding safety, AdVs of subgroup C can cause minor to mild respiratory infections sometimes associated with conjunctival compromise (Ginsberg et al., 1994).

AdV virions consist of a ~36 kb linear double-stranded DNA genome encased within a non-enveloped icosahedral particle (Figure 3).

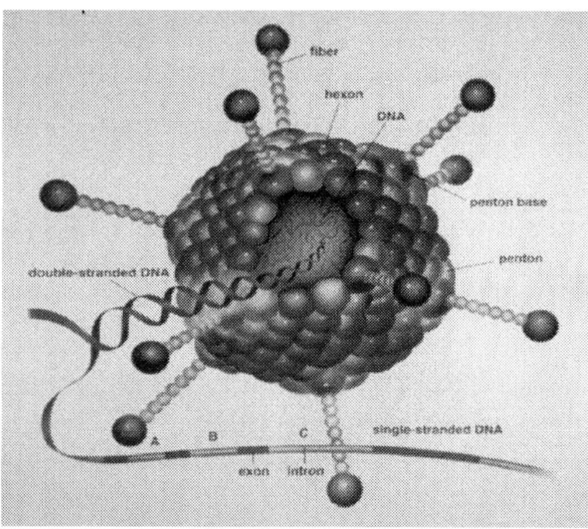

Figure 3. Adenovirus particle structure

The experimental manipulation of its genome has allowed the production of a diversity of recombinant viral particles where most of the replication genes are removed. The deletion of the E1 and E3 regions in first-generation AdV allows ~7 kb of foreign DNA to be inserted into the vector genome (Volpers and Kochanek, 2004). Another feature of these vectors is that they can be grown to extremely high titers in the HEK 293 cell line, with burst sizes typically between 10^3-10^4 viral particles (VP) per cell and final concentrations reaching 10^{13} VP/ml, after CsCl density gradient centrifugation. Whereas AdV vectors can achieve high expression in many target organs when used *in vivo*, expression of the transgene is limited to days or weeks, mainly because innate and adaptive immune host defenses against the virus. For applications where persistent expression is required to achieve a therapeutic goal, the modern, third-generation, high capacity AdV vectors have become the most efficient alternative (Hackett NR and Crystal RG, 2009). Other important obstacle in the use of RAd for gene transfer is the process of cell attachment and internalization used by the viral particles. The target cell must express the cell membrane receptor CAR (Coxsackie-Adenovirus Receptor) in order to be susceptible to the adenoviral infection (Figure 4).

For those transduction-refractory tissues, modern virology has developed modified-tropism RAds with modifications in fiber/high affinity receptor or the penton–integrin of the capsid. In an extensive survey of the tropism of AdV5-derived vectors but with fibers derived from different serotypes, the fiber genes of AdV16 were found to be better at targeting fibroblasts and chondrocytes, AdV35 at targeting dendritic cells and melanocytes and AdV50 better at targeting myoblasts and hematopoietic stem cells (Havenga et al., 2002). The addition of an oligolysine motif to the C-terminus of the fiber protein, giving the virus an affinity for polyanions such as heparin sulphate, profoundly affects the range of cell types that can be

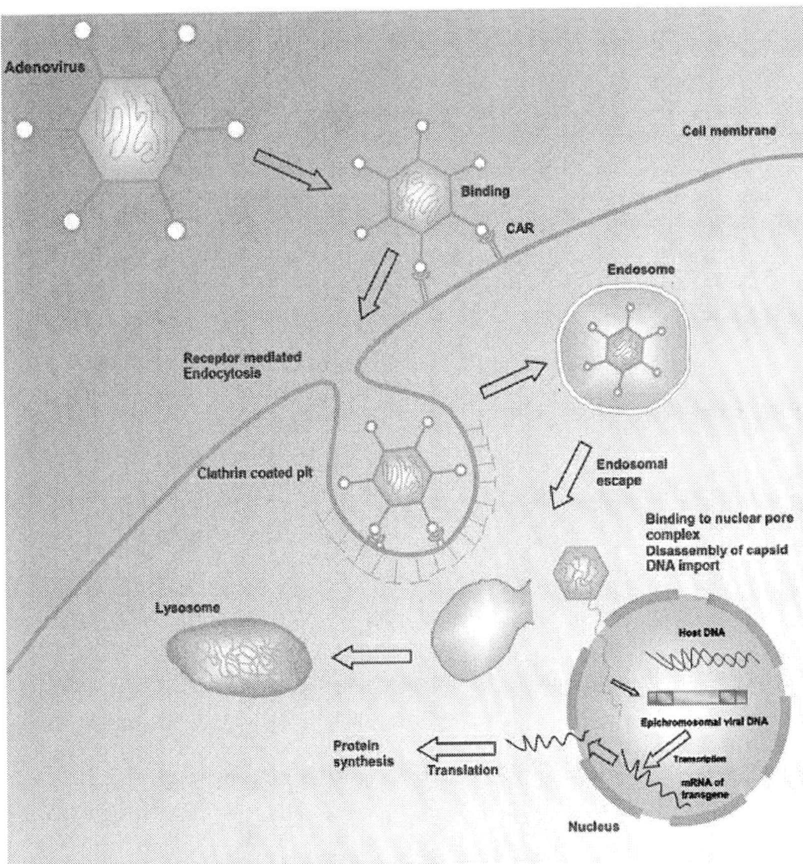

Figure 4. Recombinant adenovirus enters cells via CAR-mediated binding allowing internalization via receptor-mediated endocytosis through clathrin-coated vesicles. Inside the cytoplasm, the endocytosed adenoviral vector escapes from the endosomes, disassembles the capsid and the viral DNA enter into the nucleus through the nuclear envelope pore complex. The viral DNA is not incorporated into the host cell genome, but rather assumes an epichromosomal location, where it can still use the transcriptional and translational machinery of the host cell to synthesize recombinant protein (From: Pankajakshan Divya and Devendra K. Agrawal, 2013).

infected *in vitro* allowing cells lacking CAR to be transduced by RAdVs (Bouri et al., 1999). For a more detailed description of Adenoviral vectors see section 3.3 of the present chapter.

3.2.2. Adeno-associated virus Vectors (AAV)

AAV is a human parvovirus within the genus *Dependovirus*. It was originally observed as a contaminant of laboratory preparations of adenovirus (Carter et al., 2009). Viral particles are small (20–30 nm) and non-enveloped, containing single-stranded DNA molecules with plus and minus strands packaged with equal efficiency (Daly, 2003). No human disease has been associated with AAV infection which is an important feature when thinking in AAV as gene vectors. Six serotypes of AAV have been described with AAV2 being the most widely used for gene-transfer studies (Hermonat et al., 1984). AAV2 cell entry is mediated by binding to heparin sulfate proteoglycans and αvβ5 integrin; fibroblast growth factor receptor-1 (FGFR-1) may also be involved. The distribution of these molecules on many different cell types can explain the prolonged *in vivo* expression following AAV treatment seen in the liver, brain, skeletal muscle, lung, and hematopoietic stem cells of animal models (Daly, 2003). AAV vectors contain no viral genes that could elicit undesirable immune or inflammatory responses. The primary host reaction that might have an unwanted impact is the production of neutralizing antibodies against the viral particles.

One major concern when developing AAV vectors for gene delivery is that DNA constructs larger than the wild-type 4.7 kb sequence do not package well and vector titers decrease sharply thus constituting an insertional limitation for the cDNAs. Another important issue is the frequently seen integration of AAV genome in the host cell chromosomes. However, the available evidence indicates that integration of wild type AAV *in vivo* does not reflect the experimental *in vitro* observations, but appears to be a rather rare event and AAV genomes mostly persist as episomes, as has also been demonstrated for AAV vectors (Carter et al., 2009).

3.2.3. Retroviral vectors (RV)

Retroviruses are lipid-enveloped viruses; with nucleocapsids containing two copies of a linear, positive-stranded 7–11 kb RNA genome. The family *Retroviridae* contains various viruses that have shown potential utility for gene therapy, such as the **gammaretrovirus** *(or simple retrovirus), spumaviruses* and **lentiviruses** *(or complex retrovirus)*. Following attachment and receptor-mediated entry into host cells, viral reverse transcriptase and integrase enzymes mediate reverse transcription and integration of the virus genome into the host-cell chromatin. Retroviral vectors have the ability for stable integration and allow long-term expression so that theoretically a single administration could have a sustained, potentially even, lifelong curative effect (Schambach et al., 2009). As for any viral vector, replication-deficiency is a condition. To achieve this goal, the retroviral coding sequences have to be removed, which creates at least 6 kb space for the transgene of interest. Since neither structural proteins nor replication enzymes are encoded by the target cell, the generation of replication-competent virus is prevented. The gammaretroviruses cannot infect quiescent, non-dividing cells, which is a handicap of the vectors derived from these retroviruses. However, this can be overcome by the use of lentiviral vectors.

For most RV, taking advantage of the insertional mechanism, the simplest application is in the production of cell lines that express a transgene introduced on a retroviral vector. For modified, transgenic animals, the lentivirus group must be used because gammaretrovectors are silenced during embryonic development. RV vectors can also be used in the delivery of toxic genes to cancer cells, which are actively dividing. Another area of application is gene discovery. The integration of the viral genome can reveal function by insertional inactivation of a gene in the host cell chromosome (Somia, 2003).

3.2.4. Herpes Simplex Virus Vectors (HSV)

HSV-1 is a double-stranded DNA virus,with a capsid surrounded by a dense layer of proteins -the tegument- enveloped in a lipid bilayer with surface proteins. It has evolved to persist in a lifelong nonintegrated latent state without causing disease in the immune-competent host. Among the herpes family Herpes Simplex Virus type 1 (HSV-1) is an attractive vehicle because in natural infection, the virus establishes latency in neurons, a state in which viral genomes may persist for the life of the host as intranuclear episomal elements. Although the wild-type virus may be reactivated from latency under the influence of a variety of stresses, completely replication-defective and non-lytic viruses can be design (Goins et al., 2003). HSV-1 has a broad host range and does not require cell division for infection and gene expression. Accordingly, HSV may be generally useful for gene transfer to a variety of normal and disease tissues. The overall size of the HSV-1 genome (152 kb) represents an attractive feature for employing the vector for the transfer of large amounts of exogenous genetic sequences. Approximately one-half of the HSV-1 coding sequences are nonessential for virus replication in cell culture. At least 44 kb of HSV sequence can potentially be removed in order to accommodate a transgene (Wolfe et al., 2009).

The obstacles that need to be addressed in order to take advantage of the full potential of these vectors include elimination of residual vector toxicity, design of promoter cassettes that provide sufficient level and duration of transgene expression, and targeting of transgene expression to specific cell populations through the use of tissue-specific promoters, or by altering the virus host range through modifying receptor utilization for attachment and entry (Wolfe et al., 2009).

3.2.5. Sendai Virus Vectors (SeV)

Since its isolation in 1953 in Japan, Sendai virus (SeV) has been widely used as a research tool in cell biology and in the industry, but the application of SeV as a recombinant viral vector has been investigated only recently. Sendai virus (SeV) is a nonsegmented negative-strand RNA virus belonging to the *Paramyxoviridae* family. As SeV can infect various animal cells with an exceptionally broad host range and is not pathogenic to humans, various applications have been explored for SeV as a recombinant viral vector capable of transient but strong gene expression (Nakanishi and Otsu, 2012). Its RNA nature is advantageous for applications in which chromosomal integration of exogenous genes can be undesirable. These viral vectors are currently being tested in regenerative medicine to reprogram cell genomes to a pluripotent state with a surprisingly high efficiency (Nishimura et al., 2011; MacArthur et al., 2012) and as recombinant viral vaccines for influenza prevention (Le et al., 2011).

3.3. Recombinant adenoviral vectors

Although the pathologies associated with wild-type Adenovirus (AdV) infections are generally mild, there is a potential risk of using fully replication-competent AdV for gene transfer because the inflammatory host responses may alter organ function. There is also the possibility of overwhelming infection if AdV replication is allowed to progress when there are deficiencies in the host defense system. These situations became then a major reason to develop recombinant, non-pathogenic viral particles.

Regardless experimentation with different viral gene delivery systems, adenoviral vectors continue to be widely used for gene transfer strategies (Kovesdi et al., 1997; De Gruijl et al, 2012; Youngjoo et al, 2013; Fishbein I et al, 2013). They were also the first viral vector system to be developed. However this was more by chance than intentionally while working in the production of live adenoviral vaccines propagated in monkey cell lines. Infection of tissue culture cells with AdV vaccine accidentally contaminated with simian virus 40 (SV40) resulted in the production of the SV40 T antigen, even after removal of the SV40 virions from the AdV stocks by immunodepletion. Analysis of this adenovirus revealed that the T-antigen gene from SV40 had recombined into the E3 region of the Ad genome. This demonstrated the possibility that AdV could carry foreign genes and express them as well as demonstrating the dispensability of the E3 genes for *in vitro* replication (Roy-Chowdhury and Horwitz, 2002; Campos et al., 2007).

The first wild type adenoviruses subjected to "vectoring" process were AdVs derived from the human serotypes 5 (Ad5) and 2 (Ad2). First-generation replication-deficient Ad5 vectors were developed by deleting the E1 genes, necessary for expression of E2 and late genes required for AdV DNA synthesis, capsid protein expression, and viral replication. Further deletions included the E3 genes which are involved in the evasion of host immune defenses but dispensable for replication of the virus *in vitro*. Therefore, because of this experimental manipulation leading to viral replication impairment, there was the need to develop a biological system capable of providing the genes required for the virus propagation. This led to the creation of the HEK 293 cell line, which was transfected with sheared adenovirus-type 5 (Ad5) genomic DNA and stably expresses the E1 genes (Graham et al., 1977).

Although the first-generation AdV vectors are generally considered replication defective, there is some low level expression of viral antigens that limits the duration of transgene expression *in vivo*, due to elimination of transduced cells by the cellular immune system (McConnell and Imperiale, 2004). To avoid this response and allow long-term episomal expression a new recombinant vector was developed. They are referred as "gutless" or high-capacity Adenovirus (HC-AdV) lack all viral coding sequences except the *cis*-elements required for the genome replication and encapsidation. Therefore, they need to be assisted by Helper Adenoviruses that provide all necessary replication genes in *trans*; therefore, they are also called Helper-Dependent adeno vectors (HD-AdV) (Amalfitano, 1999). In addition, these vectors have a much higher packaging capacity of ~35 kb of foreign DNA, enabling the expression of large transgenes or the inclusion of human genomic regulatory elements (Palmer and Ng, 2005). See Table 1 for comparison of the profile of different vector generations.

A variety of applications can be found for these vectors. The most widely explored are Gene Therapy for prostate, colon, cervix, ovary and CNS tumors, for genetic diseases such as Hemophilia, Duchene Muscular Dystrophy, Familial Hypercholesterolemia, Autoimmune Diabetes; viral vaccines; supplementary therapy for degenerative conditions as Parkinson, Alzheimer and Rheumatic Arthritis and production of recombinant proteins among others.

Ad vector generation	AdV genome deletions	foreign DNA insertion capacity	propagation in cell lines	type of transgene expression	host immune response	risks
first	E1 and/or E3	6.5 kb	HEK293 (express E1)	transient	significant	Possibility of the production of replication-competent adenovirus (RCA)
second	E2 (complete or partial)	8kb	293-C2 (express E2) AE1-2a	transient	important; especially after repeated administration	No production of RCA
third	most of the genome. ITR and packaging seq. retained ("gutless", HC or HD-Ad)	35-37 kb	293 cells expressing Cre; require helper virus	long-lasting	low or none	No production of RCA

Table 1. Comparison of adenoviral vectors generations (Russell, 2000).

One of the most spectacular results in gene therapy using HD-AdV was published by Kim et al., 2001. They used a gutless AdV harboring the apoE gene to treat apoE-deficient C57BL/6 mice which display spontaneous hypercholesterolemia. These mice received a single intravenous (i.v) injection of HD-AdV-ApoE at the 12th week of age. This isolated intervention managed to normalized plasma apoE concentrations and therefore diminish the plasma cholesterol level to values found in wild-type mice. However, the most outstanding finding involves the duration of the protective effect of this gene therapy, being of at least 2.5 years, the lifespan of these mice.

Many of these AdV vectors produced in accordance with regulated quality standards, are now being used for human clinical trials. For detailed information the reader is referred to the following reviews: Russell, 2000; Józkowicz and Dulak, 2005; Campos, 2007.

3.4. Adenoviral vectors in tissue-specific gene transfer: The skeletal muscle

The skeletal muscle gene transfer approach using Adenoviral Vectors has created controversy. There are several studies with heterogeneous efficiency rates and, in some cases, divergent outcomes. On this section we will intend to present and discuss these results as well as to introduce a novel technology that might overcome the difficulties experienced in the transduction of this tissue.

Direct gene transfer into skeletal muscle cells *in vitro* and *in vivo* using either plasmid DNA or recombinant viruses has medical applications in vaccination and gene therapy and also has been widely used in studies of developmental and physiological regulation of muscle gene expression (Hallauer et al., 2000). Many factors regarding muscle tissue barriers, immune response, systemic dissemination, potential toxicity and specific properties of each viral system need to be taken into account when selecting the proper approach for skeletal muscle. It is well known that skeletal muscle is a highly developed and organized tissue in which the constituent myofibers become post-mitotic in fetal life. The mononucleated myogenic precursor cells (satellite cells) that are located between the extracellular matrix and the plasma membrane of myofibers are known to be capable of fusing together or with preexisting myofibers in response to various types of stimuli, mainly to injury (Chargé et al., 2004). These satellite cells are relatively easy to isolate and cultivate *in vitro* and can also be efficiently transduced using virtually any viral vector in contrast with the mature multinucleated myofibers. Thus, is believed that viral transduction during skeletal muscle maturation might require mitotically active myoblasts (van Deutekom et al., 1998). However, because some of the viral vectors have been shown to transduce post-mitotic, immature myofibers *in vitro* and *in vivo* other factors are also likely to be involved in the poor level of viral transduction of mature myofibers (Wolff et al., 1990; Acsadi et al., 1994 and Huard et al., 1996;).

Retroviral vectors (RV) can infect dividing myoblasts with a high efficiency although they remain incapable of infecting post mitotic myotubes or myofibers (Miller et al., 1990 and Salvatori et al., 1993). In addition, the ability to become stably integrated into the host cell genome, which can provide long-term, stable expression of the delivered gene, may also represent a risk for insertional mutagenesis. As mentioned in Section 3.2 of the present chapter, other limitations to the use of retroviruses are the gene insert capacity (less than 7 kb) and the relatively low production titers (10^5–10^6 plaque-forming units per milliliter (pfu/mL)). The recombinant vectors obtained from the *Herpes Simplex Virus type 1 (HSV-1)* can persist in the host cell in a nonintegrated state and be prepared at adequately high titers (10^7–10^9 pfu / ml). They are capable of transducing muscle cells in most maturation stages while carrying large DNA fragments. However, are still unable to highly penetrate and transduce mature myofibers (Huard et al., 1997). *Adeno-associated derived viral vectors (AAV)* have also been used to approach muscle cells. Although a long-term gene expression (up to 18 months) and a high efficiency of mature myofibers transduction have been observed in mouse skeletal muscle, the application of adeno-associated viral vectors for gene therapy may be limited by their restrictive gene insert capacity (Pruchnic et al., 2000). In the last few years a novel gene transfer viral vector has been used into skeletal muscle, the *recombinant Sendai virus (SeV) vectors*. As explained before, the wild SeV is a non-segmented negative-strand RNA virus belonging to the *Paramyxoviridae* family that can infect various animal cells with an exceptionally broad host range. Shiotani et al., 2001 accomplished a significant overexpression of hIGF-1 in the adult rat Tibialis Anterior muscle when injected the tissue with a recombinant SeV vector (hIGF-1/SeV). They indicate a favorable gene delivery to mitotic myoblasts, post-mitotic immature and mature myofibers.

Despite this wide range of viral delivery systems, *Adenoviral (AdV) vectors* are probably the most prominent ones in this matter and have been extensible use to deliver genes into skeletal muscle. However, several obstacles have been identified in the application of adenovirus as gene delivery vehicles to skeletal muscle (Acsadi et al., 1994 and van Deutekom et al., 1998). The major limitations facing first generation adenoviral gene transfer to skeletal muscle are (1) the lack of transgene persistence due to the immune rejection of transduced myofibers; (2) the relatively low insert carrying capacity; (3) the reduced viral transducibility during muscle maturation; and (4) repeated administration associated with the production of neutralizing antibodies is limited to the viral capsid (Cao et al., 2001). During experiments of AdV gene transfer in animals of different ages it became clear that the transduction efficiency was related with the maturation state of the muscle. While the skeletal muscle of newborn mice achieved high levels of AdV infection, the mature muscle from adult animals was significantly less susceptible to infection under the same conditions (Huard et al., 1995). Here the high adeno-viral transduction of newborn myofibers could be explained due partly to transduction of myoblasts and partly to the higher levels of CAR in these myofibers (Nalbantoglu et al., 1999). Several studies have shown that in developing human, mouse and rat muscle, expres-sion of the primary AdV membrane receptor CAR is severely downregulated even at early ages with CAR mRNA being barely detectable in adult myofibers (Nalbantoglu et al., 1999). Furthermore, it has been demonstrated that forced expression of CAR in myotubes by different approaches, such as RAdV encoding hCAR or transgenic mice overexpressing the receptor, overcomes the poor AdV mediated transducibility of these cells (Nalbantoglu et al., 2001 and Kimura et al., 2001). On the other hand, basal lamina and glycocalyx surrounding mature skeletal muscle cells appear to be an anatomical barrier that may limit the access and distri-bution of exogenously introduced virus. (van Deutekom et al., 1998 and Cao et al., 2001). It has been reported that the extracellular matrix of mature myofibers may form a physical barrier and prevent the passage of some viral particles that are too big to pass through its pores, which are estimated at 40 nm in size. Adenoviral particles are about 70 nm and 100 nm in diameter and appear to be too large to penetrate the pores of the basal lamina (Cao et al., 2001). Re-gardless these difficulties some authors have published high rates of skeletal muscle trans-duction using AdV. In 2002, Sapru et al., was able to achieve nearly 100% of transduced fibers in the adult rat Soleus and more than 80% in the Tibialis Anterior muscle when infected with an adenoviral vector harbouring the cDNA of the GFP under the control of the CMV promoter (AdVCMV-GFP). These authors claim that the viral titer used was an important factor since they could increase the number of transduced fiber when the viral dose was doubled. Other major factor regarding skeletal muscle infection with AdV and AAV seems to be the fiber composition of the muscle, with suggested preferential transduction of slow fibers (Pruchnic et al., 2000 and Sapru et al., 2002). New AdV vectors lacking all viral genes, the *Helper-Dependent AdV vectors*, show a markedly decreased immunogenicity and hence, an improved persistence of transgene expression in muscle *in vivo* (Bilbao et al., 2005). These observations suggest that the limitations regarding the immunogenicity with the use of adenoviral vectors are being overcome. However, the inability of adenoviral vectors to efficiently transduce mature myofibers remains a major hurdle facing the widespread application of adenoviral gene transfer to skeletal muscle (van Deutekom et al., 1998).

On the next section we will introduce the novel combination of Magnetic Nanoparticles and Recombinant Adenoviral Vectors as an efficient alternative for gene delivery in transduction-resistant differentiated skeletal muscle cells.

3.5. Use of magnetic nanoparticles and magnetic fields to enhance viral vector-based gene delivery

Nowadays, the novel association of non-viral or viral vector-based gene delivery with nanotechnology offers the possibility to develop more efficient gene transfer strategies for a number of applications. In 2002 the concept of Magnetofection was first published by Scherer et al. Here, the Magnetic Drug Targeting (MDT) approach (Widder et al., 1978) classically used to concentrate magnetically responsive therapeutic complexes in target areas of the body by means of external gradient magnetic fields was applied for gene delivery. Therefore, Magnetofection is based on the association of Magnetic Nanoparticles (MNPs) with non-viral or viral vectors in order to optimize gene delivery when exposed to a magnetic field (Scherer et al., 2002).

There are currently several synthetic formulations of MNPs commercially available for biomedical applications such as cell separation, drug/gene delivery, magnetic resonance imaging (MRI) and hyperthermia (Gupta et al., 2005). Despite the differences, they all need to comprise some basic functionality to allow them to be associated with a gene delivery vector. Furthermore, the magnetic properties of these particles have to be sufficient to concentrate the vector at the target cells under a magnetic force and the formulation has to be biocompatible enough for application in living cells or organisms (Plank et al., 2011). Their general structure is based on a magnetic core of magnetite (Fe3O4) or maghemite (g-Fe2O3) coated with synthetic polymers that provides both protection and biological functionality. Occasionally specific organic linkers are added to this structure to generate new attachment sites for drugs or gene vectors (Yallapu et al., 2010). (Figure 5).

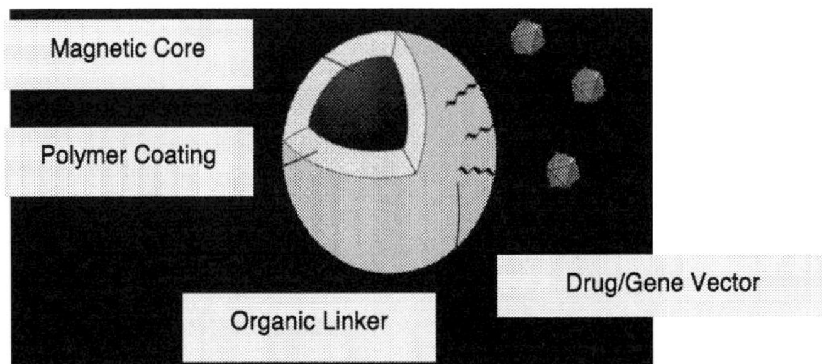

Figure 5. General Structure of a Magnetic Nanoparticle

For association with nucleic acids and/or viral particles, coatings comprised of cationic polymers such as polyethylenimine (PEI) are particularly useful (Mykhaylyk et al., 2007). The negatively charged phosphate backbone of nucleic acids as well as the negative electrokinetic (or zeta-) potential of all types of viral particles in aqueous media allow their assembling with cationic species and particles due to electrostatically induced aggregation (Scherer et al., 2002). The resulting magnetic-viral vectors complexes are then forced into sedimentation over the cell monolayer when a magnetic field is applied.

As detailed in Section 2.3, one of the most remarkable aspects of magnetofection enhancing viral gene transfer is the lack of need for virus interaction with cellular receptors. This has a particular significance for adenoviral vectors. It is known that for all the Wild-Type (except Subgroup B) and Recombinant Adenovirus cell entry takes place through receptor-mediated endocytosis via the Coxsackie–Adenovirus Receptor (CAR) (Meier and Greber, 2003). Various cell types differ widely in their level of CAR expression, and this may be a limiting factor for the transduction efficiency achievable with Adenovirus (Chorny et al., 2006). When AdVs are complex with MNPs all the viral surface proteins become hidden and unreachable for the CAR receptor. The cell entry in then achieved by unspecific clathrin-mediated endocytosis (Plank et al., 2011).

In order to find the optimal magnetic vector formulations for plasmid, siRNA and viral vector delivery, Plank and co-workers have aimed at maximal association with the magnetic component but avoiding an excess of magnetic particles (Plank et al., 2011). They have published that an excess of magnetic nanoparticles can inhibit transfection/transduction efficiency and cause toxicity (Tresilwised et al., 2010 and Sanchez-Antequera et al., 2011). Therefore, finding the optimal MNPs-to-nucleic acid ratios (about 0.5–1 units of iron weight per unit of the nucleic acid weight for triplexes with an enhancer) as well as MNPs-to-virus ratios (2.5–10 fg iron per virus particle further referred to as fg Fe/VP) have turned out useful for a variety of magnetic nanoparticle types (Sanchez-Antequera et al., 2011). The complexes formulated in this way were efficient and hardly toxic in delivery of DNA and siRNA as well in delivery of adenoviral and lentiviral vectors *in vitro* and *ex vivo*. Particularly to this regard there are several studies that have demonstrated the higher efficiency of this method when compare with traditional viral transduction using Adenovirus, Adeno-associated Virus, Baculovirus, Lentivirus and Retrovirus (Chan et al., 2005; Chorny et al., 2009; Hughes et al., 2001; Kaikkonen et al., 2008; Mah et al., 2000 and 2002; Morizono et al., 2009; Raty et al., 2004; Tresilwised et al., 2010 and 2012).

Regarding our experimental field, development of reliable techniques for manipulation of gene expression in mature skeletal muscle fibers is critical for understanding molecular mechanisms involved in both physiology and physiopathology. As explained before, differentiated skeletal muscle myotubes and myofibers are refractory to most standard protocols for gene transfer in vitro and in vivo and the use of adenoviral vectors offers relatively low efficiency. It is believed that a maturation-dependent loss of the CAR receptor together with structural and biochemical changes are responsible for these decreased transduction efficiencies (Nalbantoglu et al., 1999). It has been proposed that these limitations can be overcome by achieving adenoviral cellular uptake via a CAR independent pathway using genetic modifications of the capsid proteins or chemical modifications of the virus. However, these strategies are not sufficient for rapid infection of the cells at the target site, as the delivery process itself is diffusion-limited (Haim et al., 2005 and Schillinger et al., 2005). Here, magnetofection provided us a powerful, accessible and efficient tool for transducing differentiated myotubes of the C2C12 cell line (Pereyra A et al., 2011-Posters Sessions). A first generation (E1/E3-deleted), serotype 5, Recombinant Adenoviral vector harboring de cDNA of the Green Fluorescent Protein (RAdV-GFP) under the control of the CMV promoter was constructed in our laboratory. This vector was incubated with Atto550PEI-Mag2, a magnetic nanoparticle conjugated with a red fluorescent dye that allows particle tracking during the cellular uptake and internalization. Then the [RAdV-GFP-Atto550PEI-Mag2] complexes were incorporated to the supernatant of the mature myotubes cultures. The magnetic field required for sedimentation was provided by a commercial plate (Oz Biosciences®, Marseille, France) composed of cylindrical-permanent-Nd-Fe-B magnets. The same protocol was tested in undifferentiated C2C12 myoblast cultures and conventional RAdV transduction experiments using the same viral multiplicity of infection (MOI) were also performed for efficiency comparisons. (Figure 6)

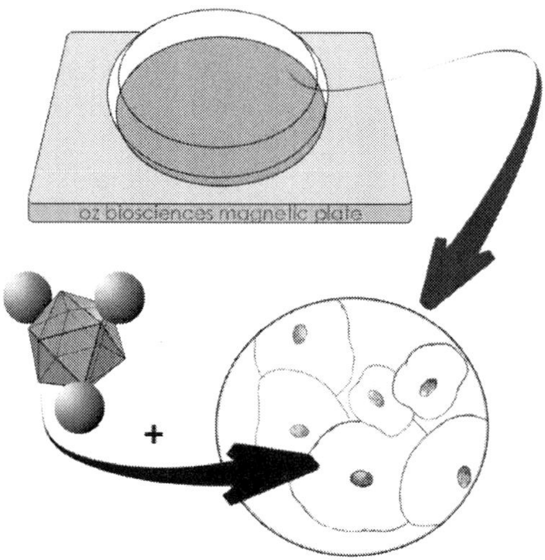

Figure 6. Magnetofection in cell culture. This basic scheme shows the process of magnetofection in cultured cells. The pre-incubated RAdV-MNPs complexes are introduced to the cell culture. Then the culture plate is exposed to a magnetic field created by the magnetic plate placed under the cells.

As showed in Figure 7, the poor RadV-GFP transduction of mature myotubes was overcome by magnetofection. In myoblasts, were the conventional transduction protocols show an acceptable efficiency, the magnetofection method displayed an enhancer effect. The intracellular localization of the magnetic nanoparticles can be seen in Figure 8.

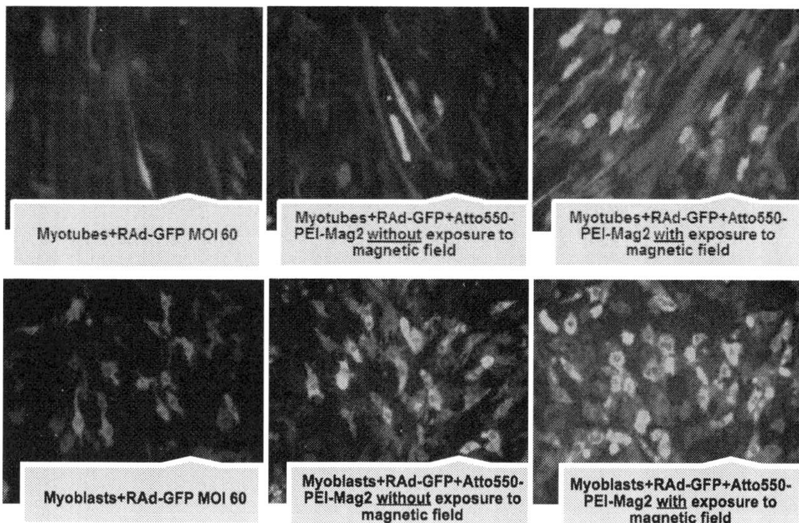

Figure 7. Magnetofection in mature C2C12 myotubes and myoblasts. The conventional RAdV transduction system was compared against magnetofection. The images were obtained 48 hs after incubation. The green fluorescence corresponds to the expression of the GFP protein encoded by the viral genome. Magnification 40X (Pereyra A et al., 2011-Posters Sessions).

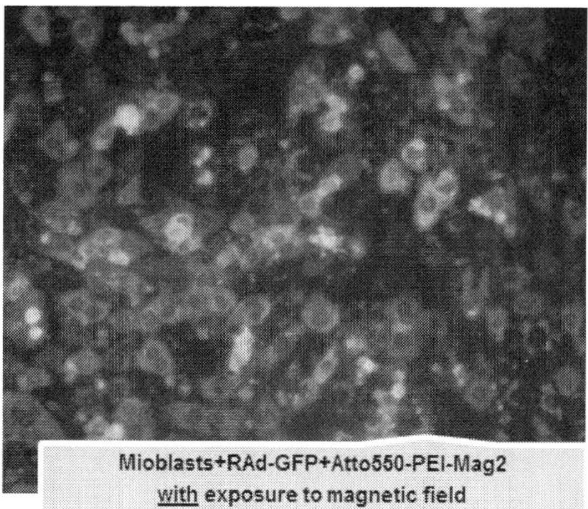

Figure 8. This picture was taken with a Double Band filter in order to appreciate the cellular co-localization of the green and red (Atto550-PEI-Mag2) fluorescence. Magnification 40X (Pereyra A et al., 2011-Posters Sessions).

4. Final remarks

There is no doubt that genetic manipulation of cells, tissues and whole individuals has become a fundamental tool for both basic and clinical research. The information that can be retrieved from experimentation using gene transfer techniques is highly significant and the therapeutic interventions that can be made by gene therapy are positioned as a promissory future for medicine. In this chapter we intended to describe the pros and cons of the most commonly used viral and non-viral gene delivery systems as well as to introduce the novelties in this field such as magnetic nanoparticles and magnetofection technology. The current applications for all of these systems seems endless; from the traditional recombinant protein production to the cutting edge cell reprograming. It is certain then that is of major importance to continue working in the pursuit of the ideal gene delivery system with high efficiency, selective tissue-tropism, non-toxic, and long-lasting expression.

Acknowledgements

The authors are grateful to Ms. Romina Pereyra for some of the illustrations displayed in this chapter. This work was supported by the National Institute of Health (NIH) grant R03TW008091 from the Fogarty International Center to Dr. Hereñu and by the National Agency of Scientific and Technological Promotion (ANPCYT) grant PICT08-2006 to Dr. Hereñu.

Author details

Andrea Pereyra and Claudia Hereñu

*Address all correspondence to: c_herenu@yahoo.com

Biochemistry Research Institute of La Plata (INIBIOLP-CONICET), Departments of Biology, Histology B and Pathology B, School of Medicine, National University of La Plata, Argentina

References

[1] Acsadi G, Jani A, Massie B, Simoneau M, Holland P, Blaschuk K, Karpati G. A differential efficiency of adenovirus-mediated *in vivo* gene transfer into skeletal muscle cells of different maturity. *Hum Mol Genet* 1994; 3: 579–584.

[2] Amalfitano A. Next-generation adenoviral vectors: new and improved. *Gene. Ther* 1999; 6: 1643-1645

[3] Bangham AD. Surrogate cells or Trojan horses. The discovery of liposomes. *Bioessays* 1995; 17(12): 1081-8.

[4] Bilbao R, Reay DP, Wu E, Zheng H, Biermann V, Kochanek S, Clemens PR. Comparison of high-capacity and first-generation adenoviral vector gene delivery to murine muscle in utero. Gene Ther. 2005;12(1): 39-47.

[5] Bhavsar D, Subramanian K, Sethuraman S, Krishnan UM. Translational siRNA therapeutics using liposomal carriers: prospects & challenges. Curr Gene Ther 2012; 12(4): 315-32.

[6] Bouri K, Feero WG, Myerburg MM, Wickham TJ, Kovesdi I, Hoffman EP, and Clemens PR. Polylysine modification of adenoviral fiber protein enhances muscle cell transduction. *Hum Gene Ther* 1999; 10: 1633–1640.

[7] Campos SK, Barry MA. Current advances and future challenges in Adenoviral vector biology and targeting. *Curr Gene Ther 2007*; 7(3):189-204.

[8] Cao B, Pruchnic R, Ikezawa M, Xiao X, Li J, Wickham TJ, Kovesdi I, Rudert WA, Huard J. The role of receptors in the maturation-dependent adenoviral transduction of myofibers. *Gene Therapy* 2001; 8: 627–637.

[9] Carter BJ, Burstein H, Peluso RW. Adeno-Associated Virus and AAV Vectors for Gene Delivery; In: Smyth Templeton N, Ed. Gene and Cell Therapy: Therapeutic mechanisms and strategies; 3ʳᵈ edition. Boca Raton, FL. CRC Press 2009; p 116-143.

[10] Chan L, Nesbeth D, Mackey T, Galea-Lauri J, Gaken J, Martin F, Collins M, Mufti G, Farzaneh F, Darling D. Conjugation of lentivirus to paramagnetic particles via nonviral proteins allows efficient concentration and infection of primary acute myeloid leukemia cells. *J Virol* 2005; 79: 13190–13194.

[11] Chargé S, Rudnicki M. Cellular and Molecular Regulation of Muscle Regeneration. *Physiol Rev* 2004; 84: 209-238.

[12] Chorny M, Fishbein I, Alferiev I, Levy RJ. Magnetically responsive biodegradable nanoparticles enhance adenoviral gene transfer in cultured smooth muscle and endothelial cells. *Mol Pharm* 2009; 6: 1380–1387.

[13] Daly TM. Overview of Adeno-Associated Viral Vectors; In: Methods in Molecular Biology, vol. 246: Gene Delivery to Mammalian Cells: Vol. 2: Viral Gene Transfer Techniques. Edited by: WC Heiser © Humana Press Inc., Totowa, NJ. 2003; p 157-165.

[14] De Gruijl TD, van de Ven R. Chapter six--Adenovirus-based immunotherapy of cancer: promises to keep. dv Cancer Res. 2012;115:147-220. doi: 10.1016/ B978-0-12-398342-8.00006-9.

[15] Dimmock N, Easton A, Leppard K. Horizons in Human Virology. In: Dimmock N, Easton A, Leppard K. Introduction to Modern Virology. UK: Blackwell Publishing; 6th edition 2007, p. 416-442.

[16] Faraji AH, Wipf P. Nanoparticles in cellular drug delivery. *Bioorganic & Medicinal Chemistry* 2009; 17: 2950–2962

[17] Feynman RP. There's plenty of room at the bottom. Eng Sci 1960; 23:22–36.

[18] Fishbein I, Forbes SP, Chorny M, Connolly JM, Adamo RF, Corrales RA, Alferiev IS, Levy RJ. Adenoviral vector tethering to metal surfaces via hydrolyzable cross-linkers for the modulation of vector release and transduction. Biomaterials. 2013 Sep;34(28): 6938-48. doi: 10.1016/j.biomaterials.2013.05.047. Epub 2013 Jun 15

[19] Gabizon A. and Papahadjopoulos D. Liposome formulations with prolonged circulation time in blood and enhanced uptake by tumors. Proc. Natl. Acad. Sci. USA. Vol. 85, pp. 6949-6953, September 1988

[20] Gao X, Huang L. Cationic liposome-mediated gene transfer. *Gene Ther* 1995; 2: 710–722.

[21] Ginsberg HS and Prince GA. The molecular basis of adenovirus pathogenesis. *Infect Agents Dis* 1994; 3: 1–8.

[22] Gobe M, Kon-no K, Kandori K, Kitahara K. Preparation and characterization of m 1983onodisperse magnetite sols in WO microemulsion. *J Colloid Interface Sci* 1983; 93(1): 293–295

[23] Goins WF, Wolfe D, Krisky DM, Bai Q, Burton EA, Fink DJ, Glorioso JC. Delivery Using Herpes Simplex Virus. An Overview; In: Methods in Molecular Biology, vol. 246: Gene Delivery to Mammalian Cells: Vol. 2: Viral Gene Transfer Techniques. Edited by: WC Heiser © Humana Press Inc., Totowa, NJ. 2003; p 257-299.

[24] Graham FL, Smiley J, Russell WC and Nairn R. Characteristics of a human cell line transformed by DNA from adenovirus type 5. *J Gen Virol* 1997; 36: 59-72.

[25] Gupta AK, Gupta M. Synthesis and surface engineering of iron oxide nanoparticles for biomedical applications. *Biomaterials* 2005; 26(18): 3995-4021.

[26] Hackett NR, Crystal RG. Adenovirus Vectors for Gene Therapy; In: Smyth Templeton N, Ed. Gene and Cell Therapy: Therapeutic mechanisms and strategies; 3rd edition. Boca Raton, FL. CRC Press 2009; p 39-56.

[27] Haim H, Steiner I, Panet A. Synchronized infection of cell cultures by magnetically controlled virus. *J Virol* 2005; 79(1): 622-5.

[28] Hallauer P, Karpati G, Hastings K. Skeletal muscle gene transfer: regeneration-associated deregulation of fast troponin I fiber type specificity. *Am J Physiol Cell Physiol* 2000; 278: C1266–C1274.

[29] Havenga MJ, Lemckert AA, Ophorst OJ, van Meijer M, Germeraad WT, Grimbergen J, van Den Doel MA, Vogels R, van Deutekom J, Janson AA, de Bruijn JD, Uytdehaag F, Quax PH, Logtenberg T, Mehtali M, and Bout A. Exploiting the natural diversity in adenovirus tropism for therapy and prevention of disease. *J Virol* 2002; 76: 4612–4620.

[30] Hayes F. Transposon-based strategies for microbial functional genomics and proteomics. *Annu Rev Genet* 2003; 37: 3-29.

[31] He C, Hu Y, Yin L, Tang C, Yin C. Effects of particle size and surface charge on cellular uptake and biodistribution of polymeric nanoparticles. *Biomaterials* 2010; 31(13): 3657–3666.

[32] Hermonat PL, Muzyczka N. Use of adeno-associated virus as a mammalian DNA cloning vector: transduction of neomycin resistance into mammalian tissue culture cells. *Proc Natl Acad Sci USA* 1984; 81: 6466–6470.

[33] Herweijer H, Zhang G, Subbotin VM, Budker V, Williams P, Wolff JA. Time course of gene expression after plasmid DNA gene transfer to the liver. *J Gene Med* 2001; 3(3): 280-91.

[34] Huard J, Lochmüller H, Acsadi G, Jani A, Holland P, Guérin C, Massie B, Karpati G. Differential short-term transduction efficiency of adult versus newborn mouse tissues by adenoviral recombinants. *Exp Mol Pathol* 1995; 62: 131–143.

[35] Huard J, Feero WG, Watkins SC, Hoffman EP, Rosenblatt DJ, Glorioso JC. The basal lamina is a physical barrier to herpes simplex virus-mediated gene delivery to mature muscle fibers. *J Virol* 1996; 70(11): 8117-23.

[36] Huard J, Krisky D, Oligino T, Marconi P, Day CS, Watkins SC, Glorioso JC. Gene transfer to muscle using herpes simplex virus-based vectors. *Neuromuscul Disord* 1997; 7(5): 299-313.

[37] Hughes C, Galea-Lauri J, Farzaneh J, Darling D. Streptavidin paramagnetic particles provide a choice of three affinity-based capture and magnetic concentration strategies for retroviral vectors. *Mol Ther* 2001; 3: 623–630.

[38] Jiao S, Acsadi G, Jani A, Felgner PL, Wolff JA. Persistence of plasmid DNA and expression in rat brain cells in vivo. *Exp Neurol* 1992; 115(3): 400-13.

[39] Jiao S, Williams P, Berg RK, Hodgeman BA, Liu L, Repetto G, Wolff JA. Direct gene transfer into nonhuman primate myofibers in vivo. *Hum Gene Ther* 1992; 3(1): 21-33.

[40] Józkowicz A, Dulak J. Helper-dependent adenoviral vectors in experimental gene therapy. *Acta Biochim Pol* 2005; 52(3): 589–599

[41] Kaikkonen MU, Viholainen JI, Narvanen A, Yla-Herttuala S, Airenne KJ. Targeting and purification of metabolically biotinylated baculovirus. *Hum Gene Ther* 2008; 19: 589–600.

[42] Kim IH, Józkowicz A, Piedra PA, Oka K, Chan L. Lifetime correction of genetic deficiency in mice with a single injection of helper-dependent adenoviral vector. Proc Natl Acad Sci USA 2001; 98(23): 13282-7.

[43] Kimura E, Maeda Y, Arima T, Nishida Y, Yamashita S, Hara A, Uyama E, Mita S, Uchino M. Efficient repetitive gene delivery to skeletal muscle using recombinant adenovirus vector containing the Coxsackievirus and adenovirus receptor cDNA. *Gene Ther* 2001; 8: 20–27.

[44] Kovesdi I, Brough DE, Bruder JT, Wickham TJ. Adenoviral vectors for gene transfer. *Curr Opin Biotechnol* 1997;8:583–9.

[45] Le TV, Mironova E, Garcin D, Compans RW. Induction of influenza-specific mucosal immunity by an attenuated recombinant Sendai virus. *PLoS One* 2011; 6(4): e18780.

[46] Lemieux P, Guerin N, Paradis G, Proulx R, Chistyakova L Kabanov A, Alakhov V. A combination of poloxamers increases gene expression of plasmid DNA in skeletal muscle. *Gene Ther* 2000; 7: 986–991.

[47] Longmuir KJ, Robertson RT, Haynes SM, Baratta JL, Waring AJ. Effective targeting of liposomes to liver and hepatocytes in vivo by incorporation of a Plasmodium amino acid sequence. *Pharm Res.* 2006; 23(4): 759-69.

[48] MacArthur CC, Fontes A, Ravinder N, Kuninger D, Kaur J, Bailey M, Taliana A, Vemuri MC, Lieu PT. Generation of human-induced pluripotent stem cells by a nonintegrating RNA Sendai virus vector in feeder-free or xeno-free conditions. Stem Cells Int. 2012; 2012:564612.

[49] Mah C, Zolotukhin I, Fraites TJ, Dobson J, Batich C, Byrne BJ. Microsphere-mediated delivery of recombinant AAV vectors in vitro and in vivo. *Mol Ther* 2000; 1(239)

[50] Mah C, Fraites TJ, Zolotukhin I, Song S, Flotte TR, Dobson J, Batich C, Byrne BJ. Improved method of recombinant AAV2 delivery for systemic targeted gene therapy. *Mol Ther* 2002; 6: 106–112.

[51] Matsunaga T, Takeyama H. Biomagnetic nanoparticle formation and application. *Supramol Sci.*1998; 5: 391–394.

[52] McConnell MJ, Imperiale MJ. Biology of adenovirus and its use as a vector for gene therapy. Hum Gene Ther. 2004;15(11): 1022-33.

[53] Meier O, and Greber UF. Adenovirus endocytosis. *J Gene Med.* 2003; 5: 451 – 462.

[54] Meir YJ, Wu SC. Transposon-based vector systems for gene therapy clinical trials: challenges and considerations. *Chang Gung Med J* 2011; 34(6): 565-79.

[55] Miller DG, Adam MA, Miller AD. Gene transfer by retrovirus vectors occurs only in cells that are actively replicating at the time of infection. *Mol Cell Biol* 1990; 10: 4239–4242.

[56] Morizono K, Xie YM, Helguera G, Daniels TR, Lane TF, Penichet ML, Chen ISY. A versatile targeting system with lentiviral vectors bearing the biotinadaptor peptide. *J Gene Med* 2009; 11: 655–663.

[57] Mykhaylyk O, Antequera YS, Vlaskou D, Plank C. Generation of magnetic nonviral gene transfer agents and magnetofection in vitro. *Nat Protoc* 2007; 2(10): 2391-411.

[58] Nakanishi M, Otsu M. Development of Sendai Virus Vectors and Their Potential Applications in Gene Therapy and Regenerative Medicine. *Curr Gene Ther.* 2012; Aug 23 [Epub ahead of print]

[59] Nalbantoglu J, Pari G, Karpati G, Holland PC. Expression of the primary coxsackie and adenovirus receptor is downregulated during skeletal muscle maturation and limits the efficacy of adenovirus-mediated gene delivery to muscle cells. *Hum Gene Ther* 1999; 10: 1009–1019.

[60] Nalbantoglu J, Larochelle N, Wolf E, Karpati G, Lochmuller H, Holland PC. Muscle-specific overexpression of the adenovirus primary receptor CAR overcomes low efficiency of gene transfer to mature skeletal muscle. *J Virol* 2001; 75: 4276–4282.

[61] Nishimura K, Sano M, Ohtaka M, Furuta B, Umemura Y, Nakajima Y, Ikehara Y, Kobayashi T, Segawa H, Takayasu S, Sato H, Motomura K, Uchida E, Kanayasu-Toyoda T, Asashima M, Nakauchi H, Yamaguchi T, Nakanishi M. Development of Defective and Persistent Sendai Virus Vector. A unique gene delivery/expression system ideal for cell reprogramming. *J Biol Chem* 2011; 286(6): 4760-71.

[62] Palmer DJ and Ng P. Helper-dependent adenoviral vectors for gene therapy. *Hum. Gene Ther* 2005; 1: 1-16.

[63] Pankajakshan Divya and Devendra K. Agrawal. Clinical and Translational Challenges in Gene Therapy of Cardiovascular Diseases. Chapter 27. Biochemistry, Genetics and Molecular Biology » "Gene Therapy - Tools and Potential Applications", book edited by Francisco Martin Molina, ISBN 978-953-51-1014-9, Published: February 27, 2013

[64] Park S, Kim YS, Kim WB, Jon S. Carbon nanosyringe array as a platform for intracellular delivery. *Nano Lett* 2009; 9(4): 1325–1329.

[65] Pereyra A, Taylor J, Morel G, Goya R, Mykhaylyk O, Delbono O, Hereñu C. Potential of Magnetic Field-Assisted Gene Therapy on Skeletal Muscle C2C12 Cell Line. *LVI Anual Meeting of the Sociedad Argentina de Investigación Clínica.* November 16-19, 2011, Mar del Plata, Buenos Aires, Argentina. Poster Session.

[66] Pereyra A, Taylor J, Morel G, Goya R, Mykhaylyk O, Delbono O, Hereñu C. Magnetic field-assisted gene delivery into skeletal muscle derived cells. Use of the Green Fluorescent Protein as a reporter. *The Third South American Workshop and International Gregorio Weber Conference on New Trends in Advanced Fluorescence Microscopy Techni-*

ques. December 12-17, 2011, Ciudad Universitaria, Universidad de Buenos Aires, Buenos Aires, Argentina. Poster Session.

[67] Plank C, Zelphati O, Mykhaylyk O. Magnetically enhanced nucleic acid delivery. Ten years of magnetofection—Progress and prospects. *Adv Drug Deliv Rev* 2011; 63 (14-15): 1300-31

[68] Probst CE, Zrazhevskiy P, Bagalkot V, Gao X . Quantum dots as a platform for nano-particle drug delivery vehicle design. *Advanced Drug Delivery Reviews* (2012). doi: 10.1016/j.addr.2012.09.036. [Epub ahead of print]

[69] Pruchnic R, Cao B, Peterson ZQ, Xiao X, Li J, Samulski RJ, Epperly M, Huard J. The use of adeno-associated virus to circumvent the maturation-dependent viral trans-duction of muscle fibers. *Hum Gene Ther* 2000; 11(4): 521-36.

[70] Prud'homme G, Lawson BR, Chang Y, Theofilopoulos AN. Immunotherapeutic gene transfer into muscle. *Trends Immunol* 2001; 22: 149–155.

[71] Raty JK, Airenne KJ, Marttila AT, Marjomaki V, Hytonen VP, Lehtolainen P, Laitinen OH, Mahonen AJ, Kulomaa MS, Yla-Herttuala S. Enhanced gene delivery by avidin-displaying baculovirus. *Mol Ther* 2004; 9: 282–291.

[72] Rockenberger J, Scher EC Alivisatos AP. A New Nonhydrolytic Single-Precursor Ap-proach to Surfactant-Capped Nanocrystals of Transition Metal Oxides. *J. Am. Chem. Soc.* 1999; 121 (49): 11595–11596

[73] Rowe WP, Huebner RJ, Gilmore LK, Parrott RH, Ward TG. Isolation of a cytopatho-genic agent from human adenoids undergoing spontaneous degeneration in tissue culture. *Proc Soc Exp Biol* Med 1953; 84(3):570- 3.

[74] Roy-Chowdhury J and Horwitz M. Evolution of adenoviruses as gene therapy vec-tors. *Mol Ther* 2002; 5: 340-344.

[75] Russell WC. Update on adenovirus and its vectors. *Journal of General Virology* 2000; 81: 2573–2604.

[76] Salvatori G, Ferrari G, Mezzogiorno A, Servidei S, Coletta M, Tonali P, Giavazzi R, Cossu G, Mavilio F. Retroviral vector-mediated gene transfer into human primary myogenic cells leads to expression in muscle fibers in vivo. *Hum Gene Ther* 1993; 4(6): 713-23.

[77] Sanchez-Antequera Y, Mykhaylyk O, van Til N. P, Cengizeroglu A, de Jong J. H, Huston M. W, Anton M, Johnston I. C, Pojda Z, Wagemaker G and Plank C. Magse-lectofection: an integrated method of nanomagnetic separation and genetic modifica-tion of target cells. *Blood* 2011; 117: 171–181.

[78] Sandri M, Bortoloso E, Nori A, Volpe P. Electrotransfer in differentiated myotubes: a novel, efficient procedure for functional gene transfer. *Exp Cell Res* 2003; 286: 87–95.

[79] Sapru MK, McCormick KM, Thimmapaya B. High-efficiency adenovirus-mediated in vivo gene transfer into neonatal and adult rodent skeletal muscle. *J Neurosc Methods* 2002; 114: 99-106.

[80] Schambach A, Maetzig T, Baum C. Retroviral Vectors for Cell and Gene Therapy; In: Smyth Templeton N, Ed. Gene and Cell Therapy: Therapeutic mechanisms and strategies; 3rd edition. Boca Raton, FL. CRC Press 2009; p 3-12.

[81] Scherer F, Anton M, Schillinger U, Henke J, Bergemann C, Kruger A, Gansbacher B, Plank C. Magnetofection: enhancing and targeting gene delivery by magnetic force in vitro and in vivo. *Gene Ther* 2002; 9: 102–109.

[82] Schillinger U, Brill T, Rudolph C, Huth S, Gersting S, Krotz F, et al. Advances in magnetofection - magnetically guided nucleic acid delivery. *J Magn Magn Mat* 2005; 293(1): 501-8.

[83] Shanmugam T, Banerjee R. Nanostructured self assembled lipid materials for drug delivery and tissue engineering. Ther Deliv 2011; 2(11): 1485-516.

[84] Shiotani A, Fukumura M, Maeda M, Hou X, Inoue M, Kanamori T, Komaba S, Washizawa K, Fujikawa S, Yamamoto T, Kadono C, Watabe K, Fukuda H, Saito K,Sakai Y, Nagai Y, Kanzaki J, Hasegawa M. Skeletal muscle regeneration after insulin-like growth factor I gene transfer by recombinant Sendai virus vector. *Gene Ther* 2001; 8(14): 1043-50.

[85] Somia N. Gene Transfer by Retroviral Vectors. An Overview; In: Methods in Molecular Biology, vol. 246: Gene Delivery to Mammalian Cells: Vol. 2: Viral Gene Transfer Techniques. Edited by: WC Heiser © Humana Press Inc., Totowa, NJ. 2003; p 463-490.

[86] Spradling AC, Rubin GM. Transposition of cloned P elements into Drosophila germ line chromosomes. *Science* 1982; 218: 341-7.

[87] Sun S, Zeng H. Electrochemically Driven Clathration/Declathration of Ferrocene and Its Derivatives by a Nanometer-Sized Coordination Cage. *J Am. Chem. Soc.* 2002; 124 (39): 11570–11571.

[88] Tolmachov O. Self-entanglement of long linear DNA vectors using transient non-B-DNA attachment points: A new concept for improvement of non-viral therapeutic gene delivery. *Medical Hypotheses* 2012; 78: 632–635.

[89] Tresilwised N, Pithayanukul P, Mykhaylyk O, Holm PS, Holzmuller R, Anton M, Thalhammer S, Adiguzel D, Doblinger M, Plank C. Boosting oncolytic adenovirus potency with magnetic nanoparticles and magnetic force. *Mol Pharm* 2010; 7: 1069–1089.

[90] Tresilwised N, Pithayanukul P, Holm PS, Schillinger U, Plank C, and Mykhaylyk O. Effects of nanoparticle coatings on the activity of oncolytic adenovirus-magnetic nanoparticle complexes. *Biomaterials* 2012; 33: 256-269.

[91] van Deutekom JC, Floyd SS, Booth DK, Oligino T, Krisky D, Marconi P, Glorioso JC, Huard J. Implications of maturation for viral gene delivery to skeletal muscle. *Neuromuscul Disord* 1998; 8(3-4): 135-148.

[92] Volpers C and Kochanek S. Adenoviral vectors for gene transfer and therapy. *J Gen Virol* 2004; 6: S164-S171.

[93] Widder, KJ, Senyel, AE, Scarpelli, GD. Magnetic microspheres: A model system of site specific drug delivery in vivo. *Proc Soc Exp Biol Med* 1978; 158: 141-146.

[94] Wolfe D, Goins WF, Frampton AR, Mata M, Fink DJ, Glorioso JC. Therapeutic Gene Transfer with Replication-Defective Herpes Simplex Viral Vectors; In: Smyth Templeton N, Ed. Gene and Cell Therapy: Therapeutic mechanisms and strategies; 3rd edition. Boca Raton, FL. CRC Press 2009; p 159-180

[95] Wolff J, Malone RW, Williams P, Chong W, Acsadi G, Jani A, Felgner PL. Direct gene transfer into mouse muscle in vivo. *Science* 1990; 247: 1465–1468.

[96] Yallapu MM, Foy SP, Jain TK, Labhasetwar V. PEG-functionalized magnetic nanoparticles for drug delivery and magnetic resonance imaging applications. Pharm Res 2010; 27(11): 2283-95.

[97] Youngjoo Choi and Jun Chang. Viral vectors for vaccine applications. Clin Exp Vaccine Res. 2013 July; 2(2): 97–105

[98] Yum K, Na S, Xiang Y, Wang N, Yu MF. Mechanochemical delivery and dynamic tracking of fluorescent quantum dots in the cytoplasm and nucleus of living cells. Nano Lett 2009; 9(5): 2193–2198.

[99] Yurchenco PD. Assembly of basement membranes. *Ann NY Acad Sci* 1990; 580: 195–213.

Chapter 8

Use of RNA Domains in the Viral Genome as Innate Immunity Inducers for Antiviral Strategies and Vaccine Improvement

Miguel R. Rodríguez Pulido, Francisco Sobrino,
Belén Borrego and Margarita Sáiz

Additional information is available at the end of the chapter

http://dx.doi.org/10.5772/56099

1. Introduction

This chapter will focus on the role of innate immunity induction on antiviral responses with an emphasis on nucleic acids as type-I interferon (IFN) inducers and their use as antiviral compounds and vaccine adjuvants. A general and up-to-date view of the different mechanisms operating in the host cell for sensing viral genomes will be given, as well as viral strategies counteracting this response through immune evasion or specifically targeted antagonism. Our own recent data describing the ability to induce IFN and mediate protection against viral infection in vivo of synthetic RNA transcripts enclosing structural domains present in the 5′- and 3′-terminal regions of the foot-and-mouth disease virus (FMDV) genome will be summarized and discussed in this context. New vaccine formulations including innate immunity inducers are being developed for improvement of current vaccines. The potential of exogenous nucleic acids as modulators of immune response outcomes and vaccine adjuvants will be reviewed and discussed. A schematic summary of the interrelated topics addressed in this chapter is shown in Figure 1. Additionally, a glossary of all the acronyms and abbreviations used in the text and figures is shown in Table 1.

2. Innate immune response against viral infection

The mammalian immune system is composed of the innate and the adaptive arms which work in combination to battle against a large variety of pathogens such as bacteria, fungi, parasites

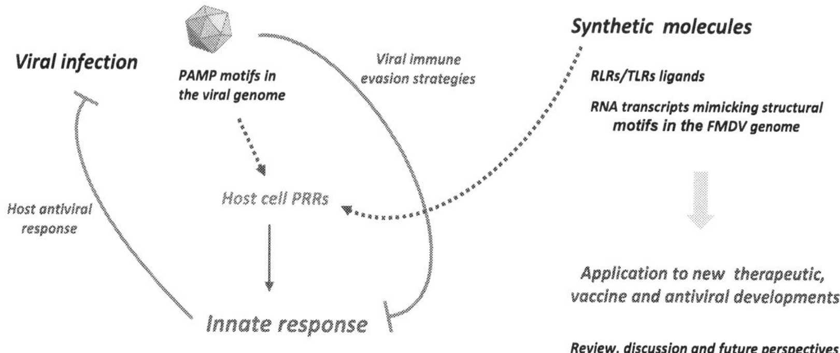

Figure 1. Schematic summary of the topics reviewed and discussed in this chapter

and viruses. Both systems have the molecular task to distinguish "self" from "non-self" components in the organisms in a sensitive and faithful manner. The innate immune system is hence the first line of defense against infection by pathogens. The existence of pattern-recognition receptors (PRRs) expressed in cells of the innate immune system that are capable to specifically sense pathogen-associated molecular patterns (PAMPs) was first proposed by Charles Janeway in 1989 [1]. PAMPs are chemical or structural features present in pathogens but not in host cells acting then as alert signals to the innate immune system of the host. Lipids, polysaccharides, nucleic acids and CpG DNA are among the basic PAMPs recognized by PRRs. Their recognition triggers cellular responses aimed to counteract the pathogen and initiate and promote other responses such as inflammation and adaptive immune responses.

In 1996, the gene/protein Toll, initially described as a transmembrane protein required for dorsal-ventral polarity in the Drosophila embryo, was found to play also a role in immunity against fungal infections [2]. One year later, their mammalian orthologs, the Toll-like receptors (TLRs), were reported to mediate recognition of pathogens by the innate immune system [3]. In 2004, a new and TLR-independent pathway was described for recognition of viral nucleic acids in the cytoplasm of the infected cells, the retinoic acid-inducible gene-1 (RIG-I) [4]. Four different families of PRRs have been found to date, including TLRs [5], RIG-I-like receptors (RLRs), C-type lectin receptors (CLRs) [6] and nucleotide oligomerization domain (NOD)-like receptors, (NLRs) [7], as well as unidentified proteins that mediate sensing of cytosolic DNA or retroviral infection [8]. Among them, TLRs, RLRs and NLRs are involved in the recognition of viral nucleic acids [9]. NOD2, a member of the NLR family, typically involved in antimicrobial immune defenses, and highly expressed in antigen presenting cells (APCs) such as monocytes, macrophages and dendritic cells (DCs), has been shown to bind viral single-stranded (ss) RNA [10].

TLRs are a family of type-I transmembrane proteins that traffic between the plasma membrane and endosomal vesicles, expressed on various immune cells, including dendritic cells, which recognize a wide range of PAMPs including double-stranded (ds) RNA (TLR3), ssRNA (TLR7 and TLR8) and DNA (TLR9). All TLRs signal as dimmers and share a common architecture of extracellular leucine-rich repeats and intracellular Toll/Interleukin-1 receptor (TIR) domain (Figure 2). Nucleic acid sensing-TLRs localize in intracellular vesicles, including endoplasmic reticulum (ER), endosomes, lysosomes and endolysosomes of dendritic cells and other innate immune cells. Intracellular localization enables TLRs to recognize nucleic acids delivered to intracellular compartments after the uptake of viruses or infected cells. Unc93b1 is a membrane protein which anchors TLRs to the ER and regulates their trafficking to the endosomal compartments. TLR signaling pathway depends on the recruitment of a TIR-domain containing adaptor, MyD88, for all TLRs (with the exception of TLR3) and culminates with NF-κB and MAP kinase activation and induction of inflammatory cytokines (Figure 2). TLR3 uses TRIF to activate NF-κB and IRF-3 through an alternative pathway and the induction of type-I IFN and inflammatory cytokines. Activation of the TLR signaling leads to maturation of DCs, contributing to the induction of adaptive immunity.

The RLRs are a family of ubiquitous cytosolic helicases consisting of the three members: RIG-I, melanoma differentiation-associated gene 5 (MDA5) and laboratory of genetics and physiology-2 (LGP2) (Figure 2). RIG-I and MDA5 have tandem caspase activation and recruitment domains (CARD) followed by a DExD/H box RNA helicase domain and a repressor domain. LGP2 lacks the N-terminal CARD domains and may function to regulate RIG-I and MDA5 as a repressor [11]. It has been reported that RIG-I recognizes ssRNA bearing a 5′-ppp and short dsRNA, while MDA5 senses long dsRNA [12-14]. When the inactive forms of RIG-I or MDA5 bind viral RNA, the helicases undergo a conformational change, multimerization and then, interaction with the adaptor molecule MAVS (also called IPS1, VISA or CARDIF), localised to the outer mitochondrial membrane via CARD-CARD interaction. Then, MAVS induces activation of IRF3/7 resulting in the transcription of type-I IFNs and also activates NF-κB (Figure 2). It has been recently shown that MAVS resides on peroxisomes also and can induce antiviral signaling from this organelle acting with mitochondrial MAVS sequentially to create an antiviral cellular state [15]. Upon viral infection, peroxisomal MAVS induces the rapid interferon-independent expression of defense factors for short-term protection, whereas mitochondrial MAVS activates an interferon-dependent signaling pathway with delayed kinetics, amplifying and stabilizing the antiviral response.

In addition to PRRs, which inhibit viral infections indirectly by activating signaling cascades that result in the transcription of IFN and other antiviral molecules, there are intrinsic antiviral factors which act blocking viral replication immediately and directly, often before the onset of IFN response, like PKR, MxA, TRIM5α or the IFIT and IFITM families [16]. Intrinsic innate factors preexist in certain cell types though they can be further induced by IFNs to amplify their antiviral activity.

Recent work supports the non-redundant functional requirement for TLRs and RLRs [17]. On the contrary, the cooperation and crosstalk between different PRRs mediates activation of an effective immune response and host defense against viral infections [8, 18]. Unique links between NLRs and RLRs signaling responses have also been identified [10, 19]. Polymicrobial infection involve complex host interactions that are likely to engage a variety of response pathways including different PRRs.

IFNs exert auto- and paracrine actions within a few hours in response to a viral infection. Their protective effect is dual: they induce an antiviral cellular state and promote the clearance of infected cells in synergy with other proapoptotic agents as tumor necrosis factor (TNF). Through the secretion of IFN, triggered by activation and translocation to the nucleus of NF-κB, IRF3 and IRF7, the antiviral response can be amplified and spread to surrounding uninfected cells by binding to the IFN-α/β receptor (IFNAR) in the cell surface. Binding of the cytokine triggers a Jak-STAT signaling pathway and subsequently activates hundreds of IFN-stimulated genes (ISGs), most of them encoding proteins with antiviral functions such as inhibition of viral gene expression or degradation of the viral genome [20].

In addition to its antiviral properties, IFNs exhibit potent immunomodulatory properties that contribute to their antiviral effect such as stimulation of the effector function of natural killer (NK) cells, cytotoxic T lymphocytes and macrophages, upregulation of MHC class I and II molecules, induction of immunoglobulin production by B cells and stimulation of proliferation of memory T cells [21]. This enables several ways to control viral replication by modulating of the innate and adaptive immune responses [22]. Type-I IFNs act through activation and maturation of dendritic cells leading to MHC upregulation. They can also regulate certain chemokines, chemokine receptors and costimulatory molecules, which, in turn, stimulate CD4- and CD8-positive T cell responses and promote Th1 differentiation, modulating T lymphocyte responses [23].

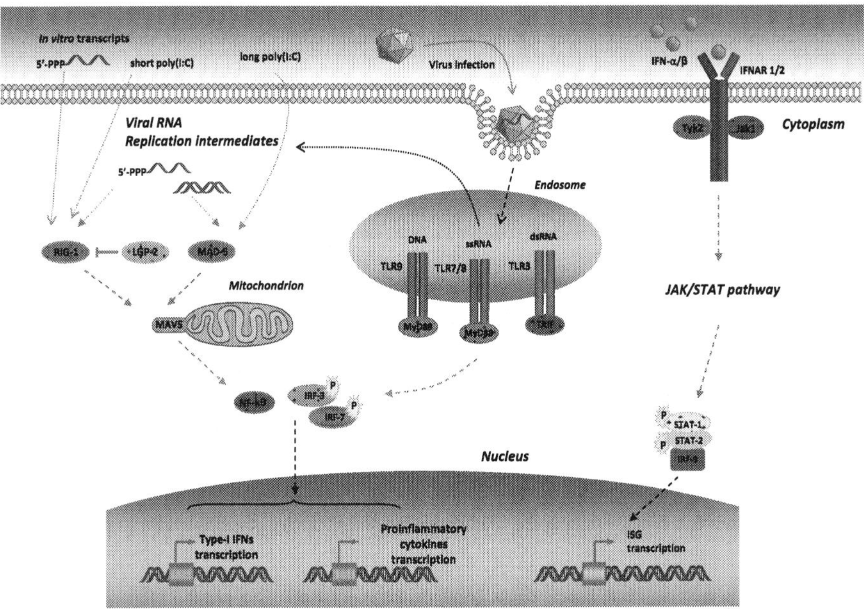

Figure 2. Schematic representation of viral RNA detection by TLRs and RLRs and signalling pathways

Induction of innate immune signaling pathways through PRRs is a crucial step antagonised by many viruses [24, 25]. Over 170 different virus-encoded IFN antagonists from 93 different viruses had been reported by 2010 (reviewed in [26]), and the list keeps constantly growing, indicating that most viruses interfere with multiple stages of the IFN response. Four main mechanisms are used to circumvent host innate responses: general inhibition of gene expression, and sequestration, proteoltytic cleavage or proteasome degradation of key factors of the IFN circuit such as RLRs, MAVS, IRFs, Jak/STAT, PKR... Several IFN antagonists are conserved within different RNA virus families, while that seems not to be the case for DNA viruses [26]. This can be explained by the multi-functionality of RNA virus proteins, imposed by restriction in genome size, unlike large dsDNA viruses which might have a higher coding capacity for new viral proteins displaying a wider range of antagonistic activities.

The potential of IFN antagonists knockout viruses as promising candidates for live virus vaccines has been suggested based on studies with Influenza A/B viruses, Japanese encephalitis virus, human respiratory syncytial virus and coronaviruses [26-29]. These severely attenuated viruses are rapidly cleared in vivo by a potent IFN response, while inducing long-lasting immune memory due to their replication competent nature. Viral miRNAs may also function in evasion of the host antiviral response (reviewed in [30]). The contribution to viral evasion of type-I IFN response of small non-coding subgenomic flavivirus RNAs generated as degradation products by a host exoribonuclease, has been recently shown [31]. Hence, IFN antagonists are good targets for antiviral drugs development.

3. RNA motifs in the viral genome can trigger innate responses

Accurate discrimination of self from non-self is critical to avoid immune triggering against self that leads to autoimmunity [32]. In that sense, it has been proposed for hepatitis C virus (HCV) that a combinatorial non-self signature in the viral genome for PRR binding may lead to accurate PAMP discrimination [33].

TLRs involved in recognition of the viral genomes are TLR3, TLR7/8 and TLR9, all of them localized to the endosomal compartment [34]. TLR3 is widely expressed in innate immune cells with the exception of neutrophils and pDCs and responds to dsRNA, a common viral PAMP, and its synthetic analog polyriboinosinic-polyribocytidylic acid (poly I:C) [35]. TLR7 and TLR8 are closely related receptors that recognize nearly any long ssRNA with some differences between them. Short ssRNA containing certain motifs preferentially activate TLR7, and activation with synthetic agonists specific to TLR7 or TLR8 trigger different cytokine profiles [36]. TLR9 is highly expressed in pDCs and responds to the unmethylated deoxycytidylate-phosphate-deoxyguanylate (CpG) motifs in viral and bacterial DNA [37].

Different features have been defined for RIG-I recognition as RNA PAMPs, including the presence of a free 5′-triphosphate, absent from eukaryotic cytoplasm due to RNA metabolism in the nucleus, length (longer than 19 nt), secondary structure characteristics (a base-pairing region of 10-20 nt near the 5′-ppp) [38] and nucleotide sequence motifs (such as a 3′-poly U/UC tract in the HCV genome) [33]. Panhandle structures adopted by Sendai virus DI-genomes or self-complementary influenza virus genome have been described as potent PAMPs sensed by RIG-I [39]. Data on MDA5 ligands are scarce. MDA5 seems to sense dsRNA analog poly I:C in mice [40] and higher-order RNA structures present in infected cells have been found to activate MDA5 [41]. A recent report shows the direct interaction of MDA5 with dsRNA replicative intermediate forms of positive strand RNA viruses [14]. RLRs have evolved to sense the presence of largely different sets of viruses but not always acting in a mutually exclusive way [13, 42].

In addition to the direct antiviral function of RNase L degrading ssRNA, RNase L can generate viral or host-derived small RNAs that amplify the IFN response by generating PAMPs that activate the RLR pathway. RNase L mediated cleavage of HCV RNA generates svRNA that activates RIG-I, thus propagating innate immune signaling to the IFN-β gene [43, 44].

Given the ability of RLRs to sense viral RNAs and activate IFN signaling cascades that eliminate viral infections, many viruses have developed immune evasion strategies to overcome detection by RLRs. This is carried out through RNA modification of viral RNA genomes to prevent host detection [24]. For example, some viruses engage cap snatching (e.g. influenza virus), modification of 5′-ppp to monophosphate through virus encoded enzymes (e.g. Borna disease virus, Lassa virus), 2′O-methylation of viral mRNA cap structure by virus encoded methyltransferases, exploiting nucleotide modifications found at higher frequency in eukaryotic versus prokaryotic/viral RNA, and the use of proteins to protect the 5′ends (picornavirus have a virus encoded protein, VPg, covalently linked to the 5′end of their genome) or overhangs (e.g. arenavirus) [24].

In 2008, Saito et al. showed that the 3′ non-coding region (NCR) of HCV (a flavivirus) encoded PAMP motifs triggering innate immune signaling in the host cell. Thus, the 100 nt-polyuridine motif (poly U/UC) within the 3′ NCR was identified as a potent PAMP, substrate of RIG-I recognition and immune triggering in human and murine cells [45]. In contrast, the structured 3′-terminal X region failed to trigger signaling. The entire HCV 5′NCR, containing four major stem-loop structures including the internal ribosome entry site (IRES), was a weak inducer of IFN promoter signaling. However, prior treatment of cells with IFN-β to increase RIG-I levels rendered them responsive to signaling induced by the 5′NCR or the X region [45], suggesting that dsRNA regions of the HCV genome are not potent PAMPs but may confer signaling during the IFN response. Some studies on the 5′- and 3′-NCRs of other flaviviruses show remarkable differences in their IFN-inducing capacity. The 5′and 3′NCRs of dengue virus (DEN) elicited low but measurable stimulation of innate immune signaling, while the smaller highly structurally conserved 3′-terminal stem-loop RNAs of DEN, West Nile virus (WNV) and yellow fever viruses were minimally active [46]. Additionally, the base-pairing extent of the 5′-triphosphate of the RNAs may have an enhancing effect on RLR recognition and signaling [38, 47]. Therefore, the ability of different RNAs as IFN inducers must be tested independently, being difficult to predict their behaviour/potency by their sequence, secondary structure or homology with analog molecules. In this sense, we have recently shown that FMDV (a picornavirus) full-length transcripts with the 3′NCR deleted induce lower levels of IFN-β than complete RNA transcripts in cell culture [48]. These results are equivalent to those reported for HCV transcripts lacking the PAMP motif poly-U/UC. In this case, it has been proposed that, as the virus must maintain this motif in the 3′NCR for its viability, the host takes advantage of this and targets this region as a discriminator of PAMP RNA through RIG-I interaction [45]. Thus, HCV infection seems to be regulated by hepatic immune defenses triggered by the cellular RIG-I helicase. For FMDV, we also found that RNA transcripts corresponding to structural domains predicted to enclose stable dsRNA regions in the 5′and 3′NCRs of the viral genome were able to trigger an IFN-α/β response in epithelial porcine kidney cultured cells and induce an antiviral state [48] (Figure 3). A direct link between antiviral activity induced by FMDV NCR transcripts and IFN could be established in cultured cells, as treatment with monoclonal antibodies against IFN-α/β effectively blocked the antiviral activity induced by the RNAs [48]. Different levels of IFN-β mRNA induction were observed for the different RNAs assayed, being the one mimicking the complete 3′NCR, enclosing two predicted stem-loop structures, the best inducer. The in vitro RNA transcripts corresponding to the complete 5′ NCR, the IRES and the S hairpin (Figure 3), were also able to induce IFN-β transcription, though at lower levels than the 3′NCR transcript. The removal of the poly A tail within the 3′NCR RNA had a detrimental effect on IFN-β induction, but milder than removal of the 5′-ppp by treatment with alkaline phosphatase, which strongly reduced but did not completely abolish induction. However, deletion of any of the 3′NCR stem-loop (SL) structures rendered RNAs minimally active for IFN-β signaling, suggesting a relevant role for RNA structure in this region for its recognition as a PAMP. Unlike the FMDV NCR transcripts, the 5′-end of the viral genome is linked to the viral protein VPg lacking a 5′-ppp, making difficult to draw conclusions on the putative role of these structural regions in viral pathogenesis.

Encouraged by the results of IFN-β induction in swine cultured cells transfected with the NCR transcripts, we aimed to address the potential of such molecules as type-I IFN inducers in vivo. For that, the FMDV NCRs were inoculated intraperitoneally into Swiss suckling mice and the levels of IFN-α/β proteins and antiviral activity in sera were measured [48]. Newborn mice are a suitable model for innate immune responses while their adaptive immunity is still immature. All the FMDV NCRs were able to induce a peak of IFN-α/β in sera of the inoculated animals at 8 h after injection, remarkably higher than those observed for poly I:C-transfected mice. This peak was maintained up to 24 h in the case of the S RNA. The presence of antiviral activity in sera from NCR-transfected mice was also detected and measured, and a good correlation with IFN-α/β levels tested by ELISA was found. Interestingly, even those transcripts showing a lower capacity for IFN-β induction in porcine cultured cells were able to induce an innate immune response in mice. On one hand, this suggests that the effect of low level-inductions of type-I IFN observed in cultured cells can be magnified in vivo. On the other hand, the action of other viral sensors in vivo, mainly TLRs, may account for the enhancing effect observed. Thus, the specific immunostimulatory activity of each NCR RNA may be different depending on the host cell context assayed. This was the case for the IRES: despite of its complex structure, it was a poor inducer in cultured cells. However, the IRES acted as a strong IFN inducer in suckling mice. We further showed that the innate immune responses triggered by the NCRs in suckling mice resulted in a reduced susceptibility to FMDV infection in all cases, being remarkable the antiviral effect of inoculation with the IRES RNA [48, 49].

The antiviral effect exerted in vivo by these small synthetic non-infectious RNA molecules was analyzed extensively, using a wide range of viral doses and different serotype isolates [49]. The time course of resistance to FMDV of the RNA-transfected mice was also studied. Inoculation with all RNAs remarkably increased the 50% lethal dose (LD_{50}) of the virus, determined for the control group. Mice inoculated with IRES or S transcripts 24 h before challenge became at least 10000-fold less susceptible to the virus than PBS-inoculated mice. Interestingly, 90% of the IRES-transfected mice survived after infection with a viral dose of 7×10^6 plaque forming units (PFU) (undiluted viral stock), showing the outstanding protective effect of these RNA molecules. The level of protection against viral infection was dose-dependent. Complete or very high protection was achieved when IRES RNA was inoculated 8 or 24 h prior to FMDV infection with 7×10^4 PFU, with 100 and 86% survival, respectively. Inoculation of the transcripts at longer times pre-infection strongly decreased their protective effect against viral infection. Co-inoculation of S or IRES transcripts and the virus induced high levels of protection (about 90%), and the IRES RNA had a higher protective effect inoculated at 8 h than at 4 h before infection, suggesting that a fine balance between the routes activating the innate immune response by the RNAs and the viral replication kinetics or antagonistic mechanisms triggered by the virus, might determine either the outcome of disease or the viral clearance. Additionally, high survival percentages were observed for those groups inoculated with the RNAs at short times after infection (89 and 87% of mice inoculated with the IRES at 4 h and 8 h post-infection survived, respectively), and complete protection (100% survival) was achieved when mice were inoculated with the S transcripts at 4 or 8 h post-infection [49]. No protective effect was observed for mice inoculated with the RNAs 24 h after viral infection. These results suggest that the antiviral response induced by the RNAs is rapidly established and effective to counteract the viral replication if administered shortly after infection, while 24 h later it was too late to restrain the progress of infection. Our data support the potential use of this RNAs as both prophylactic as well as therapeutic molecules in a certain time window. These small non-infectious RNAs could be useful to induce a rapid antiviral state in combination with effective FMD vaccines to overcome the problem of the susceptibility window until protective levels of antibodies are produced by vaccinated animals. These results provide, as well, a new insight into broad-spectrum antiviral development strategies (Figure 3).

4. Exploiting innate responses for antiviral, therapeutic and adjuvation strategies

The example described above illustrates the potential of RNA regions in the viral genome, known to elicit innate responses, for antiviral and therapeutic applications.

Viral pattern recognition system may offer unique translational implications in medical approaches, taking advantage of the innate immune function of PRRs to trigger cell autonomous responses in tumour cells along with cytotoxic T-cell responses that target tumour cells. Tumour cells are much more sensitive to cytotoxic effects after RLR ligation than are untransformed cells, allowing for tumour-specific effects despite systemic application of the ligands in mouse models [50, 51]. The concept of using targeted application of PAMPs to mimic a situation of viral infection for clinical application like immunotherapy is being extensively explored [52]. In addition to being interesting targets for the immunotherapy of cancer, RLRs have been found to play a role in other disease conditions like type-I diabetes and other autoimmune diseases like psoriasis or selective IgA-deficiency [53-55]. RLRs ligands have been shown to have a therapeutic effect for autoimmune inflammatory disease of the central nervous system in mice [56].

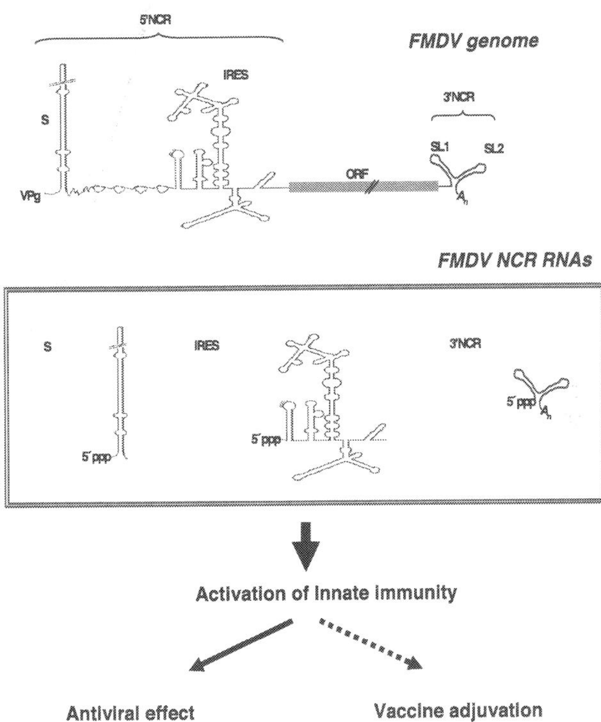

Figure 3. Synthetic RNA transcripts derived from structural domains present in the 5′ and 3′NCRs of the FMDV genome as innate immunity elicitors

Other than its role in driving innate immune defenses, IFN plays a major role in modulating the adaptive immune response [57]. IFN is required to promote T cell survival and clonal expansion after antigen presentation. IFN also induces the cytolytic activity of NK cells and cytotoxic lymphocytes and promotes B cell differentiation and antibody production, as well as expression of MHC class I molecules [58, 59]. The specific role of RLR signaling in regulating IFN production and its regulation of the adaptive immune response is less clear and appears to vary from virus to virus [60].

Luke et al. showed that the potency of a DNA vaccine against influenza virus could be augmented by the incorporation of a RIG-I activating immunostimulatory RNA into the vector backbone [61]. Mice receiving the vaccine exhibited increased virus-specific serum antibody response as compared to those receiving the DNA vaccine alone. These results suggest that RLR signaling can enhance antibody development after vaccination by activation of innate immunity and improved adaptive immune responses. However, in a previous study Koyama et al. found a defect in antigen-specific B and T cell activation in MyD88-defficient mice, unlike MAVS-deficient mice, suggesting that adaptive immune responses against influenza A virus are governed by the TLR pathway [62]. On the contrary, MAVS-deficient mice infected with WNV displayed uncontrolled inflammation including elevated systemic IFN, proinflammatory cytokine and chemokine levels, and enhanced humoral responses marked by complete loss of virus neutralization activity with a failure to protect against WNV infection [63]. This work defined an innate/adaptive immune interface mediated through MAVS-dependent RLR signaling that regulates the quantity, quality and balance of the immune response to WNV. Using MDA5-defficient mice, Anz et al. showed that the loss of T regulatory cell function on infection with encephalomyocarditis virus (EMCV) is strictly dependent on RLR signaling [64]. In a different study, a normal and effective adaptive immune response against respiratory syncytial virus (RSV) infection was reported in the absence of both MAVS and MyD88, RLR- and TLR-adaptor molecules, respectively [65]. All together, the results suggest that, in the case of flavivirus and picornavirus infections, RLR signaling is important in the control of the quantity, quality and balance of the adaptive immune response, with specific and differential mechanisms of regulation operating depending on the specific virus infections. The understanding of the molecular features underlying these processes could offer new strategies for immune and antiviral therapy targeting the RLR pathway for therapeutic control of the viral infection and enhancement of the induced immune response.

The vaccine development field is evolving from traditional whole cell vaccines to more defined and safer subunit vaccines. In the idea of exploiting the innate responses in vaccine adjuvant design, a growing demand for the use of immunopotentiators in poorly immunogenic subunit vaccines is arising with the development of a new generation of vaccine adjuvants. New vaccine adjuvants are designed to improve the recruitment and activation of dendritic cells, then enabling transition from the innate to adaptive immune system for priming of B- and T-cell responses. Endogenous or therapeutically induced early type-I IFN responses may confer protection until adaptive immunity is activated to an extent that the pathogen can be eliminated. In that context, PRRs come into sight as targets of new vaccine adjuvants beside their role as sentinels in innate immunity.

Evidence is now emerging that many empiric vaccines and adjuvants inherently stimulate PRRs, like the yellow fever vaccine 17D, one of the most effective vaccines available, shown to activate multiple DC subsets through stimulation of several TLRs (including TLR-7, -8 and -9) [66], highlighting the potential of vaccination strategies that use combinations of different PRRs ligands to stimulate polyvalent immune responses.

The current vaccine adjuvants licensed for use in human vaccines are limited [67], but other PRR agonists in clinical stages of development are emerging as potential vaccine adjuvant candidates [68], such as the TLR3 and MDA5 agonist poly I:C, a promising mucosal adjuvant for intranasal H5N1 influenza vaccination [69]. Clearly, TLR agonists are the most clinically advanced in adjuvant development, but require additional considerations, as variation in TLR expression and influence of age in responsiveness or the risk of autoimmunity by induction of excessive inflammatory responses [68]. Other TLR agonists currently in clinical trials of a malaria vaccine are the TLR-9 agonists CpG oligodeoxynucleotides (ODNs).

In addition to recognition of viral ssRNA like that derived from HIV (Human immunodeficiency virus) or influrenza viruses, TLR7/8 can also be activated by certain synthetic agonists such as the imidazoquinoline derivatives imiquimod and resiquimod (R-848), as well as the guanine nucleotide analog loxoribine. TLR7/8 imidazoquinolines can activate appropriate immune cells and modulate cellular and humoral immunity and have been found to be excellent vaccine adjuvants [70]. Most TLR agonists induce antibody and Th1 responses, although some can induce Th2 and possibly Th17 responses. Knowledge of the response outcomes in terms of cytokines, chemokines and T-cell subtypes generated by activation of combination of PRRs would help in the design of vaccine formulations including the appropriate combination of adjuvants in the future that can contribute to develop new vaccines against infectious diseases.

In the case of FMDV, many efforts are being invested on development of new vaccine formulations aimed to improve currently used vaccines [71]. Although FMD vaccines are available since the early 1900s, the disease still affects millions of animals around the globe and remains the main sanitary barrier to the commerce of animals and animal products. Among the limitations of the currently available inactivated antigen vaccines are the short duration of immunity and the lack of serotype-cross protection. Administration of this vaccine or an experimental vaccine based on a replication-defective human adenovirus (Ad5) vector that delivers the FMDV capsid and 3C proteins requires ~7 days to induce protective immunity in animals [72, 73]. New approaches aimed to shorten this susceptibility window and induce a more robust and long-lasting adaptive immune response are being developed, such as expression of type-I,-II and – III IFNs with Ad5 vectors with good but differential results depending on the host species assayed [74, 75]. Treatment with IFN has proved, so far, to be the best biotherapeutic approach tested against FMDV. Recent data show that poly I:C stabilized with poly-L-lysine and carboxymethyl cellulose (poly ICLC), a TLR-3 and MDA5 agonist, is a potent stimulator of IFN and ISGs in swine and at an adequate dose is sufficient to induce complete protection against FMD [77]. A different study shows that the combined application of recombinant adenoviruses expressing IFN-α or siRNA and other antiviral agents such as ribavirin may enhance their inhibitory effect on FMDV [77]. Our own data, discussed above, support the use of the FMDV NCR RNAs, mimicking structural domains in the viral genome acting as potent type-I IFN inducers, as promising non-infectious and synthetic molecules in future antiviral and vaccine developments against FMDV and likely other viral infections.

5. Conclusions

Understanding how the innate immune system senses the infection of different viruses with a variety of genome structures and signals and the crosstalk between different PRRs will help to understand the complex regulation of immunity to infection. The increasing knowledge on the nature of PAMPs and sensor specificity will surely contribute to the development of safer and more effective vaccines for infectious diseases. PRR agonists arise as promising molecules due to their synergistic effects on cytokine production and contributing to effective immune responses. The success of rationally designed vaccine formulations in the near future will likely correlate with the advances on understanding cell signalling mechanisms as well as PRR adjuvanticity and response outcomes. Targeted immunomodulatory strategies will require knowledge of the virus-specific aspects of the pathway. Viral proteins with IFN antagonistic activity are potential drug targets for antiviral strategies. Moreover, small, synthetic and non-infectious RNAs mimicking viral PAMPs can act as potent IFN inducers and exert an antiviral effect in vivo, providing new insight into broad-spectrum antiviral development strategies.

Acronyms and abbreviations

Ad5	replication-defective human adenovirus vector
APC	Antigen presenting cell
CARD	Caspase activation and recruitment domain
CLR	C-type lectin receptor
CpG	Unmethylated deoxycytidylate-phosphate-deoxyguanylate
DC	Dendritic cell
DEN	Dengue virus
DI genomes	Defective-interfering genomes
ds	Double-stranded
EMCV	Encephalomyocarditis virus
ER	Endoplasmic reticulum
FMDV	Foot-and-mouth disease virus
HCV	Hepatitis C virus
HIV	Human immunodeficiency virus
IFITM	Interferon inducible transmembrane protein
IFN	Interferon
IFNAR	IFN-α/β receptor
IRES	Internal ribosome entry site
IRF	Interferon regulatory factor
ISG	IFN-stimulated gene
JAK/STAT	Janus kinase/signal transducers and activators of transcription
LD$_{50}$	50% lethal dose

LGP2	Laboratory of genetics and physiology-2
MAP Kinase	Mitogen-activated protein kinase
MAVS	Mitochondrial antiviral signaling
MDA5	Melanoma differentiation-associated gene 5
MHC	Major histocompatibility complex
miRNA	Micro RNA
NCR	Non-coding region
NK	Natural killer
NLR	Nucleotide oligomerization domain (NOD-like receptor)
ODNs	Oligodeoxynucleotides
ORF	Open reading frame
PAMP	Pathogen-associated molecular pattern
pDC	Plasmacytoid dendritic cell
PFU	Plaque forming units
PKR	Protein kinase R
Poly I:C	Polyriboinosinic-polyribocytidylic acid
Poly ICLC	poly-L-lysine and carboxymethyl cellulose
PRR	Pattern-recognition receptor
RIG-I	Retinoic acid-inducible gene-1
RLR	RIG-I-like receptor
RSV	Respiratory syncytial virus
siRNA	Small interfering RNA
SL	Stem-loop structure
ss	Single-stranded
svRNA	Small viral RNA
TIR	Toll/Interleukin-1 receptor
TLR	Toll-like receptor
TNF	Tumor necrosis factor
TRIF	TIR-domain-containing adaptor inducing interferon
TRIM	Tripartite motif protein
VPg	Virus encoded protein
WNV	West Nile virus

Table 1. List of the acronyms and abbreviations used in this chapter

Acknowledgements

The work on FMDV discussed in this chapter was supported by grants BIO2008-04487-C03-01, AGL2011-24509 and BIO2011-24351 and by the ICTS program from the Spanish Ministry of Science and Innovation. We also thank Francisco Mateos, Nuria de la Losa and Lourdes Peña for their assistance in mice experiments. We are also grateful to Encarna Martínez-Salas for reagents and helpful discussion.

Author details

Miguel R. Rodríguez Pulido[1], Francisco Sobrino[1], Belén Borrego[2] and Margarita Sáiz[1*]

*Address all correspondence to: msaiz@cbm.uam.es

1 Centro de Biología Molecular Severo Ochoa, CSIC-UAM, Cantoblanco, Madrid, Spain

2 CISA-INIA, Valdeolmos, Madrid, Spain

References

[1] Janeway CA, Jr. Approaching the asymptote? Evolution and revolution in immunology. Cold Spring Harb Symp Quant Biol. 1989;54 Pt 1 1-13.

[2] Lemaitre B, Nicolas E, Michaut L, Reichhart JM, Hoffmann JA. The dorsoventral regulatory gene cassette spatzle/Toll/cactus controls the potent antifungal response in Drosophila adults. Cell. 1996;86(6) 973-983.

[3] Medzhitov R, Preston-Hurlburt P, Janeway CA, Jr. A human homologue of the Drosophila Toll protein signals activation of adaptive immunity. Nature. 1997;388(6640) 394-397.

[4] Yoneyama M, Kikuchi M, Natsukawa T, Shinobu N, Imaizumi T, Miyagishi M, et al. The RNA helicase RIG-I has an essential function in double-stranded RNA-induced innate antiviral responses. Nat Immunol. 2004;5(7) 730-737.

[5] Kawai T, Akira S. Toll-like receptors and their crosstalk with other innate receptors in infection and immunity. Immunity. 2012;34(5) 637-650.

[6] Osorio F, Reis e Sousa C. Myeloid C-type lectin receptors in pathogen recognition and host defense. Immunity. 2012;34(5) 651-664.

[7] Elinav E, Strowig T, Henao-Mejia J, Flavell RA. Regulation of the antimicrobial response by NLR proteins. Immunity. 2012;34(5) 665-679.

[8] Kawai T, Akira S. Toll-like receptors and their crosstalk with other innate receptors in infection and immunity. Immunity. 2011;34(5) 637-650.

[9] Yoneyama M, Fujita T. Recognition of viral nucleic acids in innate immunity. Rev Med Virol. 2010;20(1) 4-22.

[10] Sabbah A, Chang TH, Harnack R, Frohlich V, Tominaga K, Dube PH, et al. Activation of innate immune antiviral responses by Nod2. Nat Immunol. 2009;10(10) 1073-1080.

[11] Childs K, Randall R, Goodbourn S. Paramyxovirus V proteins interact with the RNA Helicase LGP2 to inhibit RIG-I-dependent interferon induction. J Virol. 2012;86(7) 3411-3421.

[12] Kato H, Takeuchi O, Mikamo-Satoh E, Hirai R, Kawai T, Matsushita K, et al. Length-dependent recognition of double-stranded ribonucleic acids by retinoic acid-inducible gene-I and melanoma differentiation-associated gene 5. J Exp Med. 2008;205(7) 1601-1610.

[13] Kato H, Takeuchi O, Sato S, Yoneyama M, Yamamoto M, Matsui K, et al. Differential roles of MDA5 and RIG-I helicases in the recognition of RNA viruses. Nature. 2006;441(7089) 101-105.

[14] Triantafilou K, Vakakis E, Kar S, Richer E, Evans GL, Triantafilou M. Visualisation of direct interaction of MDA5 and the dsRNA replicative intermediate form of positive strand RNA viruses. J Cell Sci. 2012;125(Pt 20) 4761-4769.

[15] Dixit E, Boulant S, Zhang Y, Lee AS, Odendall C, Shum B, et al. Peroxisomes are signaling platforms for antiviral innate immunity. Cell. 2010;141(4) 668-681.

[16] Yan N, Chen ZJ. Intrinsic antiviral immunity. Nat Immunol. 2012;13(3) 214-222.

[17] Clingan JM, Ostrow K, Hosiawa KA, Chen ZJ, Matloubian M. Differential roles for RIG-I-like receptors and nucleic acid-sensing TLR pathways in controlling a chronic viral infection. J Immunol. 2012;188(9) 4432-4440.

[18] Ramos HJ, Gale M, Jr. RIG-I like receptors and their signaling crosstalk in the regulation of antiviral immunity. Curr Opin Virol. 2011;1(3) 167-176.

[19] Poeck H, Bscheider M, Gross O, Finger K, Roth S, Rebsamen M, et al. Recognition of RNA virus by RIG-I results in activation of CARD9 and inflammasome signaling for interleukin 1 beta production. Nat Immunol. 2010;11(1) 63-69.

[20] Bowie AG, Unterholzner L. Viral evasion and subversion of pattern-recognition receptor signalling. Nat Rev Immunol. 2008;8(12) 911-922.

[21] Chevaliez S, Pawlotsky JM. Interferons and their use in persistent viral infections. Handb Exp Pharmacol. 2009;189 203-241.

[22] Guidotti LG, Chisari FV. Noncytolytic control of viral infections by the innate and adaptive immune response. Annu Rev Immunol. 2001;19 65-91.

[23] Solis M, Goubau D, Romieu-Mourez R, Genin P, Civas A, Hiscott J. Distinct functions of IRF-3 and IRF-7 in IFN-alpha gene regulation and control of anti-tumor activity in primary macrophages. Biochem Pharmacol. 2006;72(11) 1469-1476.

[24] Leung DW, Basler CF, Amarasinghe GK. Molecular mechanisms of viral inhibitors of RIG-I-like receptors. Trends Microbiol. 2012;20(3) 139-146.

[25] Randall RE, Goodbourn S. Interferons and viruses: an interplay between induction, signalling, antiviral responses and virus countermeasures. J Gen Virol. 2008;89(Pt 1) 1-47.

[26] Versteeg GA, Garcia-Sastre A. Viral tricks to grid-lock the type I interferon system. Curr Opin Microbiol. 2010;13(4) 508-516.

[27] Donelan NR, Basler CF, Garcia-Sastre A. A recombinant influenza A virus expressing an RNA-binding-defective NS1 protein induces high levels of beta interferon and is attenuated in mice. J Virol. 2003;77(24) 13257-13266.

[28] Liang JJ, Liao CL, Liao JT, Lee YL, Lin YL. A Japanese encephalitis virus vaccine candidate strain is attenuated by decreasing its interferon antagonistic ability. Vaccine. 2009;27(21) 2746-2754.

[29] Zust R, Cervantes-Barragan L, Kuri T, Blakqori G, Weber F, Ludewig B, et al. Coronavirus non-structural protein 1 is a major pathogenicity factor: implications for the rational design of coronavirus vaccines. PLoS Pathog. 2007;3(8) e109. doi:10.1371/journal.ppat.0030109

[30] Cullen BR. Viruses and microRNAs: RISCy interactions with serious consequences. Genes Dev. 2011;25(18) 1881-1894.

[31] Schuessler A, Funk A, Lazear HM, Cooper DA, Torres S, Daffis S, et al. West Nile virus noncoding subgenomic RNA contributes to viral evasion of the type I interferon-mediated antiviral response. J Virol. 2012;86(10) 5708-5718.

[32] Kawasaki T, Kawai T, Akira S. Recognition of nucleic acids by pattern-recognition receptors and its relevance in autoimmunity. Immunol Rev. 2011;243(1) 61-73.

[33] Schnell G, Loo YM, Marcotrigiano J, Gale M, Jr. Uridine Composition of the Poly-U/UC Tract of HCV RNA Defines Non-Self Recognition by RIG-I. PLoS Pathog. 2012;8(8) e1002839. doi: 10.1371/journal.ppat.1002839

[34] Thompson MR, Kaminski JJ, Kurt-Jones EA, Fitzgerald KA. Pattern recognition receptors and the innate immune response to viral infection. Viruses. 2011;3(6) 920-940.

[35] Edelmann KH, Richardson-Burns S, Alexopoulou L, Tyler KL, Flavell RA, Oldstone MB. Does Toll-like receptor 3 play a biological role in virus infections? Virology. 2004;322(2) 231-238.

[36] Hornung V, Barchet W, Schlee M, Hartmann G. RNA recognition via TLR7 and TLR8. Handb Exp Pharmacol. 2008;183 71-86.

[37] Hemmi H, Takeuchi O, Kawai T, Kaisho T, Sato S, Sanjo H, et al. A Toll-like receptor recognizes bacterial DNA. Nature. 2000;408(6813) 740-745.

[38] Schlee M, Roth A, Hornung V, Hagmann CA, Wimmenauer V, Barchet W, et al. Recognition of 5' triphosphate by RIG-I helicase requires short blunt double-stranded RNA as contained in panhandle of negative-strand virus. Immunity. 2009;31(1) 25-34.

[39] Baum A, Sachidanandam R, Garcia-Sastre A. Preference of RIG-I for short viral RNA molecules in infected cells revealed by next-generation sequencing. Proc Natl Acad Sci U S A. 2010;107(37) 16303-16308.

[40] Gitlin L, Barchet W, Gilfillan S, Cella M, Beutler B, Flavell RA, et al. Essential role of mda-5 in type I IFN responses to polyriboinosinic:polyribocytidylic acid and encephalomyocarditis picornavirus. Proc Natl Acad Sci U S A. 2006;103(22) 8459-8464.

[41] Pichlmair A, Schulz O, Tan CP, Rehwinkel J, Kato H, Takeuchi O, et al. Activation of MDA5 requires higher-order RNA structures generated during virus infection. J Virol. 2009;83(20) 10761-10769.

[42] Fredericksen BL, Keller BC, Fornek J, Katze MG, Gale M, Jr. Establishment and maintenance of the innate antiviral response to West Nile Virus involves both RIG-I and MDA5 signaling through IPS-1. J Virol. 2008;82(2) 609-616.

[43] Malathi K, Dong B, Gale M, Jr., Silverman RH. Small self-RNA generated by RNase L amplifies antiviral innate immunity. Nature. 2007;448(7155) 816-819.

[44] Malathi K, Saito T, Crochet N, Barton DJ, Gale M, Jr., Silverman RH. RNase L releases a small RNA from HCV RNA that refolds into a potent PAMP. RNA. 2010;16(11) 2108-2119.

[45] Saito T, Owen DM, Jiang F, Marcotrigiano J, Gale M, Jr. Innate immunity induced by composition-dependent RIG-I recognition of hepatitis C virus RNA. Nature. 2008;454(7203) 523-527.

[46] Uzri D, Gehrke L. Nucleotide sequences and modifications that determine RIG-I/RNA binding and signaling activities. J Virol. 2009;83(9) 4174-4184.

[47] Schmidt A, Schwerd T, Hamm W, Hellmuth JC, Cui S, Wenzel M, et al. 5'-triphosphate RNA requires base-paired structures to activate antiviral signaling via RIG-I. Proc Natl Acad Sci U S A. 2009;106(29) 12067-12072.

[48] Rodriguez-Pulido M, Borrego B, Sobrino F, Saiz M. RNA structural domains in noncoding regions of foot-and-mouth disease virus genome trigger innate immunity in porcine cells and mice. J Virol. 2011;85(13) 6492-6501.

[49] Rodriguez-Pulido M, Sobrino F, Borrego B, Saiz M. Inoculation of newborn mice with non-coding regions of foot-and-mouth disease virus RNA can induce a rapid,

solid and wide-range protection against viral infection. Antiviral Res. 2011;92(3) 500-504.

[50] Besch R, Poeck H, Hohenauer T, Senft D, Hacker G, Berking C, et al. Proapoptotic signaling induced by RIG-I and MDA-5 results in type I interferon-independent apoptosis in human melanoma cells. J Clin Invest. 2009;119(8) 2399-2411.

[51] Tormo D, Checinska A, Alonso-Curbelo D, Perez-Guijarro E, Canon E, Riveiro-Falkenbach E, et al. Targeted activation of innate immunity for therapeutic induction of autophagy and apoptosis in melanoma cells. Cancer Cell. 2009;16(2) 103-114.

[52] Schmidt A, Rothenfusser S, Hopfner KP. Sensing of viral nucleic acids by RIG-I: from translocation to translation. Eur J Cell Biol. 2012;91(1) 78-85.

[53] Nejentsev S, Walker N, Riches D, Egholm M, Todd JA. Rare variants of IFIH1, a gene implicated in antiviral responses, protect against type 1 diabetes. Science. 2009;324(5925) 387-389.

[54] Strange A, Capon F, Spencer CC, Knight J, Weale ME, Allen MH, et al. A genome-wide association study identifies new psoriasis susceptibility loci and an interaction between HLA-C and ERAP1. Nat Genet. 2010;42(11) 985-990.

[55] Ferreira RC, Pan-Hammarstrom Q, Graham RR, Gateva V, Fontan G, Lee AT, et al. Association of IFIH1 and other autoimmunity risk alleles with selective IgA deficiency. Nat Genet. 2010;42(9) 777-780.

[56] Kalinke U, Prinz M. Endogenous, or therapeutically induced, type I interferon responses differentially modulate Th1/Th17-mediated autoimmunity in the CNS. Immunol Cell Biol. 2012;90(5) 505-509.

[57] Mattei F, Schiavoni G, Tough DF. Regulation of immune cell homeostasis by type I interferons. Cytokine Growth Factor Rev. 2010;21(4) 227-236.

[58] Jego G, Palucka AK, Blanck JP, Chalouni C, Pascual V, Banchereau J. Plasmacytoid dendritic cells induce plasma cell differentiation through type I interferon and interleukin 6. Immunity. 2003;19(2) 225-234.

[59] Le Bon A, Schiavoni G, D'Agostino G, Gresser I, Belardelli F, Tough DF. Type I interferons potently enhance humoral immunity and can promote isotype switching by stimulating dendritic cells in vivo. Immunity. 2001;14(4) 461-470.

[60] Loo YM, Gale M, Jr. Immune signaling by RIG-I-like receptors. Immunity. 2011;34(5) 680-692.

[61] Luke JM, Simon GG, Soderholm J, Errett JS, August JT, Gale M, Jr., et al. Coexpressed RIG-I agonist enhances humoral immune response to influenza virus DNA vaccine. J Virol. 2010;85(3) 1370-1383.

[62] Koyama S, Ishii KJ, Kumar H, Tanimoto T, Coban C, Uematsu S, et al. Differential role of TLR- and RLR-signaling in the immune responses to influenza A virus infection and vaccination. J Immunol. 2007;179(7) 4711-4720.

[63] Suthar MS, Ma DY, Thomas S, Lund JM, Zhang N, Daffis S, et al. IPS-1 is essential for the control of West Nile virus infection and immunity. PLoS Pathog. 2010;6(2) e1000757. doi: 10.1371/journal.ppat.1000757.

[64] Anz D, Koelzer VH, Moder S, Thaler R, Schwerd T, Lahl K, et al. Immunostimulatory RNA blocks suppression by regulatory T cells. J Immunol. 2010;184(2) 939-946.

[65] Bhoj VG, Sun Q, Bhoj EJ, Somers C, Chen X, Torres JP, et al. MAVS and MyD88 are essential for innate immunity but not cytotoxic T lymphocyte response against respiratory syncytial virus. Proc Natl Acad Sci U S A. 2008;105(37) 14046-14051.

[66] Querec T, Bennouna S, Alkan S, Laouar Y, Gorden K, Flavell R, et al. Yellow fever vaccine YF-17D activates multiple dendritic cell subsets via TLR2, 7, 8, and 9 to stimulate polyvalent immunity. J Exp Med. 2006;203(2) 413-424.

[67] Olive C. Pattern recognition receptors: sentinels in innate immunity and targets of new vaccine adjuvants. Expert Rev Vaccines. 2012;11(2) 237-256.

[68] Duthie MS, Windish HP, Fox CB, Reed SG. Use of defined TLR ligands as adjuvants within human vaccines. Immunol Rev. 2011;239(1) 178-196.

[69] Ichinohe T, Ainai A, Tashiro M, Sata T, Hasegawa H. PolyI:polyC12U adjuvant-combined intranasal vaccine protects mice against highly pathogenic H5N1 influenza virus variants. Vaccine. 2009;27(45) 6276-6279.

[70] Tomai MA, Vasilakos JP. TLR-7 and -8 agonists as vaccine adjuvants. Expert Rev Vaccines. 2011;10(4) 405-407.

[71] Rodriguez LL, Grubman MJ. Foot and mouth disease virus vaccines. Vaccine. 2009;27 Suppl 4 90-94.

[72] Moraes MP, Mayr GA, Mason PW, Grubman MJ. Early protection against homologous challenge after a single dose of replication-defective human adenovirus type 5 expressing capsid proteins of foot-and-mouth disease virus (FMDV) strain A24. Vaccine. 2002;20(11-12) 1631-1639.

[73] Pacheco JM, Brum MC, Moraes MP, Golde WT, Grubman MJ. Rapid protection of cattle from direct challenge with foot-and-mouth disease virus (FMDV) by a single inoculation with an adenovirus-vectored FMDV subunit vaccine. Virology. 2005;337(2) 205-209.

[74] Moraes MP, de Los Santos T, Koster M, Turecek T, Wang H, Andreyev VG, et al. Enhanced antiviral activity against foot-and-mouth disease virus by a combination of type I and II porcine interferons. J Virol. 2007;81(13) 7124-7135.

[75] Perez-Martin E, Weiss M, Diaz-San Segundo F, Pacheco JM, Arzt J, Grubman MJ, et al. Bovine type III interferon significantly delays and reduces severity of foot-and-mouth disease in cattle. J Virol. 2012;86(8) 4477-4487.

[76] Dias CC, Moraes MP, Weiss M, Diaz-San Segundo F, Perez-Martin E, Salazar AM, et al. Novel antiviral therapeutics to control foot-and-mouth disease. J Interferon Cytokine Res. 2012;32(10) 462-473.

[77] Kim SM, Park JH, Lee KN, Kim SK, Ko YJ, Lee HS, et al. Enhanced inhibition of foot-and-mouth disease virus by combinations of porcine interferon-alpha and antiviral agents. Antiviral Res. 2012;96(2) 213-220.

Chapter 9

Gene Constellation of Influenza Vaccine Seed Viruses

Ewan P. Plant and Zhiping Ye

Additional information is available at the end of the chapter

http://dx.doi.org/10.5772/55289

1. Introduction

The Age of Enlightenment is the period in time when the method of reasoning known as the Scientific Method was developed. This revolution in science began with the description of the sun as the center of our solar system rather than the earth. Natural phenomena previously explained by spiritualists were now described by science. Given our still evolving understanding of influenza, it is perhaps no coincidence that we describe the combined effects of the influenza virus gene segments with the word 'constellation', which has astrological roots describing the position of the stars. Interestingly, the name influenza also has astrological roots: it was borrowed from the Italian word *influenza* in the mid-17th century which, in turn, was derived from the Medieval Latin word *influentia*, a 14th century term that refers to the influence of the stars. The scientific rational to describe the influence of the influenza gene constellation on virus phenotype is currently being resolved. Here we try to shine some light on the subject by providing the reader with background information, recent experimental results and provide a framework for questions that remain unanswered.

Influenza is a common infectious respiratory disease caused by influenza viruses. The host range of these viruses can include birds, humans and other mammals. Influenza viruses cause seasonal epidemics and are almost globally ubiquitous. They cause significant morbidity and mortality each year yet some infected persons remain asymptomatic. Influenza is typically transmitted by aerosols produced by coughing or sneezing. Although virus particles on contaminated surfaces can be easily inactivated, the virus is still able to spread easily and rapidly. Vaccination is the recommended approach to prevent disease because of the possible emergence of drug resistance.

Vaccines are produced each year to counter the currently circulating seasonal strains. The influenza vaccine seed viruses used to produce the immunogenic proteins are reassortant viruses. That is, they contain a mix of gene segments from different viruses. The genomes of

the influenza A and B viruses are made up of eight negative-strand RNA segments. The haemagglutinin (HA) and neuraminidase (NA) proteins found on the surface of the virus are found on two different segments. Usually these two segments from a seasonal virus are combined with another six gene segments from a high yield strain to make a vaccine seed virus. HA is the major immunogenic protein recognized by the host immune system. Because of influenza's high rate of mutation, and the capacity of the genome to tolerate many mutations, there is a need to update the influenza vaccine seed virus strains each year. The influenza virus is able to avoid control by the host immune system via two major types of mutation. Antigenic drift is the process of gradual genetic mutation, especially in the HA gene, that results in newer viruses not being well recognized by antibodies that recognized the progenitor virus. Antigenic shift is the replacement of one or more segments from one influenza virus with those of another. This unpredictable event can lead to a change in host range, transmission or pathogenicity. Likewise, genetic reassortment, the mixing of genomic segments from different strains, can generate undesirable characteristics in the influenza vaccine seed viruses. Here we explore possible reasons for this and describe approaches that might be beneficial to the development of influenza vaccine seed viruses.

2. The segmented influenza genome

The Orthomyxoviridae family is comprised of negative-sense, segmented RNA viruses. There are three influenza genera, influenza A, B and C viruses, that belong to the family. Eight negative-sense RNA segments make up the viral genome for the A and B viruses, one more than the influenza C virus genome which has seven segments (Figure 1). The terminal ends of each gene segment are conserved and this allows control over several aspects of the influenza lifecycle. The terminal sequences are partially complementary and can form structures that serve as regulatory signals for transcription and replication (Figure 1). The structures are dynamic and it is thought that switching between the structures allows different steps of replication to occur [1].

Each negative-sense viral RNA is encapsidated with nucleoprotein to form a ribonucleoprotein (RNP). Attached to the 5′ end of each segment is the influenza polymerase complex [2]. This arrangement allows a message RNA (mRNA) from each segment to be transcribed independently of other segments in the nucleus. Most of the segments encode a single protein but some of the RNA segments are spliced or have alternative translation mechanisms so that usually more than 10 proteins are made during an infection [3;4]. The mRNAs transcribed from all the segments are exported from the nucleus to the cytoplasm where they are translated into the viral proteins. The nucleoprotein and polymerase proteins each contain nuclear localization sequences and are imported into the nucleus where they participate in the production of new viral RNA. Some of the remaining proteins are processed in the secretory pathway and transported to the cell surface for incorporation into the virions while others remain in the cytoplasm or nucleus and modulate the host cell immune response.

The negative-sense segmented genome bestows some advantages and disadvantages to the virus. Having the genome divided up into segments creates some challenges such as ensuring that one of each segment is packaged into one virion. However, it also helps alleviate a problem faced by many RNA viruses, the high error rate inherent in RNA synthesis. The error rate for the influenza viruses has be calculated to be 2.0×10^{-6} and 0.6×10^{-6} mutations per site per infectious cycle for influenza A and B respectively [5]. Rates ranging from 3.72 to 6.77×10^{-4} substitutions per site per year have been calculated for the influenza C segments [6]. Influenza viruses can exist as a quasispecies, that is, a group of diverse viruses that collectively contribute to the characteristics of the population (reviewed in [7]). This enables mutations to exist that, by themselves, may not increase the fitness of the virus and could even be detrimental. A combination of these mutations, that together increase the fitness of the virus, may result in a virus with some selective advantage. Such a combination could occur by gene reassortment. The separation of different mutations on different virus segments facilitates this proc-

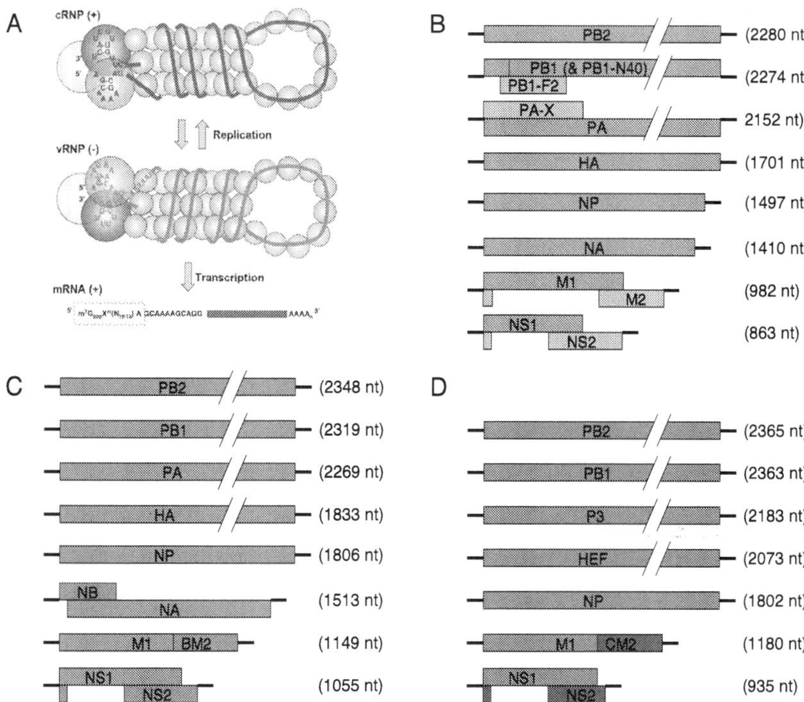

Figure 1. Genomes of influenza viruses. A) The ends of the negative strand influenza genomic RNA are complexed with the three polymerase proteins and the remaining sequence is encapsidated with nucleoprotein (vRNP (-)). The positive strand cRNP is similarly complexed. The mRNA is transcribed with a 5′ cap structure and poly-A tail (see main text for details). Figure used with permission from Resa-Infante et al., 2011. B) Schematic of the influenza A virus genome. The bold black lines represent the 3′ and 5′ untranslated regions. The blue and pink boxes represent the major protein coding regions. C) Schematic of the influenza B virus genome. The green and brown boxes represent the major protein coding regions. D) Schematic of the influenza C virus genome. The red and purple boxes represent the major protein coding regions. The protein coding regions are not to scale. Coding regions in a different reading frame are shown above or below each other, coding regions in the same frame are show as contiguous blocks.

ess and allows reassortant viruses to be made if a cell is infected with two influenza viruses at the same time.

Reassortment occurs in all three influenza virus genera [6;8;9]. But there is no evidence of reassortment between the genera. This is likely due, in part, to the level divergence between the viruses, both in the non-coding regulatory elements and in the proteins which interact with each other. Also, there is evidence that one virus may suppress another through pathways that are not well understood [10]. Reassortment has long been known to occur naturally in humans, swine and birds [11-13].

Often evidence of reassortment is based on incongruence in the phylogenetic trees of each of the segments. Although naturally occurring reassortants can be compared to previously sequenced strains, the actual strains that a particular segment is derived from, and the steps in reassortment, are only deduced. The different segments from the recent 2009 pandemic H1N1 virus were phylogenetically similar to human, avian, and classical, Eurasian and triple reassortant swine virus segments. Using all the available whole virus genome sequences Bokhari et al. [14] used a bioinformatics approach to determine which viruses were most likely homologous to the ancestors to the 2009 pandemic strain, and what reassortments needed to occur. Interestingly, they found that among 92% of the possible paths there were certain bottleneck viruses [14]. That is, these viruses contained the mutations and segment reassortments that made them most like the next virus in the reassortment path that eventually gave rise to the 2009 pandemic strain. This suggests that there are certain sequence requirements that need to be present before reassortment occurs.

Sequence analysis indicates that a disproportionate number of the naturally occurring reassortants are the result of novel haemagglutinin and/or neuraminidase genes being introduced into a previously circulating strain [15]. The introduction of the HA and NA segments into another strain is also the goal of influenza vaccine seed virus strain construction.

2.1. Making vaccine seed virus

There are two predominant ways that an influenza virus can be engineered; one, by simultaneous infection of a cell with two viruses each bearing some desired trait, or two, by reverse genetics. Both methods are commonly used to produce influenza vaccine seed viruses. A disadvantage of co-infection is that it may result in the production of viruses with unwanted combinations of the gene segments. A disadvantage to the reverse genetic approach is that it may be difficult to generate a virus if the introduced HA and NA genes cause a detrimental gene constellation effect.

In the late 1950s and throughout the 1960s Edwin Kilbourne pioneered the development of genetic recombination with influenza [16]. He recognised that an influenza vaccine seed virus should have certain desirable characteristics; good growth, low virulence, thermal stability and the proper antigenicity. At that time, development of high yield vaccine seed virus strains was via empirical methods such as mouse-lung passage. Kilbourne developed and promoted the use of "the deliberate mating of 2 or more viruses, each bearing a desired trait" so that "an appropriate progeny virus can be selected without the need for tedious" adaptation "until appropriate mutants, if any, become manifest" [16]. This was achieved by infecting an egg with a combination of a non-infective strain (with a high yield characteristic) and infective influenza strain (with the desired antigenic trait) in the presence of antiserum to suppress the antigenic proteins of the non-infective strain. This allowed for the selection of virus with the same antigenicity as the infectious strain and the same growth characteristics of the non-infectious strain [17]. This method for making vaccine seed virus strains has been widely used since. One drawback of this approach is that, because the antiserum only suppresses virus expressing the non-infective strain surface proteins, segments from the infective strain, in addition to the HA and NA segments, are often present in the resulting strains [18;19]. This may result in an undesirable trait being present in the vaccine seed virus strain and, as reassortments involving whole gene segments cannot revert like single point mutations, the traits are not as easily reverted as point mutations.

More recently, with the development of reverse genetics, it has been possible to make reassortant viruses from cloned viral segments. This allows genetically defined vaccine seed virus strains to be produced and the methodology has been employed extensively for the production of live attenuated vaccine seed virus strains [20]. The cold adaption and attenuation mutations are spread out on multiple segments. A large number of human influenza vaccine seed virus strains have been made with the cold adapted strains A/Ann Arbor/6/60-H2N2 and B/Ann Arbor/1/66 strains [20]. High yield attenuated backbone strains for vaccination of livestock such as birds or pigs have also been developed and are typically made using reverse genetics [21-23]. In addition, viruses expressing the haemagglutinin from highly pathogenic avian strains (H5N1) need to have the polybasic cleavage site removed by mutagenesis and reverse genetic viruses made to reduce pathogenicity. Viruses made by reverse genetics use the same plasmid derived internal gene segments. In constrast, vaccine seed strains made using *in ovo* reassortment are sometimes made with a recent high growth reassortant as the non-infective strain. This could result in the carrying forth of internal segments that are not from the high yield strain Puerto Rico/8/1934 (PR/8) or mutations that have appeared during the passage of the earlier reassortant strain. Although the reverse genetically engineered viruses are genetically defined, there is no avenue for reassortment to occur if there is some incompatibility between the glycoproteins from the season strain and the remaining proteins from the backbone virus.

Influenza viruses are frequently isolated and propagated in tissue culture. Madin-Darby canine kidney (MDCK) cells are widely used because they are quite susceptible to influenza virus infection. This is because the antiviral activity of MDCK cells is lacking due to inadequate interferon-induced myxovirus resistance protein 1 (Mx1) activity [24]. As noted above, the introduction of HA or NA segments into circulating strains is dominant among naturally occurring reassortants [15]. In contrast, recombinants generated in MDCK cells with no selection show a positive correlation between other segments: most of the segment pairs that segregate with each other in MDCK cells were polymerase combinations [25-27]. This suggests that for naturally occurring reassortants there is some selective pressure that is not present in laboratory-based experiments. One explanation for this bias may be due to the limited MDCK cell antiviral response; without this response the replication efficiency of the virus might be the main limiting factor. Thus, viruses with combinations of polymerase factors that are most efficient will become dominant. One report of recombinants generated in eggs without selection described a preference for cosegregation of the HA and M segments [28]. Although the number of reports regarding reassortment without selection is limited, the current data suggests that egg-based experiments may more closely reflect naturally occurring events.

The replicase proteins may play a role in reassortment via their independent interaction with each RNA segment. Each genomic segment has it's own replicase proteins associated with it when it enters the nucleus. A doubly infected cell is capable of producing each of the segments from both viruses independently of each other. The timing of the overall replication will depend on many factors such as which segments are imported into the nucleus first, translation promoter sequences and replication signals on each segment, and the induced host response. If a cell is simultaneously infected with two viruses, it is possible that early in the infection the polymerase that transcribes RNA more efficiently will control the dynamics of the infection. The resulting dynamics between the host cell and the viruses may favor the production of either virus or a reassortant. It is unknown if the dynamics of transcription and replication play a role in reassortment but, given that the transciption and replication signals on each segment can differ, it is easy to imagine the dynamics of an infection being altered when a cell is infected with two viruses. One could also imagine polymerase proteins from one strain transcribing or replicating another strain's RNA, or a polymerase protein from one virus interacting with, and altering the activity of, a polymerase protein from the other virus.

2.2. Glycoproteins

There are two major surface glycoproteins for the influenza A and B viruses. The haemagglutinin (HA) protein is a sugar-binding protein that facilitates virus entry into epithelial cells that have sialic acid sugars on the cell surface. After the HA is cleaved by a protease, the virion is imported into the cell by endocytosis. Virus replication culminates with the accumulation of new virions at the cell surface. The neuraminidase (NA) cleaves the glycosidic linkages of the sialic acids to mediate virion release. The HA protein is encoded by segment 4 and the NA protein by segment 6 in influenza A and B viruses. Influenza C has only seven segments. The haemagglutinin-esterase-fusion (HEF) glycoprotein is encoded on segment 4 and this protein performs the functions analogous to HA and NA of influenza A and B viruses.

In contrast to the influenza B and C viruses which are only sorted by type and strain, the influenza A viruses are sorted into subtypes. The HA and NA proteins are used for virus classification. There are at least 16 different HA subtypes and 9 different NA subtypes among the influenza A viruses. The HA subtypes are divided into two groups. Certain subtypes from both groups are able to infect and transmit among humans. Most human influenza A infections are from H1N1, H2N2, and H3N2 subtypes. Occasionally a strain will jump the species barrier. A limited number of avian subtypes (H5, H7 and H9) have infected humans. Sometimes the disease is much more severe than that from a human influenza strain but these strains seem to lack the ability to transmit from human to human efficiently. There is a fear that these highly virulent viruses may reassort with human viruses creating virulent viruses that spread easily among humans. New pandemic strains arising by reassortment is clearly a concern. An additional subtype was recently identified in bats [29]. The HA from the bat virus is more similar to the Group 1 HAs (subtypes H1, 2, 6, 8, 9, 11, 12, 13 and 16) than the Group 2 subtypes but the NA shows no similarity to any previously identified NA subtypes [29]. It remains to be seen if these viruses can reassort and cause disease in humans.

Infection of humans with viruses containing swine origin HA and NA is known to occur. Like other zoonoses, most of these swine viruses do not spread efficiently in humans. However, the swine origin 2009 pandemic H1N1 virus spread around the world supplanting the prior seasonal human H1N1 strains. This virus was a reassortant derived from a North American triple reassortant H1N2 swine strain and a Euroasiatic H1N1 swine strain.

2.3. Replicase proteins

Four proteins are required for influenza virus replication; the nucleocapsid protein (NP) and the three polymerase proteins PB1, PB2 and PA. The polymerase proteins are the larger influenza proteins and are encoded in the largest segments 1-3. NP is encoded in segment 5. The RNA from each viral segment form ribbon-like closed superhelical structures (reviewed in [30]). The 5′ and 3′ ends of the RNA are in close proximity to one another and are associated with the three polymerase proteins. Nucleoprotein is associated with the remaining genomic RNA and there is one NP monomer present for each 24 nucleotides. Nuclear localization sequences on NP facilitate import of each ribonucleoprotein complex (RNP) into the nucleus (reviewed in [31]). Inside the nucleus mRNA transcription and viral replication take place.

Two of the polymerase proteins, PB1 and PB2, have biochemical interactions with NP proteins. In addition, the three polymerase proteins interact with each other. The carboxyl-terminus of PA interacts with the amino-terminal end of PB1 and the carboxyl-terminal end of PB1 interacts with the amino-terminal end of PB2. The same arrangement is described for both the negative strand viral RNA (vRNP) and the positive strand copy RNA (cRNP) with the polymerase being associated with the 5′ end of the RNAs. New negative-strand viral genomes are derived from the cRNP and also have a newly synthesized polymerase complex associated with the 5′ end. The NP and three polymerase proteins all have nuclear localization signals which enable them to be imported into the nucleus after they are synthesized in the cytoplasm. There is evidence that PA and PB1 associate with each other before localizing to the nucleus (reviewed in [31]).

Both types of positive-strand RNA, cRNP and mRNA, are generated from the vRNP. In constrast to the vRNP and cRNP, the mRNA is not associated with NP. The viral mRNAs also have a capped 5′ leader sequence snatched from a cellular mRNA and a polyadenylated 3′ end. It is not yet known exactly what regulates the polymerase complex so that it makes two distinct products from one template. The cap-binding domain is in PB2 and the endonuclease domain is in PA. Together these parts of the polymerase complex capture and remove the 5′ capped region of cellular mRNA and this is used as the priming sequence for viral mRNA production. The RNA-dependent RNA polymerase domain required for all RNA production is in the PB1 protein.

Combining the polymerase proteins from different strains to produce chimeric polymerase complexes has been studied with regard to polymerase activity and pathogenicity. It is sometimes found, but not always, that increased polymerase activity leads to more virus and increased pathogenicity [32]. Most recent studies have focused on the replicase genes from the 2009 pandemic H1N1 strain and prior seasonal strains. It was found that the pandemic PB2 gene combined with seasonal PB1, PA and NP genes resulted in significantly less polymerase activity [33;34]. Conversely, inclusion of a seasonal PB2 gene in a pandemic background significantly increased polymerase activity. When the corresponding reassortant viruses, with a PR/8 backbone, were generated the growth kinetics for both types were reduced. This suggests that the level of polymerase activity needs to be optimized for the best virus production in vitro. In addition these viruses had higher mouse LD_{50} values suggesting polymerase activity and replication are important for virulence [33]. Interestingly, introduction of the pandemic NP gene into a seasonal virus also dramatically reduced the virus replication and pathogenicity demonstrating that both altered polymerase and RNP could give rise to detrimental gene constellation effects.

In an analysis of ressortant viruses with a 2009 pandemic strain background it was found that introduction of a PA, PB1 or PB2 segment from another virus typically reduced the virus titer [27;35]. This included instances when all three segments from another virus increased polymerase activity (A/swine/Korea/JNS06/04 or A/mallard/Korea/6L/07) or reduced polymerase activity (A/duck/Korea/LPM91/06 or A/aquatic bird/Korea/ma81/07). Again, this suggests that the level of polymerase activity needs to be optimized for efficient virus production in vitro. Each of these viruses were less pathogenic in mice but several other viruses were generated that were more pathogenic in mice. One, containing the just the PA segment from A/aquatic bird/Korea/ma81/07 in the 2009 pandemic backbone had a similar level of polymerase activity to the reassortant virus containing all three A/aquatic bird/Korea/ma81/07 polymerase genes indicating that polymerase activity per se is not the cause of pathogenicity [35]. It is possible that specific virulence determinants are associated with the PA segment but, in the absence of a gene constellation effect, this would not account for the lower pathogenicity of the virus containing all three polymerase segments.

One PB2 virulence marker, the amino acid at position 627, is a determinant of host range and contributes to pathogenicity in mice [36]. It has been shown that introduction of a PB2 gene from a low pathogenic H1N1 virus into the highly pathogenic 1918 strain attenuated the virus in mice but pathogenicity was restored with a E627K mutation [37]. In contrast, studies of swine influenza in pigs have shown that there is no correlation between pathogenicity and viruses with either a swine- or avian-origin PB2 gene containing the 627K or 627E mutation [38]. While it has been suggested that the 627 residue mediates an interaction with NP [39], and the strength of this interaction correlates with polymerase activity [40], recent evidence suggests that restricted activity is due to a lack of compatibility with a host cell factor [41]. Thus, although amino acid signatures of virulence may be important in the context of genetic drift, these results demonstrate that gene constellation effects can attenuate virulence in some hosts.

Clearly, although specific functions of replicase complex reside in each protein, the interaction of the replicase proteins plays a large role in several virus attributes including replication and virulence. At present there is a lot of genetic information for the replicase genes available. Unfortunately our current understanding of influenza replication does not enable the prediction of replication efficiency or associated pathogenicity based on replicase gene sequence alone. While new functional information is being generated regularly, a more complete understanding of influenza replication and its contribution toward pathogenicity will require more comprehensive structure-function information.

2.4. Other proteins

There are at least four additional proteins produced during influenza infection of a cell. The two segments not mentioned so far, segments 7 and 8, are the smallest genome segments. In both influenza A and B viruses each of these segments encodes at least two proteins. The analogous segments in influenza C viruses are segments 6 and 7 (Figure 1).

Influenza A segment 7 encodes two matrix proteins; M1 and M2. The M2 proteins from influenza B and C viruses are called BM2 and CM2 respectively. M1 binding to RNPs in the nucleus inhibits viral transcription [42;43]. M1 proteins form a continuous shell on the inner side of the lipid bilayer. M2, and the analogous BM2 and CM2 proteins, are ion-channel proteins that form as a homotetramers in the virus envelope. These small hydrophobic integral membrane proteins allow hydrogen ions to enter the viral particle from the endosome. The lower pH causes M1 to dissociate from the RNPs leading to the uncoating of the virus. Different coding strategies are used by the different influenza species (Figure 2 and [44]). M2 protein is translated from a spliced transcript while BM2 protein is translated by a coupled termination/reinitiation event [45;46]. In contrast CM1 is translated from a spliced transcript and CM2 is the produced by peptidase cleavage of a precursor protein [47;48]. Influenza B viruses encode an additional small hydrophobic integral membrane protein on segment 6. The open reading frame starts 4 nucleotides upstream from the NA ORF (Figure 2). Although NB is conserved in influenza B genomes it is apparently not essential and it's function remains unknown at present [49]. Interestingly, influenza A viruses also encode an alternative M2 protein on a splicing variant [3]. The same conserved redundancy in two different influenza families highlights the importance of the ion channels for the viruses. This diversity in coding and expression of similar functions may also be a reason why reassortant viruses containing segments from different influenza types are not readily obtained.

The smallest influenza segment encodes the non-structural protein NS1 and the nuclear export protein xlink. xlink is translated from a spliced transcript and is incorporated into the virions in small numbers (Figure 2). The major role of NS1 is to modulate the host immune response. It is a multifunctional protein that interacts with several host proteins and has an RNA binding domain (reviewed in [50]). Protein sequence features from influenza NS1 proteins indicate that there are variant types that seem to correlate with certain host species [51]. The exact nature of this relationship has not been teased out yet. The NS1 protein from the 2009 pandemic strain was less effective at blocking the innate immune response in cultured cells than other seasonal strains but attempts to make the 2009 NS1 protein more like the seasonal strain did not result in the same effect, rather the virus had reduced the virulence and was more easily cleared [52]. A better understanding the relationship between NS1 from specific virus strains and the host cell type could lead to the development of vaccine seed virus backbone strains that are more suitable for vaccine production.

The presence of segment 7 or 8 from differing viruses can alter the phenotype of another virus. For example, addition of different NS segments from an H3N2 virus or different H5N1 viruses into a PR/8 backbone could result in no attenuation or complete attenuation [53;54]. Similarly, the same gene can have different effects on different viruses. For example, replacement of the A/Korea/82 (H3/N2) M segment with the A/Ann Arbor/6/60 M segment attenuated the virus. However, introducing the same A/Ann Arbor/6/60 M segment into A/Udorn/72 (H3/N2) did not attenuate the virus [55]. This clearly demonstrates the greater impact the gene constellation has toward the virus phenotype than an individual segment in this instance.

The characterization of the laboratory generated reassortants provides useful hypothesis-driven information. Natural reassortment of the smallest influenza segments may give additional information about the role these segments play in the virus lifecycle. Some viruses isolated recently from North American pigs contain the 2009 H1N1 M segment in the context of a previously endemic H1N2 strain [56]. This suggests that this particular gene constellation may increase viral fitness. Supporting this Chou et al., [57] were able to show that inclusion of the M segment in a PR/8 backbone was essential for transmission in guinea pigs. Another group reported that the neuraminidase segment from the 2009 H1N1 strain, in addition to the M segment, was required for efficient replication and transmission in pigs [58]. Finally, Hause et al. [59] reported transmission but lower viral titers for the reassortant viruses containing the M segment from the 2009 pandemic strain in pig lung homogenate when compared to infection with the parental strains. The major difference between the viruses analyzed by these groups is that the backbone strains differed; one group used comtemporary H1N2 swine viruses to generate reassortants while the other groups used laboratory adapted strains. The different outcomes observed are based on gene constellation effects.

3. The gene constellation effect

It stands to reason that if a certain gene constellation confers some desirable attribute then viruses containing those segments should occur more often in the population over time than other reassortments. That is, the combination of gene segments should occur independently many times in a large enough population if the same parental viruses continue to co-circulate in the population. However, if the reassortant has a relative fitness that is much greater than other circulating strains then the virus may quickly replace other virus strains in the population. An example of this is the emergence and spread of the 2009 pandemic H1N1 strain. Very quickly, and during a season not typically associated with high influenza rates, the 2009 pandemic strain became the prevalent H1N1 strain and prior seasonal H1N1 strains become less common [60].

Possibly more common, but less well documented, are reassortants that have a small increase in fitness compared to the parental strains. A recent analysis of reassortant H3N2 viruses in swine demonstrated that multiple reassortants generating the same gene constellation occurred [61]. Several H3N2 swine lineage viruses were isolated from humans in 2011 and found to contain the segment 7 from the 2009 pandemic H1N1 virus. This prompted Nelson and colleagues to analyze the reassortants present in swine populations. What they discovered was that, in addition to the reassortants that transmitted to humans, reassortants in swine with a range of genetic backbones contained the 2009 pandemic segment 7 [61]. It is not known if the presence of the 2009 segment 7 in swine viruses plays a role in viral fitness in swine or if it has a role in zoonotic infection of humans, but clearly the presence of this segment gives rise to viruses from different reassortment events that are stably represented in the population.

In addition to the appearance of this particular gene constellation in North America pigs, gene constellations involving all segments from the 2009 pandemic strain are becoming dominant in other parts of the world. It has been reported that the 2009 pandemic strain is present in pigs and reassorting with H1N2 and H3N2 strains [62-65]. It is not clear how the combination of gene segments present in strains like the 2009 pandemic strain results in a greater viral fitness but study of different viral characteristics has given us some insight. Here we highlight research describing the effect different gene constellations have on viral fitness.

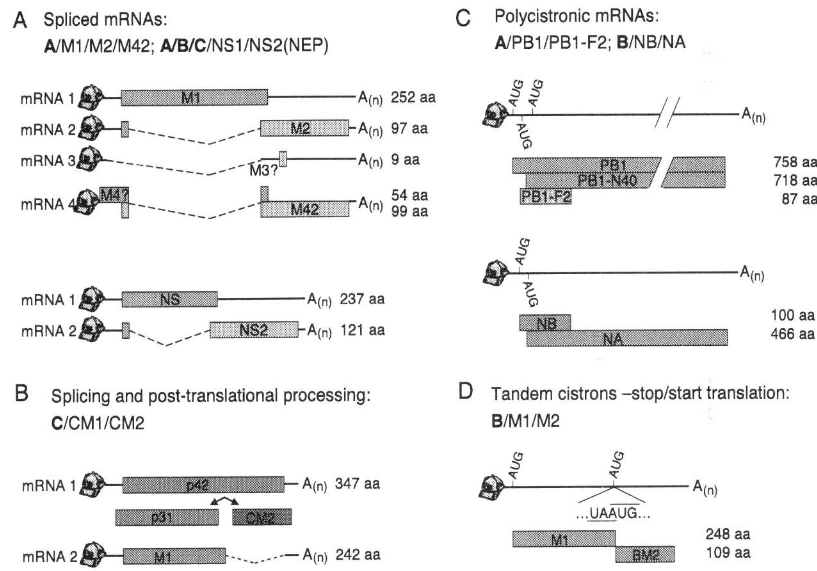

Figure 2. Coding strategies of influenza viruses (adapted from Lamb & Takeda, 2001). A) Multiple splice variants are transcribed from one segment to enable the production of multiple proteins. As many as four mRNAs (and proteins) are produced from influenza A segment 7. The mRNA from the smallest segment of all influenza species is spliced for the translation of the nuclear export protein xlink. B) A larger protein is translated and processed to produce the influenza C protein CM2. The M1 protein is produced from a spliced mRNA. C) Different AUG start codons are used to produce up to three proteins from influenza A segment 1. A second membrane protein, NB, is produced from an alternative start codon in segment 6 of influenza B. D) The termination and start codons of the respective M1 and BM2 open reading frames in segment 7 of influenza B overlap. A stop/start translation strategy is used to produce the second protein.

3.1. Altered pathogenicity

There have been many efforts to understand which gene segments contribute to pathogenicity. If a particular segment were known to contribute to pathogenicity in a vaccine strain then safeguards could be put in place to prevent the generation of the gene constellations containing the offending segment(s). The current recommendation for Institutional Biosafety Committees in the United States is that gene constellation be included in the evaluation for determining the biocontainment level for influenza work [66]. One difficulty in applying this recommendation is that the pathogenicity of a virus in one host species may differ greatly from the pathogenicity in another host species. Another difficulty is that new strains emerge and evolve faster than the pathogenicity of the gene combinations can be assessed. Here we reflect on what is known about gene constellation and pathogenicity.

In the 1970s it became clear that both the glycoproteins and the internal proteins play a role in pathogenicity. In many experiments segments from a human or animal origin virus were introduced into a pathogenic avian influenza virus and pathogenicity tested in chickens. Often the pathogenicity was reduced [67;68]. However, increased monitoring of avian influenza viruses in Hong Kong indicated that most naturally occurring reassortant H5N1 viruses were lethal to birds [69]. Serial passage of pathogenic avian influenza in MDCK cells and selection of large plaques resulted in reduced pathogenicity in mice suggesting that differences in the host cell type that the virus is propagated in play a role in pathogenicity [70]. Interestingly, the attenuated variants all had common mutations in the polymerase genes and grew to higher titers on MDCK cells than virus purified from small plaques. This suggests that an equilibrium between replication efficiency and pathogenicity is being altered when a virus is adapting to a new host. This is in contrast to the increased pathogenicity seen when faster growing viruses are compared to slower growing viruses in the same host; for example, avian viruses grown in eggs [71;72].

There are other examples of changes in the replication machinery of the influenza virus affecting pathogenicity. In one study it was found that reassortant avian H5N1 viruses containing the PB2 gene from a human H3N2 virus had increased pathogenicity in mice [73]. It was also shown that the introduction of a human PB1 segment alone did not enhance pathogenicity but had a cooperative effect when the human PB2 was present. As noted above, the segments containing the polymerase often cosegregate, which could be troubling if this enhances pathogenicity.

Pathogenicity is not only dependent on the virus, it also depends on the host species and even where in the host the virus replicates [74]. The pathogenicity caused by a virus is often due to the host response to the virus and is perhaps best exemplified by the cytokine storm. In such instances an excessive amount of proinflammatory cytokines are released and inflammation spreads from the site of infection. Acute lung injury, or the more severe acute respiratory distress syndrome, is associated with influenza infections (reviewed in [75]). As the NS-1 protein encoded on segment 8 has a role in modulating the host immune response, one would expect that segregation of segment 8 during reassortment might alter virus pathogenicity. However, while addition of an H3N2 NS segment to a 2009 H1N1 virus increased replication efficiency, the virus was not as pathogenic in mice as the parental H1N1 strain [54]. Also, the addition of an H5N1-derived NS segment to a PR/8 backbone attenuated the virus when tested in mice [53]. Because PR/8 is a high growth strain these results may not be truly representative of the effect the NS gene can have. When the NS segment from a highly pathogenic H5N1 strain was added to a highly pathogenic H7N1 strain the virus was more pathogenic in mice [76]. Here the observed increase in virulence was also associated with enhanced cell tropism. This demonstrates the potential for reassortants created in a host specific manner to gain the ability to jump the species barrier.

The 2009 H1N1 pandemic virus arose by reassortment of swine influenza viruses of different lineages. The NA and M gene segments were from the Eurasian avian-like swine virus lineage and the remaining segments were from the North American triple reassortant lineage. The triple reassortant lineage emerged in the 1990s with the PB2 and PA segments derived from an avian virus, the PB1 from a human virus and the remaining segments from the classical swine lineage. With all three swine lineages circulating, and reassorting, there is a concerted effort to characterize current and possible reassortants for their potential to infect humans [62;65]. The ferret is a widely used model for assessing pathogenicity and transmission of viruses. Unlike mice, but like humans, ferrets infected with a seasonal influenza strain present with an increase in temperature, nasal secretions, sneezing and sometimes with a cough, making them suitable for study of human viruses. Triple reassortant swine viruses isolated before 2009 did not produce clinical signs of respiratory symptoms like sneezing and nasal secretions [77]. The pathogenicity of these viruses was similar to the 2009 pandemic virus with more lung pathology than seasonal viruses [77]. All the triple reassortant viruses transmitted between ferrets by direct contact but only those with human-like HA and NA were transmitted efficiently by respiratory droplets [65;78]. Addition of a seasonal H3N2 NA to a 2009 pandemic virus gave rise to more severe pulmonary lesions in ferrets demonstrating the importance of gene constellation [79]. Further, it has been shown that the tropism of such a virus is linked to the balance between HA and the NA even when the replication competence is lower [62]. For influenza vaccine seed viruses it has been hypothesized that a lower NA content in the virion can increase HA content [80].

3.2. Altered growth rates

A high growth phenotype in a virus seed strain is beneficial for vaccine production. There are many proteins that might have an effect on the virus growth rate. The envelope proteins determine the infectivity through effects on attachment, entry and budding. The replicase proteins affect the speed of transcription and replication. And finally, NS1 protein modulates the host response. Thus, differing combinations can result in changes in growth rates. Many vaccine seed strains have been made using PR/8 as the backbone, that is, replacement of the HA and NA segments whilst retaining the six remaining PR/8 segments. Usually the introduced HA and NA segments come from the same virus so one would assume that the encoded proteins are compatible with each other. Thus, any reduction in growth rate is due to the change in interactions between the HA and NA combination and other proteins. Increased growth rates are often achieved by passaging the virus but the enhanced growth sometimes results in antigenic changes to HA rather than adaptation of the other proteins.

Many insights about virus growth have come from analysis of changes on one segment. It has been observed that culture of many influenza viruses in eggs results in amino acid changes in HA as the virus adapts to the new host. Similar observations have been made when viruses are grown in different cell types. For example, most viruses have an asparagine residue in the haemagglutinin at position 117 (H1 subtype) or 116 (H3 subtype). Substitution of this residue with aspartic acid does not alter the growth in MDCK cells but enhances growth in Vero cells [81]. This mutation was shown to alter the pH range for virus membrane fusion indicating that this is an important factor for optimal growth [81]. However, HA does not act in isolation and the best growth of vaccine seed strains will depend on how differences in HA affect interactions with other proteins, in particular, the other major envelope proteins NA and M.

The introduction of HA from a seasonal H1N1 virus into the 2009 pandemic strain backbone resulted in larger plaque size and higher viral titers in cell culture [82]. It was further shown that, in contrast to the predominantly filamentous parental strains, the reassortant virus was predominantly spherical and enhanced yields could be obtained by introducing the same seasonal HA into other swine-origin backbones [82]. By using the opposite approach, introducing swine-origin segments into the seasonal virus backbone Octaviani et al. were able to show that the high yield was primarily due to the presence of swine-origin HA and M segments [82].

The HA and NA segments from A/Vietnam/1194/2004 and the PR/8 backbone were combined to make a reassortant prepandemic vaccine seed virus but it gave low antigen yield and did not grow well. Incorporation of the M gene segment from the A/Panama/2007/1999 H3N2 strain or from the A/Vietnam/1203/2004 H5N1 strain enhanced growth [83]. The M1 proteins differed at positions T167A, R174K, I219V, A227T and A239K, while the M2 proteins differed at positions N31S, R54L, Y57H, S82N and G89S from the PR/8 segment 7 proteins. It is unknown how the synergy between these segments works but, with structures available for the major envelope proteins and advances in electron tomography, these interactions may be revealed in the near future [84].

Several vaccine seed strains have included the seasonal PB1 gene [18;19]. Growth rates were also shown to improve when the indigenous PB1 was included in the 2009 H1N1 reassortant [85] and with a H5 reassortant [86]. However, this result did not extend to a different H5 reassortant [83]. We postulated that certain residues within the PB1 protein might be important for growth and yield. Rather than target those amino acids known to be involved enzyme activity or protein interactions with the other polymerase proteins, we made changes based on sequence similarity between diverse PB1 proteins that reassorted *in ovo* during the production of influenza vaccine seed viruses. Using this impartial approach, we made changes to the PR/8 PB1 gene that, when combined with the HA and NA from the low yield H3N2 Wyoming/03 strain, resulted in faster growth in both egg and cell culture [87].

3.3. Altered protein production

The 5′ untranslated regions of the influenza A genomic segments contain signals that stimulate translation. These signals regulate the amount of protein produced from each segment and differ between segments. The non-coding regions also differ in length among the different segments and virus types [88-90]. The sequence motifs AGGGU and GGUAGAUA that are recognized respectively by the host protein G-rich RNA sequence binding factor 1 (GRSF-1) and the viral NS1 protein may also be present in the non-coding regions (reviewed in [1]). Both of these proteins have been shown to stimulate translation [91;92]. In addition, there are changes in translation that most likely due to changes in the RNA structure. Single nucleotide changes in the 5′ and 3′ non-coding regions of PA were shown to have no effect on translation individually, but together these changes almost completely abolished protein expression [90]. It is not known if the changes altered mRNA production or affected translation itself, but one would expect that the resultant loss of viral protein would affect virus replication and possibly virulence. Indeed, the mutations that resulted in low protein production were based on the sequence of a low pathogenic avian influenza virus.

Very similar viruses may produce quite different amounts of viral protein and several groups have tried to find the underlying reasons. It is common for several reassortants to be made using different seasonal isolates in an effort to create a suitable high yield vaccine seed virus. As noted previously the PB1 segment from the seasonal strain was found in many high yield reassortants made for vaccine manufacture [18]. Analysis of protein production from reassortants with or without the seasonal PB1 segment and the 2009 H1N1 HA and NA as been performed. One group found that the presence of the PB1 from the 2009 strain increased protein yield while another found that it decreased protein yield [85;93]. Both groups used PR/8 as the high yield donor strain but they each used different 2009 pandemic isolates. This highlights one difficulty in predicting what gene constellations may be beneficial for protein production; minor strain variations may have major translational effects. Comparison of both the PR/8 strains and both 2009 pandemic strains used in these studies may provide interesting information.

3.4. Incomplete genomes and RNA structure

The in vitro production of ressortant viruses using the classical reassortant method involves high multiplicities of infection to increase the chance of both viruses infecting cell. Early studies demonstrated that passage of influenza at high multiplicities of infection could result in the production of non-infectious particles [94;95]. These particles are now known as defective interfering, or DI, particles. Subsequently it was discovered that the polymerase genes frequently contained deletions [96;97]. Thus, although eight genomic segments were packaged, the viruses could not make proteins essential for replication. In addition, non-infectious particles lacking the glycoproteins have also been described [98]. This explains the loss of infectivity even though many particles are detected by haemagglutination [99]. Kaverin et al., [99] were able to show that one fraction of a DI population could complement another fraction of the DI population. More recently, Odagiri and Tashiro [100] were able to show that non-coding sequences were responsible for the preferential packaging of DI RNA rather than the full length RNA segment. It is not clear when or how the portions of the RNA are deleted but it is likely that the structure and the sequence of the RNA play a significant role. Using a bioinformatics approach, Priore et al. [101] analyzed the extensive base pairing that exists throughout the genomic segments of avian, swine and human influenza A viruses. The results indicated that there were significant differences between species in the PB2, NP, M and NS-containing segments. These differences were only on the positive strand which could indicate a role in either protein production or negative strand synthesis. Given that these segments do not reassort as frequently as the other polymerase segments or glycoprotein segments, it would seem that genome wide RNA structural organization does not contribute to reassortment. Although, it is postulated that global organizational RNA structure could be a mechanism by which the virus adapts to the host environment [101] leaving open the possibility that the RNA structure of a particular segment affects the chances of it being involved in reassortment.

DI particles represent an evolutionary dead end with regard to a natural infection. However, particles that lack a complete genome could be either detrimental or beneficial in vaccine production. Particles that are defective in the polymerase will alter growth characteristics and would not function in a live attenuated vaccine. In contrast, particles with incomplete genomes represent an abundance of antigen with no pathogenicity, which could be viewed as desirable in an inactivated vaccine.

4. Conclusion

Understanding the gene constellation effect in influenza is important, especially for vaccine production. The mixing and matching of influenza genomic segments in nature and in the laboratory gives rise to new viruses with phenotypes that differ from the ancestral viruses. In nature, this may be a more pathogenic virus or one that has an expanded host range. In the laboratory, attenuated viruses with good growth characteristics and high protein yield are desirable for study and vaccine production. A greater understanding of what contributes to the gene constellation effect may enable researchers to produce influenza vaccine seed viruses that facilitate production with reduced risk of infection.

As we have described in this chapter, current research has provided some insight into the genomic features that contribute to the gene constellation effect, but more work needs to be done. Some segments, such as those encoding the glycoproteins and the polymerase proteins, appear to be more frequently involved in reassortments. The reassortment of the polymerase proteins is more common in laboratory manipulation whereas the reassortment of glycoproteins is more common in nature. The beneficial effects of certain protein:protein interactions may be the underlying impetus behind some of these reassortments. For example, certain combinations of HA, NA and M can lead to changes in transmission and growth. Likewise, certain combinations of PB1, PB2 and PA can affect polymerase activity and growth. Also, the two smallest segments have effects on cell tropism and viral fitness. The amount of polymerase activity is not directly associated with virus titer suggesting other factors affecting replication must be balanced with replication efficiency. Changes in the polymerase segments can also affect pathogenicity, especially when the virus is adapting to a new host cell. Changes in the glycoproteins have also been shoen to affect pathogenicity. While much work has focused on either the glycoproteins or the replicase proteins independently, some of the work described here demonstrates that these two groups of proteins have an effect on each other. The interaction between these two groups of proteins at a functional level needs to be elucidated.

While several groups have analyzed the genomes of reassortant viruses there is still a great need for better understanding what features contribute to genomic reassortment. With more whole virus genome sequences available for analysis there is a better chance that the features important for reassortment can be determined. Retrospective analysis of reassortant viruses can illuminate which genomic features are compatible. In vitro construction of reassortant viruses can highlight which segments, or parts of segments, are not compatible. After a reassortment event is detected there needs to be more analyses of the mutations that occurred in each segment as they may have facilitated the reassortment event. Mutations necessary for reassortment would occur prior to reassortment and perhaps be present in bottleneck viruses. Mutations that occur with passage are those that increase the fitness of the reassortant. Description of both types of mutations would enhance our understanding of the network of interactions between viral proteins. In addition to the changes in coding sequences, analysis of the untranslated regions of the genomes is also important. There is no available information about the compatibility of segments with the replication and translation machinery, or how this contributes to the gene constellation.

Finally, an understanding of the gene constellation effect will allow for the selection of better reassortant viruses for vaccine production. Currently both the *in ovo* and reverse genetic methods use an impirical approach to get the best viruses that express the desired HA and NA proteins. Knowing how the different segments contribute to the network of interactions that result in high yield will enable researchers to produce strains that will provide the best backbone for an influenza vaccine seed virus. The optimal backbones may be universal, or differ for the different virus subtypes, or differ according to the host species that the virus providing the HA and NA infects. But without optimal virus backbones, the production of high yield reassortant influenza vaccine seed viruses will remain inefficient.

Acknowledgements

The findings and conclusions in this chapter have not been formally disseminated by the Food and Drug Administration and should not be construed to represent an Agency determination or policy.

Author details

Ewan P. Plant and Zhiping Ye

*Address all correspondence to: ewan.plant@fda.hhs.gov

Food and Drug Administration, USA

References

[1] Gultyaev AP, Fouchier RA, Olsthoorn RC. Influenza virus RNA structure: unique and common features. Int Rev Immunol 2010 Dec;29(6):533-56.

[2] Tiley LS, Hagen M, Matthews JT, Krystal M. Sequence-specific binding of the influenza virus RNA polymerase to sequences located at the 5' ends of the viral RNAs. J Virol 1994 Aug;68(8):5108-16.

[3] Wise HM, Hutchinson EC, Jagger BW, Stuart AD, Kang ZH, Robb N, et al. Identification of a novel splice variant form of the influenza a virus m2 ion channel with an antigenically distinct ectodomain. PLoS Pathog 2012 Nov;8(11):e1002998.

[4] Jagger BW, Wise HM, Kash JC, Walters KA, Wills NM, Xiao YL, et al. An Overlapping Protein-Coding Region in Influenza A Virus Segment 3 Modulates the Host Response. Science 2012 Jun 28;337(6091):199-204.

[5] Nobusawa E, Sato K. Comparison of the mutation rates of human influenza A and B viruses. J Virol 2006 Apr;80(7):3675-8.

[6] Gatherer D. Tempo and mode in the molecular evolution of influenza C. PLoS Curr 2010;2:RRN1199.

[7] Lauring AS, Andino R. Quasispecies theory and the behavior of RNA viruses. PLoS Pathog 2010;6(7):e1001005.

[8] McCullers JA, Wang GC, He S, Webster RG. Reassortment and insertion-deletion are strategies for the evolution of influenza B viruses in nature. J Virol 1999 Sep;73(9): 7343-8.

[9] Palese P, Young JF. Variation of influenza A, B, and C viruses. Science 1982 Mar 19;215(4539):1468-74.

[10] Wanitchang A, Narkpuk J, Jaru-Ampornpan P, Jengarn J, Jongkaewwattana A. Inhibition of influenza A virus replication by influenza B virus nucleoprotein: An insight into interference between influenza A and B viruses. Virology 2012 Jul 5;432(1): 194-203.

[11] Arikawa J, Yamane N, Ishida N. Serological evidence of natural recombinant influenza virus (Hsw1N2) infection among pigs in Japan. Acta Virol 1981 Jul;25(4):225-9.

[12] Desselberger U, Nakajima K, Alfino P, Pedersen FS, Haseltine WA, Hannoun C, et al. Biochemical evidence that "new" influenza virus strains in nature may arise by recombination (reassortment). Proc Natl Acad Sci U S A 1978 Jul;75(7):3341-5.

[13] Nishikawa F, Sugiyama T. Direct isolation of H1N2 recombinant virus from a throat swab of a patient simultaneously infected with H1N1 and H3N2 influenza A viruses. J Clin Microbiol 1983 Aug;18(2):425-7.

[14] Bokhari SH, Pomeroy LW, Janies DA. Reassortment Networks and the evolution of pandemic H1N1 swine-origin influenza. IEEE/ACM Trans Comput Biol Bioinform 2012 Jan;9(1):214-27.

[15] de Silva UC, Tanaka H, Nakamura S, Goto N, Yasunaga T. A comprehensive analysis of reassortment in influenza A virus. Biology Open 2012 Apr 15;1(4):385-90.

[16] Kilbourne ED. Future influenza vaccines and the use of genetic recombinants. Bull World Health Organ 1969;41(3):643-5.

[17] Kilbourne ED, Murphy JS. Genetic studies of influenza viruses. I. Viral morphology and growth capacity as exchangeable genetic traits. Rapid in ovo adaptation of early passage Asian strain isolates by combination with PR8. J Exp Med 1960 Mar 1;111:387-406.

[18] Fulvini AA, Ramanunninair M, Le J, Pokorny BA, Arroyo JM, Silverman J, et al. Gene constellation of influenza A virus reassortants with high growth phenotype prepared as seed candidates for vaccine production. PLoS One 2011;6(6):e20823.

[19] Ottmann M, Duchamp MB, Casalegno JS, Frobert E, Moules V, Ferraris O, et al. Novel influenza A(H1N1) 2009 in vitro reassortant viruses with oseltamivir resistance. Antivir Ther 2010;15(5):721-6.

[20] Jin H, Chen Z, Liu J, Kemble G. Genetic engineering of live attenuated influenza viruses. Methods Mol Biol 2012;865:163-74.

[21] Liu M, Liu CG, Zhang Y, Shi WL, Wang W, Liu YY. Efficacy of a high-yield attenuated vaccine strain wholly derived from avian influenza viruses by use of reverse genetics. Vet Microbiol 2012 Jul 10.

[22] Paillot R, Hannant D, Kydd JH, Daly JM. Vaccination against equine influenza: quid novi? Vaccine 2006 May 8;24(19):4047-61.

[23] Pena L, Vincent AL, Ye J, Ciacci-Zanella JR, Angel M, Lorusso A, et al. Modifications in the polymerase genes of a swine-like triple-reassortant influenza virus to generate live attenuated vaccines against 2009 pandemic H1N1 viruses. J Virol 2011 Jan;85(1): 456-69.

[24] Seitz C, Frensing T, Hoper D, Kochs G, Reichl U. High yields of influenza A virus in Madin-Darby canine kidney cells are promoted by an insufficient interferon-induced antiviral state. J Gen Virol 2010 Jul;91(Pt 7):1754-63.

[25] Greenbaum BD, Li OT, Poon LL, Levine AJ, Rabadan R. Viral reassortment as an information exchange between viral segments. Proc Natl Acad Sci U S A 2012 Feb 28;109(9):3341-6.

[26] Lubeck MD, Palese P, Schulman JL. Nonrandom association of parental genes in influenza A virus recombinants. Virology 1979 May;95(1):269-74.

[27] Octaviani CP, Ozawa M, Yamada S, Goto H, Kawaoka Y. High level of genetic compatibility between swine-origin H1N1 and highly pathogenic avian H5N1 influenza viruses. J Virol 2010 Oct;84(20):10918-22.

[28] Varich NL, Gitelman AK, Shilov AA, Smirnov YA, Kaverin NV. Deviation from the random distribution pattern of influenza A virus gene segments in reassortants produced under non-selective conditions. Arch Virol 2008;153(6):1149-54.

[29] Tong S, Li Y, Rivailler P, Conrardy C, Castillo DA, Chen LM, et al. A distinct lineage of influenza A virus from bats. Proc Natl Acad Sci U S A 2012 Mar 13;109(11): 4269-74.

[30] Resa-Infante P, Jorba N, Coloma R, Ortin J. The influenza virus RNA synthesis machine: advances in its structure and function. RNA Biol 2011 Mar;8(2):207-15.

[31] Hutchinson EC, Fodor E. Nuclear import of the influenza A virus transcriptional machinery. Vaccine 2012 May 29.

[32] Mehle A, Dugan VG, Taubenberger JK, Doudna JA. Reassortment and mutation of the avian influenza virus polymerase PA subunit overcome species barriers. J Virol 2012 Feb;86(3):1750-7.

[33] Hsieh EF, Lin SJ, Mok CK, Chen GW, Huang CH, Wang YC, et al. Altered pathogenicity for seasonal influenza virus by single reassortment of the RNP genes derived from the 2009 pandemic influenza virus. J Infect Dis 2011 Sep 15;204(6):864-72.

[34] Octaviani CP, Goto H, Kawaoka Y. Reassortment between seasonal H1N1 and pandemic (H1N1) 2009 influenza viruses is restricted by limited compatibility among polymerase subunits. J Virol 2011 Aug;85(16):8449-52.

[35] Song MS, Pascua PN, Lee JH, Baek YH, Park KJ, Kwon HI, et al. Virulence and genetic compatibility of polymerase reassortant viruses derived from the pandemic (H1N1) 2009 influenza virus and circulating influenza A viruses. J Virol 2011 Jul; 85(13):6275-86.

[36] Subbarao EK, London W, Murphy BR. A single amino acid in the PB2 gene of influenza A virus is a determinant of host range. J Virol 1993 Apr;67(4):1761-4.

[37] Qi L, Davis AS, Jagger BW, Schwartzman LM, Dunham EJ, Kash JC, et al. Analysis by single gene reassortment demonstrates that the 1918 influenza virus is functionally compatible with a low pathogenicity avian influenza virus in mice. J Virol 2012 Jun 20;86(17):9211-20.

[38] Ma W, Lager KM, Li X, Janke BH, Mosier DA, Painter LE, et al. Pathogenicity of swine influenza viruses possessing an avian or swine-origin PB2 polymerase gene evaluated in mouse and pig models. Virology 2011 Feb 5;410(1):1-6.

[39] Labadie K, Dos Santos AE, Rameix-Welti MA, van der Werf S, Naffakh N. Host-range determinants on the PB2 protein of influenza A viruses control the interaction

between the viral polymerase and nucleoprotein in human cells. Virology 2007 Jun 5;362(2):271-82.

[40] Ng AK, Chan WH, Choi ST, Lam MK, Lau KF, Chan PK, et al. Influenza polymerase activity correlates with the strength of interaction between nucleoprotein and PB2 through the host-specific residue K/E627. PLoS One 2012;7(5):e36415.

[41] Cauldwell AV, Moncorge O, Barclay WS. Unstable polymerase-NP interaction is not responsible for avian influenza polymerase restriction in human cells. J Virol 2012 Oct 31.

[42] Noton SL, Medcalf E, Fisher D, Mullin AE, Elton D, Digard P. Identification of the domains of the influenza A virus M1 matrix protein required for NP binding, oligo-merization and incorporation into virions. J Gen Virol 2007 Aug;88(Pt 8):2280-90.

[43] Ye ZP, Baylor NW, Wagner RR. Transcription-inhibition and RNA-binding domains of influenza A virus matrix protein mapped with anti-idiotypic antibodies and syn-thetic peptides. J Virol 1989 Sep;63(9):3586-94.

[44] Lamb RA, Takeda M. Death by influenza virus protein. Nat Med 2001 Dec;7(12): 1286-8.

[45] Horvath CM, Williams MA, Lamb RA. Eukaryotic coupled translation of tandem cis-trons: identification of the influenza B virus BM2 polypeptide. EMBO J 1990 Aug; 9(8):2639-47.

[46] Powell ML, Napthine S, Jackson RJ, Brierley I, Brown TD. Characterization of the ter-mination-reinitiation strategy employed in the expression of influenza B virus BM2 protein. RNA 2008 Nov;14(11):2394-406.

[47] Hongo S, Sugawara K, Muraki Y, Matsuzaki Y, Takashita E, Kitame F, et al. Influenza C virus CM2 protein is produced from a 374-amino-acid protein (P42) by signal pep-tidase cleavage. J Virol 1999 Jan;73(1):46-50.

[48] Pekosz A, Lamb RA. Influenza C virus CM2 integral membrane glycoprotein is pro-duced from a polypeptide precursor by cleavage of an internal signal sequence. Proc Natl Acad Sci U S A 1998 Oct 27;95(22):13233-8.

[49] Hatta M, Kawaoka Y. The NB protein of influenza B virus is not necessary for virus replication in vitro. J Virol 2003 May;77(10):6050-4.

[50] Hale BG, Randall RE, Ortin J, Jackson D. The multifunctional NS1 protein of influen-za A viruses. J Gen Virol 2008 Oct;89(Pt 10):2359-76.

[51] Noronha JM, Liu M, Squires RB, Pickett BE, Hale BG, Air GM, et al. Influenza virus sequence feature variant type analysis: evidence of a role for NS1 in influenza virus host range restriction. J Virol 2012 May;86(10):5857-66.

[52] Hale BG, Steel J, Manicassamy B, Medina RA, Ye J, Hickman D, et al. Mutations in the NS1 C-terminal tail do not enhance replication or virulence of the 2009 pandemic H1N1 influenza A virus. J Gen Virol 2010 Jul;91(Pt 7):1737-42.

[53] Lipatov AS, Andreansky S, Webby RJ, Hulse DJ, Rehg JE, Krauss S, et al. Pathogenesis of Hong Kong H5N1 influenza virus NS gene reassortants in mice: the role of cytokines and B- and T-cell responses. J Gen Virol 2005 Apr;86(Pt 4):1121-30.

[54] Shelton H, Smith M, Hartgroves L, Stilwell P, Roberts K, Johnson B, et al. An influenza reassortant with polymerase of pH1N1 and NS gene of H3N2 influenza A virus is attenuated in vivo. J Gen Virol 2012 May;93(Pt 5):998-1006.

[55] Subbarao EK, Perkins M, Treanor JJ, Murphy BR. The attenuation phenotype conferred by the M gene of the influenza A/Ann Arbor/6/60 cold-adapted virus (H2N2) on the A/Korea/82 (H3N2) reassortant virus results from a gene constellation effect. Virus Res 1992 Sep 1;25(1-2):37-50.

[56] Ducatez MF, Hause B, Stigger-Rosser E, Darnell D, Corzo C, Juleen K, et al. Multiple reassortment between pandemic (H1N1) 2009 and endemic influenza viruses in pigs, United States. Emerg Infect Dis 2011 Sep;17(9):1624-9.

[57] Chou YY, Albrecht RA, Pica N, Lowen AC, Richt JA, Garcia-Sastre A, et al. The M segment of the 2009 new pandemic H1N1 influenza virus is critical for its high transmission efficiency in the guinea pig model. J Virol 2011 Nov;85(21):11235-41.

[58] Ma W, Liu Q, Bawa B, Qiao C, Qi W, Shen H, et al. The neuraminidase and matrix genes of the 2009 pandemic influenza H1N1 virus cooperate functionally to facilitate efficient replication and transmissibility in pigs. J Gen Virol 2012 Jun;93(Pt 6):1261-8.

[59] Hause BM, Collin EA, Ran Z, Zhu L, Webby RJ, Simonson RR, et al. In Vitro Reassortment between Endemic H1N2 and 2009 H1N1 Pandemic Swine Influenza Viruses Generates Attenuated Viruses. PLoS One 2012;7(6):e39177.

[60] Blyth CC, Kelso A, McPhie KA, Ratnamohan VM, Catton M, Druce JD, et al. The impact of the pandemic influenza A(H1N1) 2009 virus on seasonal influenza A viruses in the southern hemisphere, 2009. Euro Surveill 2010;15(31).

[61] Nelson MI, Vincent AL, Kitikoon P, Holmes EC, Gramer MR. The evolution of novel reassortant A/H3N2 influenza viruses in North American swine and humans, 2009-2011. J Virol 2012 Jun 13;86(16):8872-8.

[62] Chan RW, Kang SS, Yen HL, Li AC, Tang LL, Yu WC, et al. Tissue tropism of swine influenza viruses and reassortants in ex vivo cultures of the human respiratory tract and conjunctiva. J Virol 2011 Nov;85(22):11581-7.

[63] Hiromoto Y, Parchariyanon S, Ketusing N, Netrabukkana P, Hayashi T, Kobayashi T, et al. Isolation of the Pandemic (H1N1) 2009 virus and its reassortant with an H3N2 swine influenza virus from healthy weaning pigs in Thailand in 2011. Virus Res 2012 Aug 10;169(1):175-81.

[64] Matsuu A, Uchida Y, Takemae N, Mawatari T, Yoneyama S, Kasai T, et al. Genetic characterization of swine influenza viruses isolated in Japan between 2009 and 2012. Microbiol Immunol 2012 Aug 25;56(11):792-803.

[65] Yen HL, Liang CH, Wu CY, Forrest HL, Ferguson A, Choy KT, et al. Hemagglutinin-neuraminidase balance confers respiratory-droplet transmissibility of the pandemic H1N1 influenza virus in ferrets. Proc Natl Acad Sci U S A 2011 Aug 23;108(34): 14264-9.

[66] Biosafety in Microbiological and Biomedical Laboratories. 5 ed. Atlanta: Centers for Disease Control and Prevention; 2009.

[67] Rott R, Orlich M, Scholtissek C. Attenuation of pathogenicity of fowl plague virus by recombination with other influenza A viruses nonpathogenic for fowl: nonexculsive dependence of pathogenicity on hemagglutinin and neuraminidase of the virus. J Virol 1976 Jul;19(1):54-60.

[68] Scholtissek C, Rott R, Orlich M, Harms E, Rohde W. Correlation of pathogenicity and gene constellation of an influenza A virus (fowl plague). I. Exchange of a single gene. Virology 1977 Aug;81(1):74-80.

[69] Guan Y, Peiris JS, Lipatov AS, Ellis TM, Dyrting KC, Krauss S, et al. Emergence of multiple genotypes of H5N1 avian influenza viruses in Hong Kong SAR. Proc Natl Acad Sci U S A 2002 Jun 25;99(13):8950-5.

[70] Li J, Liu B, Chang G, Hu Y, Zhan D, Xia Y, et al. Virulence of H5N1 virus in mice attenuates after in vitro serial passages. Virol J 2011;8:93.

[71] Suzuki K, Okada H, Itoh T, Tada T, Mase M, Nakamura K, et al. Association of increased pathogenicity of Asian H5N1 highly pathogenic avian influenza viruses in chickens with highly efficient viral replication accompanied by early destruction of innate immune responses. J Virol 2009 Aug;83(15):7475-86.

[72] Tada T, Suzuki K, Sakurai Y, Kubo M, Okada H, Itoh T, et al. NP body domain and PB2 contribute to increased virulence of H5N1 highly pathogenic avian influenza viruses in chickens. J Virol 2011 Feb;85(4):1834-46.

[73] Li C, Hatta M, Nidom CA, Muramoto Y, Watanabe S, Neumann G, et al. Reassortment between avian H5N1 and human H3N2 influenza viruses creates hybrid viruses with substantial virulence. Proc Natl Acad Sci U S A 2010 Mar 9;107(10):4687-92.

[74] Rebel JM, Peeters B, Fijten H, Post J, Cornelissen J, Vervelde L. Highly pathogenic or low pathogenic avian influenza virus subtype H7N1 infection in chicken lungs: small differences in general acute responses. Vet Res 2011 Jan;42(1):10.

[75] Tisoncik JR, Korth MJ, Simmons CP, Farrar J, Martin TR, Katze MG. Into the eye of the cytokine storm. Microbiol Mol Biol Rev 2012 Mar;76(1):16-32.

[76] Ma W, Brenner D, Wang Z, Dauber B, Ehrhardt C, Hogner K, et al. The NS segment of an H5N1 highly pathogenic avian influenza virus (HPAIV) is sufficient to alter

replication efficiency, cell tropism, and host range of an H7N1 HPAIV. J Virol 2010 Feb;84(4):2122-33.

[77] Belser JA, Gustin KM, Maines TR, Blau DM, Zaki SR, Katz JM, et al. Pathogenesis and transmission of triple-reassortant swine H1N1 influenza viruses isolated before the 2009 H1N1 pandemic. J Virol 2011 Feb;85(4):1563-72.

[78] Barman S, Krylov PS, Fabrizio TP, Franks J, Turner JC, Seiler P, et al. Pathogenicity and transmissibility of north american triple reassortant Swine influenza a viruses in ferrets. PLoS Pathog 2012 Jul;8(7):e1002791.

[79] Schrauwen EJ, Herfst S, Chutinimitkul S, Bestebroer TM, Rimmelzwaan GF, Osterhaus AD, et al. Possible increased pathogenicity of pandemic (H1N1) 2009 influenza virus upon reassortment. Emerg Infect Dis 2011 Feb;17(2):200-8.

[80] Jing X, Phy K, Li X, Ye Z. Increased hemagglutinin content in a reassortant 2009 pandemic H1N1 influenza virus with chimeric neuraminidase containing donor A/Puerto Rico/8/34 virus transmembrane and stalk domains. Vaccine 2012 Jun 13;30(28): 4144-52.

[81] Murakami S, Horimoto T, Ito M, Takano R, Katsura H, Shimojima M, et al. Enhanced growth of influenza vaccine seed viruses in vero cells mediated by broadening the optimal pH range for virus membrane fusion. J Virol 2012 Feb;86(3):1405-10.

[82] Octaviani CP, Li C, Noda T, Kawaoka Y. Reassortment between seasonal and swine-origin H1N1 influenza viruses generates viruses with enhanced growth capability in cell culture. Virus Res 2011 Mar;156(1-2):147-50.

[83] Abt M, de JJ, Laue M, Wolff T. Improvement of H5N1 influenza vaccine viruses: influence of internal gene segments of avian and human origin on production and hemagglutinin content. Vaccine 2011 Jul 18;29(32):5153-62.

[84] Calder LJ, Wasilewski S, Berriman JA, Rosenthal PB. Structural organization of a filamentous influenza A virus. Proc Natl Acad Sci U S A 2010 Jun 8;107(23):10685-90.

[85] Wanitchang A, Kramyu J, Jongkaewwattana A. Enhancement of reverse genetics-derived swine-origin H1N1 influenza virus seed vaccine growth by inclusion of indigenous polymerase PB1 protein. Virus Res 2010 Jan;147(1):145-8.

[86] Rudneva IA, Timofeeva TA, Shilov AA, Kochergin-Nikitsky KS, Varich NL, Ilyushina NA, et al. Effect of gene constellation and postreassortment amino acid change on the phenotypic features of H5 influenza virus reassortants. Arch Virol 2007;152(6): 1139-45.

[87] Plant EP, Liu TM, Xie H, Ye Z. Mutations to A/Puerto Rico/8/34 PB1 gene improves seasonal reassortant influenza A virus growth kinetics. Vaccine 2012;31(1):207-212

[88] Crescenzo-Chaigne B, Barbezange C, van der Werf S. Non coding extremities of the seven influenza virus type C vRNA segments: effect on transcription and replication by the type C and type A polymerase complexes. Virol J 2008;5:132.

[89] Stoeckle MY, Shaw MW, Choppin PW. Segment-specific and common nucleotide sequences in the noncoding regions of influenza B virus genome RNAs. Proc Natl Acad Sci U S A 1987 May;84(9):2703-7.

[90] Wang L, Lee CW. Sequencing and mutational analysis of the non-coding regions of influenza A virus. Vet Microbiol 2009 Mar 30;135(3-4):239-47.

[91] de la Luna S, Fortes P, Beloso A, Ortin J. Influenza virus NS1 protein enhances the rate of translation initiation of viral mRNAs. J Virol 1995 Apr;69(4):2427-33.

[92] Kash JC, Cunningham DM, Smit MW, Park Y, Fritz D, Wilusz J, et al. Selective translation of eukaryotic mRNAs: functional molecular analysis of GRSF-1, a positive regulator of influenza virus protein synthesis. J Virol 2002 Oct;76(20):10417-26.

[93] Timofeeva TA, Ignatieva AV, Rudneva IA, Shilov AA, Kaverin NV. Yields of virus reassortants containing the HA gene of pandemic influenza 2009 virus. Acta Virol 2012;56(2):149-51.

[94] Duesberg PH. The RNA of influenza virus. Proc Natl Acad Sci U S A 1968 Mar;59(3): 930-7.

[95] Pons M, Hirst GK. The single- and double-stranded RNA's and the proteins of incomplete influenza virus. Virology 1969 May;38(1):68-72.

[96] Bean WJ, Jr., Simpson RW. Transcriptase activity and genome composition of defective influenza virus. J Virol 1976 Apr;18(1):365-9.

[97] Crumpton WM, Dimmock NJ, Minor PD, Avery RJ. The RNAs of defective-interfering influenza virus. Virology 1978 Oct 15;90(2):370-3.

[98] Rohde W, Boschek CB, Harms E, Rott R, Scholtissek C. Characterization of virus-like particles produced by an influenza A virus. Arch Virol 1979;62(4):291-302.

[99] Kaverin NV, Kolomietz LI, Rudneva IA. Incomplete influenza virus: partial functional complementation as revealed by hemadsorbing cell count test. J Virol 1980 May; 34(2):506-11.

[100] Odagiri T, Tashiro M. Segment-specific noncoding sequences of the influenza virus genome RNA are involved in the specific competition between defective interfering RNA and its progenitor RNA segment at the virion assembly step. J Virol 1997 Mar; 71(3):2138-45.

[101] Priore SF, Moss WN, Turner DH. Influenza A virus coding regions exhibit host-specific global ordered RNA structure. PLoS One 2012;7(4):e35989.

Chapter 10

Vaccines and Antiviral Agents

Hongxuan He

Additional information is available at the end of the chapter

http://dx.doi.org/10.5772/56866

1. Introduction

There has been a long history of the battle between viral diseases and the mankind. The arms at our disposal against the virus invasion are continuously expanding its inventory. Most of them fall into the category of vaccines and antiviral and each of the two kinds of viral diseases intervention agents has its own advantages and limitations.

2. Vaccine

A vaccine is a biological preparation that improves immunity to a particular disease. A vaccine typically contains an agent that resembles a disease-causing microorganism, and is often made from weakened or killed forms of the microbe, its toxins or one of its surface proteins. The agent stimulates the body's immune system to recognize the agent as foreign, destroy it, and "remember" it, so that the immune system can more easily recognize and destroy any of these microorganisms that it later encounters.

Vaccines are dead or inactivated organisms or purified products derived from them.

There are several types of vaccines in use. These represent different strategies used to try to reduce risk of illness, while retaining the ability to induce a beneficial immune response.

2.1. Inactivated

Some vaccines contain killed, but previously virulent, micro-organisms that have been destroyed with chemicals, heat, radioactivity or antibiotics. Examples are the influenza vaccine, cholera vaccine,bubonic plague vaccine, polio vaccine, hepatitis A vaccine, and rabies vaccine.

2.2. Attenuated

Some vaccines contain live, attenuated microorganisms. Many of these are live viruses that have been cultivated under conditions that disable their virulent properties, or which use closely related but less dangerous organisms to produce a broad immune response [1]. They typically provoke more durable immunological responses and are the preferred type for healthy adults. Examples include the viral diseases yellow fever, measles, rubella, and mumps. Attenuated vaccines have some advantages and disadvantages. They have the capacity of transient growth so they give prolonged protection, and no booster dose is required. But they may get reverted to the virulent form and cause the disease.

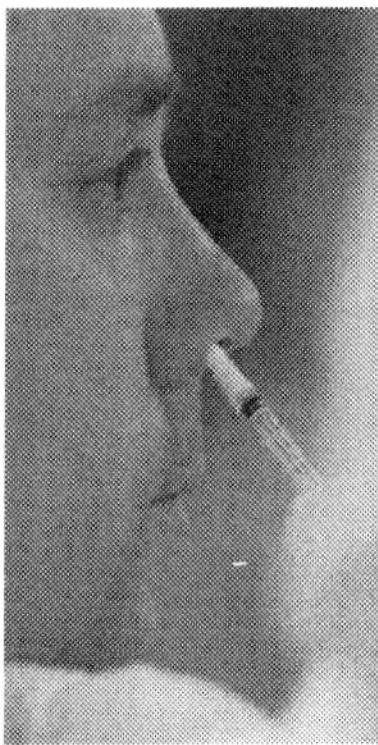

Figure 1. H1N1 flu nasal spray as an example of attenuated vaccine

2.3. Subunit

Protein subunit- rather than introducing an inactivated or attenuated micro-organism to an immune system (which would constitute a "whole-agent" vaccine), a fragment of it can create an immune response. Examples include the subunit vaccine against Hepatitis B virus that is composed of only the surface proteins of the virus (previously extracted from the blood stream of chronically infected patients, but now produced by recombination of the viral genes into yeast), the virus like particle (VLP) vaccine against human papillomavirus (HPV) that is composed of the viral major capsidprotein, and the hemagglutinin and neuraminidase subunits of the influenza virus. One method of production involves isolation of a specific protein from a virus and administering this by itself. A weakness of this technique is that isolated proteins can be denatured and will then bind to different antibodies than the proteins in the virus. A second method of subunit vaccine is the recombinant vaccine, which involves putting a protein gene from the targeted virus into another virus. The second virus will express the protein, but will not present a risk to the injector. This is the type of vaccine currently in use for hepatitis, and it is experimentally popular, being used to try to develop new vaccines for difficult to vaccinate viruses such as Ebola and HIV.

2.4. DNA Vaccine

In the past decade and a half, the DNA vaccine concept has been tested and applied against various pathogens and tumor antigens [2]. The optimized gene sequence of interest is delivered to the skin, subcutaneum or muscle by one of several delivery methods [3]. The expression of plasmid-encoded genes will produce foreign antigens and elicits immunological response. Until now, four DNA vaccine products have been approved, all in the area of veterinary medicine [4].

Vaccines are very effective on stable viruses, but are of limited use in treating a patient who has already been infected. It is also difficult to successfully deploy them against rapidly mutating viruses, such as influenza (the vaccine for which is updated every year) and HIV [5]. Antiviral drugs are particularly useful in these cases.

Vaccine Target	Product Name	Company involved	Date licensed and country	Target organism	Benefits
West Nile virus	West Nile Innovator	Centers for Disease Control and Prevention and Fort Dodge Laboratories	2005 USA	Horses	Protects against West Nile virus infection
Infectious Haematopoietic necrosis virus	Apex-IHN	Novartis	2005 Canada	Salmon	Improves animal welfare, increase food quality and quantity
Growth hormone releasing hormone	LifeTide-SWS	VGX Animal Health	2007 Austrilia	Swine and food Animals	Increases the number of piglets weaned in breeding sows
Melanoma	Canine Melanoma Vaccine	Merial, Memorial Sloan-Kettering Cancer Center and The Animal Medical Center of New York	2007 USA, conditional license	Dogs	Treats aggressive forms of cancer of the mouth, nail bed, foot pad or other areas as an alternative to radiation and surgery

Table 1. Current licensed DNA therapies (Adapted from Kutzler MA & Weiner et. al)

3. Antiviral agent

Antiviral drugs are a class of medication used specifically for treating viral infections. Like antibiotics for bacteria, specific antivirals are used for specific viruses. Unlike most antibiotics, antiviral drugs do not destroy their target pathogen; instead they inhibit their development [6].

Most of the antiviral drugs now available are designed to help deal with HIV, herpes viruses (best known for causing cold sores and genital herpes, but actually the cause of a wide
range of other diseases, such as chicken pox), the hepatitis B and C viruses, which can cause
liver cancer, and influenza A and B viruses. Researchers are working to extend the range of
antivirals to other families of pathogens.

Designing safe and effective antiviral drugs is difficult, because viruses use the host's cells to
replicate. This makes it difficult to find targets for the drug that would interfere with the virus without harming the host organism's cells. Moreover, the major difficulty in developing
vaccines and anti-viral drugs is due to viral variation.

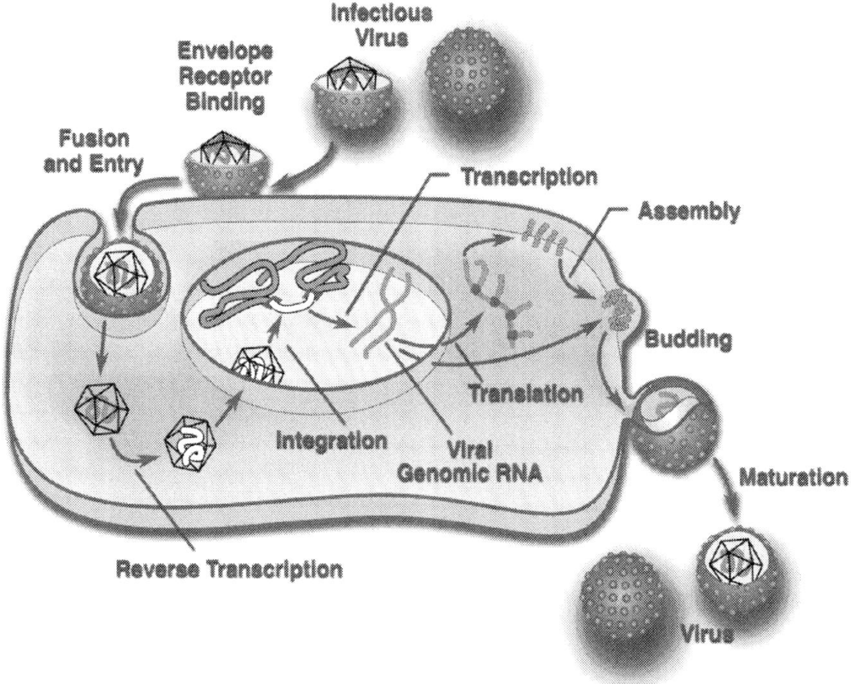

Figure 2. Virus life cycle and targets of antivirals

3.1. Before cell entry

One anti-viral strategy is to interfere with the ability of a virus to infiltrate a target cell. The
virus must go through a sequence of steps to do this, beginning with binding to a specific
"receptor" molecule on the surface of the host cell and ending with the virus "uncoating" inside the cell and releasing its contents. Viruses that have a lipid envelope must also fuse

their envelope with the target cell, or with a vesicle that transports them into the cell, before they can uncoat [7].

3.1.1. Entry inhibitor

A very early stage of viral infection is viral entry, when the virus attaches to and enters the host cell [8]. A number of "entry-inhibiting" or "entry-blocking" drugs are being developed to fight HIV. HIV most heavily targets the immune system's white blood cells known as "helper T cells", and identifies these target cells through T-cell surface receptors designated "CD4" and "CCR5". Attempts to interfere with the binding of HIV with the CD4 receptor have failed to stop HIV from infecting helper T cells, but research continues on trying to interfere with the binding of HIV to the CCR5 receptor in hopes that it will be more effective.

3.1.2. Uncoating inhibitor

Inhibitors of uncoating have also been investigated.

Amantadine and rimantadine, have been introduced to combat influenza. These agents act on penetration/uncoating. They are M2 inhibitors which block the ion channel formed by the M2 protein that spans the viral membrane. The influenza virus enters its host cell by receptor-mediated endocytosis. Thereafter, acidification of the endocytotic vesicles is required for the dissociation of the M1 protein from the ribonucleoprotein complexes. Only then are the ribonucleoprotein particles imported into the nucleus via the nuclear pores. The hydrogen ions needed for acidification pass through the M2 channel. Amantadine and rimantadine block the channel [9].

3.2. During viral synthesis

A second approach is to target the processes that synthesize virus components after a virus invades a cell.

3.2.1. Reverse transcription

One way of doing this is to develop nucleotide or nucleoside analogues that look like the building blocks of RNA or DNA, but deactivate the enzymes that synthesize the RNA or DNA once the analogue is incorporated. This approach is more commonly associated with the inhibition of reverse transcriptase (RNA to DNA) than with "normal" transcriptase (DNA to RNA).

An improved knowledge of the action of reverse transcriptase has led to better nucleoside analogues to treat HIV infections. One of these drugs, lamivudine, has been approved to treat hepatitis B, which uses reverse transcriptase as part of its replication process. Researchers have gone further and developed inhibitors that do not look like nucleosides, but can still block reverse transcriptase.

Another target being considered for HIV antivirals include RNase H-which is a component of reverse transcriptase that splits the synthesized DNA from the original viral RNA.

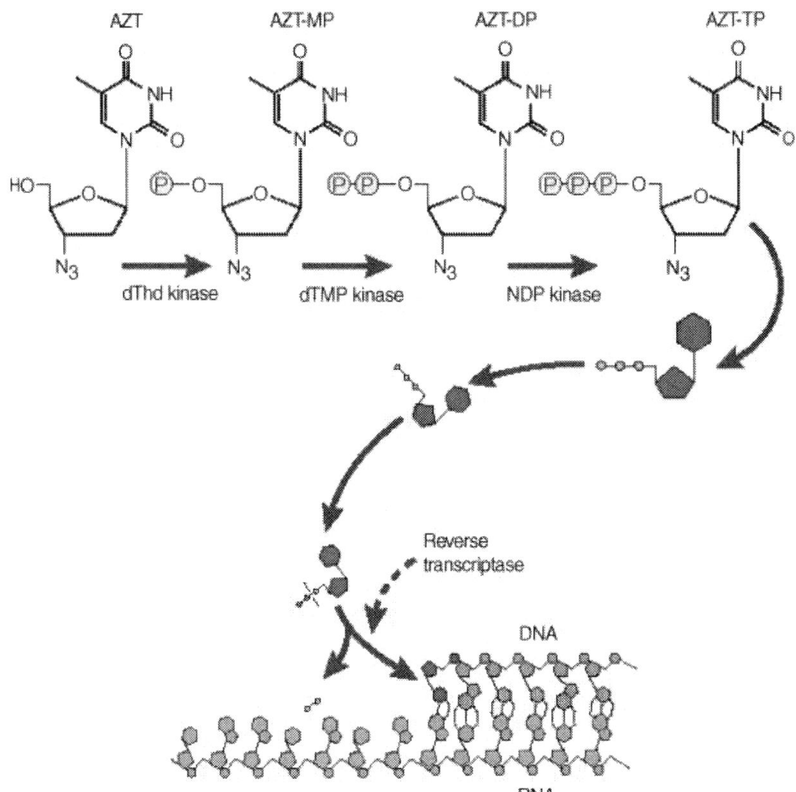

Figure 3. Example of the mechanisms of antivirals: Mechanism of action of azidothymidine (AZT). AZT needs to be phosphorylated, in three steps, to the triphospate form before it can interfere with the reverse transcriptase reaction

3.2.2. Integrase

Another target is integrase, which splices the synthesized DNA into the host cell genome. There appears to be no functional equivalent of the enzyme in human cells. The biochemical mechanism of integration of HIV DNA into the host cell genome involves a carefully defined sequence of DNA tailoring (3'-processing) and coupling (joining or integration) reactions [10]. In spite of some effort in this area targeted at the discovery of therapeutically useful inhibitors of this viral enzyme, there are no drugs for HIV/AIDS in clinical use where the mechanism of action is inhibition of HIV integrase. However there are several promising candidates in several classes of compounds, including nucleotides, dinucleotides, oligonucleotides and miscellaneous small molecules such as heterocyclic systems, natural products, diketo acids and sulfones, that have been discovered as inhibitors of HIV integrase.

3.2.3. Transcription

Once a virus genome becomes operational in a host cell, it then generates messenger RNA (mRNA) molecules that direct the synthesis of viral proteins. Production of mRNA is initiated by proteins known as transcription factors. Several antivirals are now being designed to block attachment of transcription factors to viral DNA. Kao et al. recently identified a compound called nucleozin via random screening, which was found to inhibit influenza by interacting with influenza NP. Nucleozin causes the NPs to aggregate abnormally, and consequently inhibits normal viral transcription, crippling the replication cycle by extension [11]. Examination of a nucleozin analogue revealed that the compound functions by binding to two copies of NP and forming abnormal dimers, causing the proteins to aggregate and preventing them from functioning normally. Nucleozin was also shown to inhibit influenza virus in vitro and in a mouse model, making it a promising candidate for a new antiviral drug.

3.2.4. Translation/antisense

Genomics has not only helped find targets for many antivirals, it has provided the basis for an entirely new type of drug, based on "antisense" molecules. These are segments of DNA or RNA that are designed as complementary molecule to critical sections of viral genomes, and the binding of these antisense segments to these target sections blocks the operation of those genomes. A phosphorothioate antisense drug named fomivirsen has been introduced, used to treat opportunistic eye infections in AIDS patients caused by cytomegalovirus, and other antisense antivirals are in development.

3.2.5. Translation/ribozymes

Yet another antiviral technique inspired by genomics is a set of drugs based on ribozymes, which are enzymes that will cut apart viral RNA or DNA at selected sites. In their natural course, ribozymes are used as part of the viral manufacturing sequence, but these synthetic ribozymes are designed to cut RNA and DNA at sites that will disable them.

A ribozyme antiviral to deal with hepatitis C has been suggested, and ribozyme antivirals are being developed to deal with HIV. An interesting variation of this idea is the use of genetically modified cells that can produce custom-tailored ribozymes. This is part of a broader effort to create genetically modified cells that can be injected into a host to attack pathogens by generating specialized proteins that block viral replication at various phases of the viral life cycle [12].

3.2.6. Protease inhibitors

Some viruses include an enzyme known as a protease that cuts viral protein chains apart so they can be assembled into their final configuration, such as Saquinavir (Figure 4). HIV includes a protease, and so considerable research has been performed to find "protease inhibitors" to attack HIV at that phase of its life cycle. Protease inhibitors became available in the 1990s and have proven effective, though they can have unusual side effects, for example causing fat to build up in unusual places. Improved protease inhibitors are now in development [13].

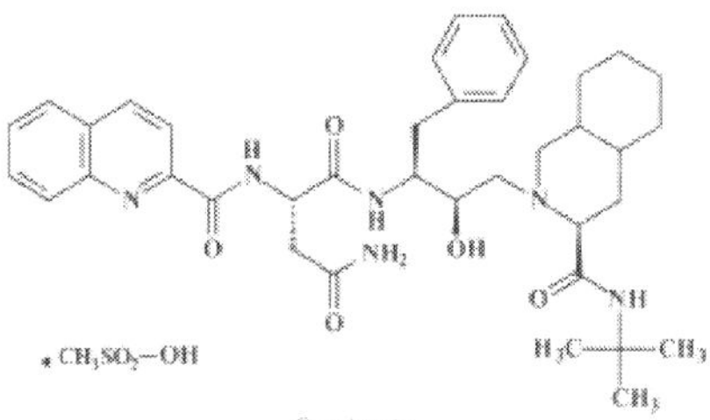

Figure 4. Protease inhibitor antiviral Saquinavir.

Structure: cis-N-*tert*-butyl-decahydro-2-[2(R)-hydroxy-4-phenyl-3(S)-[[N-2-quinolylcarbonyl-l-asparaginyl]-amino]butyl]-(4aS-8aS)-isoquinoline-3(S)-carboxamide methane sulfonate, hard gel capsules, Invirase®, also available as soft gelatin capsules (Fortovase®).

Activity spectrum: HIV (types 1 and 2).

Mechanism of action: transition-state, hydroxyethylene-based, peptidomimetic inhibitor of HIV protease.

3.3. Release phase

The final stage in the life cycle of a virus is the release of completed viruses from the host cell, and this step has also been targeted by antiviral drug developers. Two drugs named zanamivir (Relenza) and oseltamivir (Tamiflu) that have been recently introduced to treat influenza prevent the release of viral particles by blocking a molecule named neuraminidase that is found on the surface of flu viruses, and also seems to be constant across a wide range of flu strains [14].

3.4. Considerations in the clinical development of antiviral agents

A total of 37 antiviral compounds (not including interferons or immunoglobulins) have momentarily been licensed for the treatment of HIV, HBV, herpesvirus, influenza virus and/or HCV infections [15]. In the preceding sections these compounds have been discussed from the following viewpoints: chemical structure, activity spectrum, mechanism of action, principal clinical indication(s). Other points that need to be considered before the full clinical potential of any given drug could be appreciated, are: (i) duration of treatment, (ii) single-versus multiple-drug therapy, (iii) pharmacokinetics, (iv) drug interactions, (v) toxic side effects and (vi) development of resistance. A particular issue that may be important in the clinical setting is whether the listed anti-HIV agents would be equally suited for the treatment of HIV-2 and HIV-1 infections.

As to the duration of treatment, this may vary from a few days (HSV, VZV, influenza virus infections) to several months or years (HIV, HBV and HCV infections), depending on whether we are dealing with an acute (primary (i.e. influenza) or recurrent (i.e. HSV, VZV) infection or chronic, persistent (i.e. HIV, HBV, HCV) infection. For HIV infections it is still being evaluated whether long-term treatment can be interrupted, without loss of benefit (or increased benefit) to the patient (structured treatment interruption, STI) [16].

While the short-term treatment (5–7 days) of HSV, VZV and influenza virus infections, and even the more prolonged treatment of CMV infections, can be based on single-drug therapy, for the long-term treatment of HIV infections combination of several drugs in a triple-drug cocktail (also referred to as HAART for 'highly active anti-retroviral therapy') has become the standard procedure, and likewise, the long-term treatment of HBV infections may in the future also evolve from single- to dual- or triple-drug therapy [17].

Pharmacokinetic parameters to be addressed, when evaluating the therapeutic potential, include bioavailability (upon either topical, oral or parenteral administration), plasma protein binding affinity, distribution through the organism (penetration into the CNS, when this is needed), metabolism through the liver (i.e. cytochrome P-450 drug-metabolizing enzymes) and elimination through the kidney. Particularly when concocting the multiple-drug combinations for the treatment of HIV infection, possible drug–drug interactions should be taken into account: i.e. some compounds act as P-450 inhibitors and others as P-450 inducers, and this may greatly influence the plasma drug levels achieved, especially in the case of NNRTIs and PIs [18].

Toxic side effects, both short and long-term, must be considered when the drugs have to be administered for a prolonged period, as in the treatment of HIV infections. These side effects may seriously compromise compliance (adherence to drug intake), and could, at least in part, be circumvented by a reduction of the pill burden to, ideally, once-daily dosing.

Finally, resistance development may be an important issue, again for those compounds that have to be taken for a prolonged period, as is generally the case for most of the NRTIs, NNRTIs and PIs currently used in the treatment of HIV infections. Yet, the nucleoside phosphonate analogues (NtRTIs) tenofovir and adefovir do not readily or rapidly lead to resistance development, even after more than 1 year of therapy (for HIV and HBV, respectively). Resistance has been noted with HBV against lamivudine after long-term therapy (>6 months), but, if resistant to lamivudine, HBV infections remain amenable to treatment with adefovir dipivoxil. As has been occasionally observed in immunosuppressed patients, HSV may develop resistance to acyclovir, and CMV to ganciclovir, but, if based on ACV TK or CMV PK deficiency, these resistant viruses remain amenable to treatment with foscarnet and/or cidofovir [19]. In immunocompetent patients, treated for an acute or episodic HSV, VZV or influenza virus infection, short-term therapy is unlikely to engender any resistance problems

The evolution of viral vaccines from the time of Jennerian prophylaxis to today's recombinant technology has been a continuing story of success. From the relatively crude or "first generation" vaccines for smallpox, rabies, and yellow fever followed a second and third generation of improved or new viral vaccines. The application of techniques for attenuating, inactivating, and partially purifying candidate viruses yielded safe, effective vaccines against influenza, poliomyelitis, measles, mumps, and rubella. With the advent of effective national immunization programs in the United States and other areas of the world to promote wide scale use of these vaccines, we have seen a dramatic decrease in incidence of the viral infections of children. The new biotechnology serves as the cornerstone for a fourth generation of vaccines and has already provided a licensed recombinant yeast human hepatitis B vaccine. The prospects for a wide spectrum of new or improved vaccines are highly encouraging, not only because of the recent technical advances but also because vaccine development has been recognized as a priority area of research. Under the National Institute of Allergy and Infectious Diseases' Program for Accelerated Development of New Vaccines, support is being provided for developmental vaccine studies with hepatitis A and B, influenza A and B, rabies, rotavirus, varicella, and respiratory syncytial virus. The outlook for antivirals is equally optimistic. The same technologies that have provided greater insight into the genetics and molecular biology of viruses and hence the means to fashion subunit or even synthetic vaccines have yielded data that can be applied to successful development of targeted antiviral compounds.

Author details

Hongxuan He*

Address all correspondence to: hehx@ioz.ac.cn

National Research Center for Wildlife Borne Diseases, Institute of Zoology, Chinese Academy of Sciences, Beijing, PR China

References

[1] Steel J, et al. (2009) Live attenuated influenza viruses containing NS1 truncations as vaccine candidates against H5N1 highly pathogenic avian influenza. Journal of virology 83[4]:1742-1753.

[2] Kutzler MA & Weiner DB (2008) DNA vaccines: ready for prime time? Nature reviews. Genetics 9[10]:776-788.

[3] Ramakrishna L, Anand KK, Mohankumar KM, & Ranga U (2004) Codon optimization of the tat antigen of human immunodeficiency virus type 1 generates strong immune responses in mice following genetic immunization. Journal of virology 78[17]: 9174-9189.

[4]   Tacket CO, et al. (1999) Phase 1 safety and immune response studies of a DNA vaccine encoding hepatitis B surface antigen delivered by a gene delivery device. Vaccine 17[22]:2826-2829.

[5]   Roose K, Fiers W, & Saelens X (2009) Pandemic preparedness: toward a universal influenza vaccine. Drug news & perspectives 22[2]:80-92.

[6]   De Clercq E (2004) Antiviral drugs in current clinical use. Journal of clinical virology : the official publication of the Pan American Society for Clinical Virology 30[2]: 115-133.

[7]   Teissier E, Penin F, & Pecheur EI (2011) Targeting cell entry of enveloped viruses as an antiviral strategy. Molecules 16[1]:221-250.

[8]   Doms RW & Moore JP (2000) HIV-1 membrane fusion: targets of opportunity. The Journal of cell biology 151[2]:F9-14.

[9]   Hay AJ, Wolstenholme AJ, Skehel JJ, & Smith MH (1985) The molecular basis of the specific anti-influenza action of amantadine. The EMBO journal 4[11]:3021-3024.

[10]   Mathe C & Nair V (1999) Potential inhibitors of HIV integrase. Nucleosides & nucleotides 18(4-5):681-682.

[11]   Kao RY, et al. (2010) Identification of influenza A nucleoprotein as an antiviral target. Nature biotechnology 28[6]:600-605.

[12]   Olmstead AD, Knecht W, Lazarov I, Dixit SB, & Jean F (2012) Human subtilase SKI-1/S1P is a master regulator of the HCV Lifecycle and a potential host cell target for developing indirect-acting antiviral agents. PLoS pathogens 8[1]:e1002468.

[13]   Lamarre D, et al. (2003) An NS3 protease inhibitor with antiviral effects in humans infected with hepatitis C virus. Nature 426(6963):186-189.

[14]   Moscona A (2005) Neuraminidase inhibitors for influenza. The New England journal of medicine 353[13]:1363-1373.

[15]   B. Kronenberger SZ (2012) Antiviral tagets in HCV. Chronic Hepatitis C Virus, ed Shiffman ML (Springer, New York), pp 203-225.

[16]   Mancini E, Castiglione F, Bernaschi M, de Luca A, & Sloot PM (2012) HIV reservoirs and immune surveillance evasion cause the failure of structured treatment interruptions: a computational study. PloS one 7[4]:e36108.

[17]   Regidor DL, et al. (2011) Effect of highly active antiretroviral therapy on biomarkers of B-lymphocyte activation and inflammation. AIDS 25[3]:303-314.

[18]   Flentge CA, et al. (2009) Synthesis and evaluation of inhibitors of cytochrome P450 3A (CYP3A) for pharmacokinetic enhancement of drugs. Bioorganic & medicinal chemistry letters 19[18]:5444-5448.

[19] Bacon TH, Levin MJ, Leary JJ, Sarisky RT, & Sutton D (2003) Herpes simplex virus resistance to acyclovir and penciclovir after two decades of antiviral therapy. Clinical microbiology reviews 16[1]:114-128.

Chapter 11

Viral Counter Defense X Antiviral Immunity in Plants: Mechanisms for Survival

Alessandra Tenório Costa, Juliana Pereira Bravo,
Rodrigo Kazuo Makiyama,
Alessandra Vasconcellos Nunes and Ivan G. Maia

Additional information is available at the end of the chapter

http://dx.doi.org/10.5772/56253

1. Introduction

The history of plant virology dates to the late 19th century, when Iwanowski and Beijerinck, who were investigating the cause of a mysterious disease of tobacco, independently described an unusual agent that caused tobacco mosaic disease. This agent was later named *Tobacco mosaic virus* (TMV) [1]. During this period, viruses including *Potato virus X* (PVX), *Potato virus Y* (PVY) and *Lettuce mosaic virus* (LMV) were described. These viruses could be distinguished based on their transmission and method of disease induction. In addition, numerous techniques for the study of viruses were developed.

Viruses are among the most agriculturally important groups of plant pathogens, causing serious economic losses in many major crops by reducing yield and quality. A virus can be defined as a set of one or more nucleic acid template molecules, often encased in a protective coat of protein or lipoprotein, which is able to organise its own replication only within suitable host cells [1]. Because the genetic information encoded by viral genomes is limited, viruses depend entirely on host cells to replicate their genome and produce infectious progeny. Both plant and animal viruses can be classified according to the type of nucleic acid that makes up their genome. In plants, the vast majority of viruses have positive-sense (+) RNA genomes (i.e., the RNA genome has the same polarity as cellular mRNA), although negative-sense (−) RNA and double-stranded RNA viral genomes also exist. Other plant viruses have DNA genomes; the DNA can be double-stranded (caulimoviruses) or single-stranded (geminiviruses) [1, 2]. Cell-to-cell movement of plant viruses occurs through cytoplasmic "bridges" between cells

called plasmodesmata, and viruses are able to move systemically throughout plants via the phloem [1].

Viruses depend on other organisms (vectors) to transmit them from diseased to healthy plants. These vectors are often sap-sucking insects such as aphids, thrips, whiteflies, leaf-feeding beetles, plant-feeding mites, soil-inhabiting nematodes or fungal pathogens. Some viruses can be mechanically transmitted, on pruning knives or gardeners hands; or by grafting material, and a relatively small number of species can pass through infected seed [3]. Viruses use a variety of strategies that frequently induce disease in the plants they infect. Different viruses induce distinct diseases, and this can be true even for different strains of the same virus. Virus infection can profoundly alter the physiology of the host due to the interaction with cellular components. In plants, the severity of viral diseases varies considerably depending on the host genotype, the stage of infection and the environmental conditions. Diseases caused by viruses can vary broadly in intensity, from very mild symptoms observed in tolerant plants up to very severe symptoms and plant death [1, 4, 5].

Each plant virus encodes an average of 4-10 proteins necessary to coordinate the complex biochemical and intermolecular interactions required for viral infection cycles. The cycle of infection includes viral genome replication, cell-to-cell and systemic movement and transmission [6]. For efficient viral infection to occur, viral proteins must be able to interact with factors in the host cell, thereby manipulating metabolic pathways and coordinating biochemical interactions that promote infection. Thus, during co-evolution between viruses and their hosts, a variety of complex interactions have developed that involve several distinct mechanisms of plant defence and virus attack.

During evolution, plants have developed diverse defence mechanisms that are activated during viral infection. One of these is the hypersensitive response (HR), which activates initial defence responses that prevent the infection from spreading further and then kills the cells within the infected zone. The onset of a second mechanism, systemic acquired resistance (SAR), then protects the plant together with HR against new attacks by the same pathogen. SAR is induced by a variety of agents after initial infection and can provide resistance to a wide range of pathogens for days [7-10]. The HR and SAR responses are accompanied by changes in gene expression that include the production of pathogenesis-related (PR) proteins and of several proteins involved in cell signalling [11].

Plants also possess other antiviral defence mechanisms such as RNA silencing, a remarkable type of gene regulation based on sequence-specific targeting and degradation of RNA [12], and the more recently described ubiquitin/26S proteasome system (UPS), which plays a central role in the degradation of proteins. The latter system is involved in almost all phases of the defence mechanisms of plants, regardless of the type of pathogen [13]. In addition to its proteolytic activity against ubiquitinated pathogen proteins, which directs their degradation by the 20S proteasome, the degradation of viral RNA can also occur via the ribonuclease activity of the 20S proteasome [14, 15]. While the proteasome as a structure, and RNA silencing as a mechanism, are two conserved features among eukaryotes, several lines of evidence suggest that the proteasome-linked RNase activity is most likely not directly related to RNA silencing. The selective degradation of viral RNAs by the 20S complex can represent an alternative pathway of host defence that occurs in parallel to RNA silencing and reinforces cellular antiviral defence in plants [14].

Over time, the strategies used by the virus to overcome these elaborate host defences can lead to a number of fundamental changes in the plant's physiology. Such changes, including structural modifications in the host cell, may give rise to intranuclear inclusions of various types and may affect the nucleolus or the size and shape of the nucleus. Within the mitochondria, abnormal membrane systems may develop [16, 17]. In plants infected with *Turnip yellow mosaic virus* (TYMV - Tymovirus), abnormalities such as clumping of chloroplasts and abnormal size and number of starch grains in leaf cells may occur, and small vesicles near the periphery and chloroplasts may become greatly enlarged and filled with starch grains [18]. These abnormalities were also observed in squash infected by the *Zucchini yellow mosaic potyvirus* (ZYMV - Potyvirus) [17]. In plant cell walls, three types of abnormality have been observed: abnormal thickening due to the deposition of callose near the edges of virus-induced lesions; cell wall protrusions involving the plasmodesmata (these protrusions may have one or more canals and may be quite short or of considerable length); paramural bodies, which are depositions of electron-dense material between the cell wall and the plasma membrane, may appear and extend over substantial areas of the cell wall, or be limited in extent occurring in association with plasmodesmata. Moreover, in the cytoplasm of an infected cell, virus particles may accumulate in sufficient numbers to form three-dimensional crystalline arrays. The ability to form crystals within the host cell cytoplasm depends on properties of the virus itself, and is not related to the overall concentration reached in the tissue or to the ability of the purified virus to form crystals [1].

Intriguingly, in carrot plants (*Daucus carota* L.) infected by *Cucumber mosaic virus* (CMV) some cytological and physiological changes were observed due to alterations in various host metabolites. With respect to cytological changes, scattered metaphase was observed in the diseased plant cells. The mitotic index of the diseased cells was decreased, while the nucleus / cytoplasm ratio was increased. Chromatin bridges were also observed at anaphase I and II. Physiological changes resulting in decreased carbon, nitrogen and protein content and increased phosphorous content of the virally infected plants have been observed [19].

Other viral counterattack mechanisms involve changes in the plant cell cycle. In plants, as in all eukaryotes, the four phases of the mitotic cell cycle (G1, S, G2 and M) are conserved. During development, plant cells leave the cell division cycle, and in mature plants, DNA replication and the corresponding enzymes are confined to meristematic tissues [20]. Geminiviruses are good models for the study of the relationship between the cell cycle and viral DNA replication because they replicate in differentiated cells, such as mature cells of the leaves, stems and roots, in which most of the cellular factors required for viral DNA replication are normally absent. These cells have left the cell division cycle and no longer contain detectable levels of plant DNA replication enzymes necessary for geminivirus replication [21]. Due to the requirement for cellular factors, geminiviral DNA replication must be coupled to a special state of the infected cell, suggesting that the virus may have evolved mechanisms that affect the expression of cellular genes involved in S-phase progression and G1/S transition [22]. One such a mechanism involved in regulating changes of the host cell cycle appears to be the inactivation of the retinoblastoma protein (pRb), which negatively regulates the G1/S transition in cells. The Rep protein, which is encoded by all geminiviruses and is the only viral protein necessary for viral DNA replication, has been found to induce expression and also interacts with the host "proliferating cell nuclear antigen" (PCNA), an auxiliary protein of DNA polymerases required during replication and repair in non-dividing plant cells. This observation suggests that Rep protein can provide the necessary stimulus to induce the dedifferentiation process. Mechanisms other than sequestering plant pRb most likely contribute to the multiple effects of geminivirus proteins on cellular gene expression, cell growth control and cellular DNA replication [21, 22].

Ultimately, RNA silencing suppressor (RSS) proteins are able to block or attenuate plant host defence mechanisms, particularly post-transcriptional gene silencing and efficiently inhibit host antiviral responses by interacting with the key components of cellular silencing machinery, often mimicking their normal cellular functions [23]. Viral suppressors of RNA silencing have been identified from almost all plant virus genera; these VSRs are surprisingly diverse, exhibiting no obvious sequence similarities. Most identified VSRs are multifunctional; in addition to being RNA silencing suppressors, they often perform essential roles, functioning as coat proteins, replicases, movement proteins, helper components for viral transmission, proteases or transcriptional regulators. The first viral RNA silencing suppressor identified was the helper component-proteinase (HC-Pro) of potyviruses [24]; currently, many different suppressors are known.

The aim of this chapter review is to discuss the current status of knowledge regarding various components of host silencing machinery and viral suppression. We also pretend to describe how the defence response in plants is directed against the virus and, in particular, how the virus can sidetrack the host's defence response. Relevant topics on the molecular bases of the induction and suppression of the RNA silencing mechanism, as well as the applications and perspectives of the use of silencing suppression in plant biotechnology, will be emphasised.

2. The importance of silencing pathways

A major breakthrough in the history of biology was the discovery of an RNA-induced silencing response in the nematode *Caenorhabditis elegans* [25]. Prior to that discovery, this RNA-induced silencing had been interpreted as a defence mechanism against viruses and other invading nucleic acids; the discovery of endogenous small regulatory RNAs in many species led to the realisation that gene silencing is a fundamental genetic regulatory mechanism in eukaryotic organisms [26].

Several studies in genetics, biochemistry and the development of novel techniques in molecular biology have helped identify different components of the RNA silencing machinery and have confirmed that RNA interference (RNAi), co-suppression and virus-induced gene silencing share mechanistic similarities [27]. RNA silencing is a conserved eukaryotic pathway involved in the suppression of gene expression via sequence-specific interactions that are mediated by 21–23 nucleotide (nt) RNA molecules [28]. Organisms utilise RNA silencing for three purposes: 1) creating and maintaining heterochromatin at repetitive DNA and transposons; 2) regulating development, stress responses and other endogenous regulatory functions; 3) defending against viral and bacterial infections [29]. Silencing is utilised in developmental pathways and in cellular differentiation to repress genes whose products are not required at specific stages of development or in specific cell types; in plants silencing is also used to respond to internal and external stresses by changing the expression of specific genes involved in the response. In some situations, tissue- or cell-specific silencing is desirable.

Initially discovered in transgenic plants, especially in those created to acquire virus resistance, RNA silencing is now also believed to be responsible for various epigenetic effects and their maintenance; the silencing of transgenic loci in plants most likely results from the activation of defence mechanisms. A number of silencing studies with different plant systems have explored transgenes as indicators of silencing pathways; these works have received important attention in part because silencing reduces the reliability of transgenic approaches in biotechnology of several agriculturally important cultures.

RNA silencing can be classified into two major categories: transcriptional gene silencing (TGS) and post-transcriptional gene silencing (PTGS). TGS is defined as inhibition of transcription, whereas PTGS involves the post-transcriptional degradation of RNA species but does not affect the transcription rate [30]. TGS occurs when double-stranded RNA molecules (dsRNA) containing promoter sequences are present, and PTGS occurs when dsRNA comprise open reading frames (ORF). Together, TGS and PTGS depend on small interfering RNAs (siRNA) or microRNAs (miRNA) that are produced from dsRNA precursors [31]. Because RNA silencing, mainly PTGS, also contributes to antiviral immunity in plants, fungi and invertebrates, it is an important part of innate immunity [32]. The silencing may persist over many cell divisions or plant generations [33].

The basic steps in common to all RNA silencing pathways (Figure 1) include: (i) formation of a dsRNA; (ii) processing of the dsRNA by an RNase III–like enzyme named Dicer (DCL) to shorter (20-30 nucleotide) dsRNA duplexes, the so called siRNA (iii) binding of the small RNA duplexes to a protein from the Argonaute (AGO) family; and (iv) targeting of the RNA-induced complex to mRNA (or DNA) guided by strand complementary to the small dsRNA, which is called the guide [34]. At present, there is good evidence for the existence of at least four different types of RNA silencing pathways in plants. These pathways involve different types of small RNA molecules, specially siRNA and miRNA. Heterochromatin-associated siRNA (hc-siRNA), trans-acting siRNA (tasiRNA) and viral siRNA (primary and secondary) are also important in silencing [35]. A better understanding of silencing pathways is very important because of the potential usefulness of silencing as a powerful tool for gene function studies and crop improvement.

2.1. Post-transcriptional gene silencing

As already mentioned, PTGS is essential to antiviral immunity in plants, thus our focus will be concentrated in this phenomenon. PTGS was first observed in 1990, and initially referred to as 'co- suppression', it was first discovered in transgenic petunia plants in which the introduction of the gene for chalcone synthase created a block in anthocyanin biosynthesis, resulting in variegated pigmentation [36].

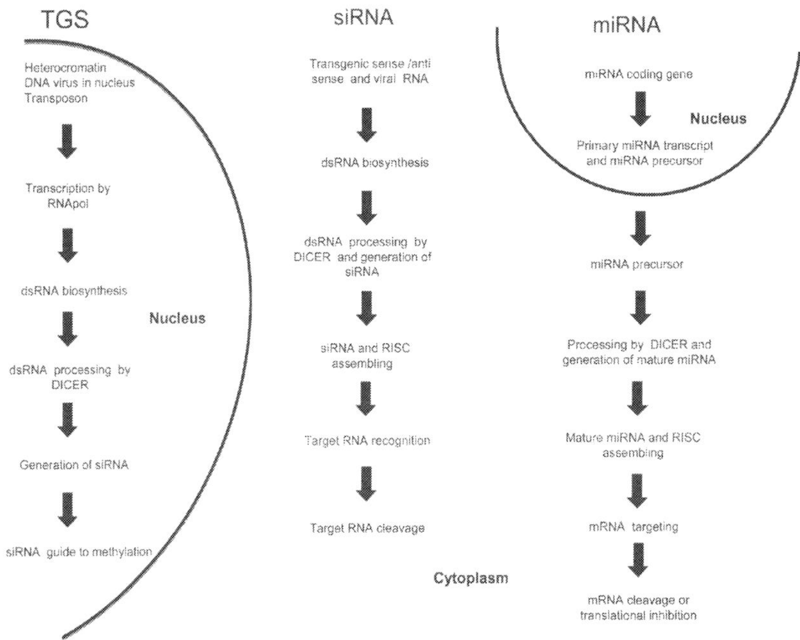

Figure 1. Overview of the transcriptional gene silencing (TGS), siRNA and miRNA pathways. The important steps of each pathway are depicted.

PTGS was also detected in transgenic plants engineered for virus resistance, and associated with the phenomenon of recovery of a host plant from viral infection. This recovery process was soon understood to be associated with the plant's inherent RNA silencing mechanism, which is an evolutionarily conserved antiviral system. The first report related to virus-induced gene silencing was published as long ago as 1929 by McKinney, who reported that tobacco plants infected with the "green" strain of TMV were protected against infection by a closely related second virus (TMV "yellow" strain). This phenomenon was later described as "cross-protection" [37]. However, a remarkable explanation of cross-protection was provided when it was shown that virus infection prevents infection by a second virus if the two viruses possess homologous sequences. Importantly, this viral-RNA-mediated cross-protection was function-ally equivalent to post-transcriptional gene silencing [38].

Viruses are not only targets of transgene induced RNA silencing but also elicit silencing themselves. Transgenic plants expressing a truncated version of the coat protein of *Tobacco etch virus* (TEV) were initially susceptible to infection and showed symptoms. However, a few weeks after the transgenic plants recovered from the TEV infection, newly developed leaves were symptomless and virus-free. Remarkably, the recovered leaves were resistant against a second TEV infection but were susceptible to infection by the heterologous PVY [39].

PTGS is a mechanism that has been preserved among eukaryote kingdoms. It is a genetic regulatory mechanism that is involved in several processes including defence of the genome against mobile DNA elements, establishment of heterochromatin, control of plant and animal development, and downregulation of gene expression by specific cleavage and translational repression of target mRNAs that contain complementary sites to miRNAs or siRNAs [40].

The involvement of short RNAs in PTGS was discovered when ~25 nt RNA molecules with sequence homology to a transgene were detected only in plants in which the corresponding transgene was silenced [41]. These molecules are also generated against replicating viruses. They are loaded in an active silencing complex called RISC (RNA-induced silencing complex). RISC shows slight variability in composition from one RNA silencing pathway to the next and from species to species; however, all RISCs contain a guide RNA strand bound to an AGO protein. AGO imparts endonuclease activity to the RISC (the so called slicer activity) being responsible for target RNA cleavage.

During PTGS, RNA is degraded predominantly in the cytoplasm. The strongest evidence in support of this comes from study of PTGS-based degradation of RNA viruses that are expressed exclusively in the cytoplasm [42]. However, recent work has demonstrated that the enzymes involved in PTGS are localised in both the cytoplasm and the nucleus. Overall, the available data suggest that PTGS activity in plant cells occurs in both the cytoplasmic and nuclear compartments. Nuclear PTGS would allow regulation of a potentially larger set of endogenous targets that cannot be regulated through cytoplasmic PTGS [31].

Some aspects of genomic silencing remain unknown. For a more complete understanding of genomic silencing, supplementary approaches are needed. This is especially important because gene silencing has the potential for use as a potent sequence-specific gene inactivation tool in functional genomics and plant biotechnology.

3. RNA silencing suppression

RNA silencing is known to serve as a mechanism for plant defence against pathogens. To counteract this mechanism, viruses have evolved the ability to avoid or suppress RNA silencing. Using this strategy, viruses protect their genomes from degradation through the production of proteins that act as suppressors of RNA silencing. These viral proteins act through a variety of molecular mechanisms including, particularly, blockage of the intercellular and systemic spread of mobile small silencing RNA molecules. The ability of viruses to infect cells can have a profound impact on host endogenous RNA silencing regulatory pathways and can result in alterations in endogenous short RNA expression profile and gene expression [43]. A general overview of the RNA silencing pathway discriminating the different steps targeted by different VSR is provided in Figure 2.

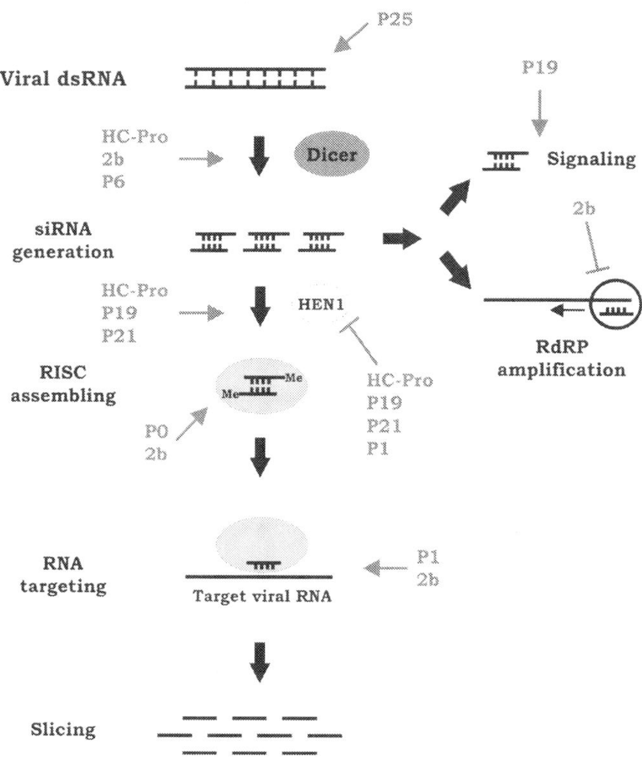

Figure 2. Schematic representation of the RNA silencing pathway triggered by double stranded RNA molecules (dsRNA) of virus origin. The steps of the pathway that are targeted by different viral suppressor proteins are depicted. Amplification of the silencing signal requires an RNA-dependent RNA polymerase (RdRP). Both strands of the small RNA duplexes (3′ termini) are methylated by HEN1 before RISC loading.

VSRs were first evident from the analysis of potyviral synergistic interactions with other viruses. It was shown that this synergism is the result of suppression of a host defence mechanism by the potyviral HC-Pro [44]. Subsequent studies established that HC-Pro is a suppressor of PTGS and provided a link between PTGS and antiviral defence [24, 45]. At the same time, analysis of a second viral protein, the 2b protein of CMV, identified this protein as a suppressor of PTGS in *Nicotiana benthamiana*. Intringuily, HC-Pro and 2b do not target the silencing mechanism in the same way; HC-Pro suppresses silencing in tissues in which it is already established, whereas the 2b protein only affects silencing initiation [45]. In 1999, a seminal study [46] showed definitively that suppression of RNA silencing is an anti-defence strategy commonly used by plant viruses with DNA or RNA genomes, such as Geminivirus (through protein AC2), Sobemovirus (through protein P1), Tombusvirus (by means of the "19K" protein) and others (Comovirus, Tobamovirus and Tobravirus).

Silencing suppression has also been documented in virus capable of infecting other organisms such as insect and fungus, including flock-house virus (FHV), cricket paralysis virus *Droso-phila* C virus, and *Cryphonectria parasitica* hypovirus [47]. In insect cells, functional similarities between the CMV 2b protein and the suppressor B2 protein from FHV were described. Deletion of the B2 ORF from FHV results in a drastic loss of virus accumulation in *Drosophila mela-nogaster* S2 cells, and this loss can be rescued by decreasing the cellular content of AGO2. B2 therefore seems to suppress the effect of the AGO2-dependent silencing response that normally restricts FHV accumulation [48].

Viral suppressors are considered to be of recent evolutionary origin, and they are often encoded by out-of-frame ORFs within more ancient genes. They are surprisingly diverse within and across kingdoms, with no obvious sequence homology [47]. VSRs are variously positioned on the viral genome and can be expressed using different strategies such as subgenomic RNAs, transcriptional read-through, ribosomal leaky-scanning or proteolytic maturation of polyproteins. Due to their molecular evolution, many of the viral suppressors identified to date are multifunctional, *i.e.*, besides being RNA silencing suppressors; they also perform essential roles in other steps of the viral life cycle [23].

Viruses have developed three efficient silencing suppression strategies to counter host antiviral immunity. The first is related to the inhibition of key components of RNA silencing pathways; the mechanisms involved in this strategy have already been well characterised for some viral proteins and will be described forward. A second silencing-suppression strategy, which will also be described later, involves the recruitment of endogenous negative regulators of RNA silencing. For example, yeast two-hybrid system results showed that HC-Pro is able to interact with the tobacco calmodulin-related rgs-CaM, a cellular suppressor of PTGS [49]. The third strategy relies on modification of the host transcriptome and is supported by studies of the geminivirus transcriptional-activator proteins (TrAPs), which have been identified as silencing suppressors (50).

VSRs are essential for viruses to replicate in host cells and to achieve systemic infection [47]. Although they do not share any obvious sequence or structural similarity across vi-ral families and groups, they have having been initially identified as pathogenicity deter-minants causing developmental defects in host plants, or as host range determinants [51, 52]. Viral silencing suppressors could cause developmental defects in plants because they act in miRNA and siRNA pathways that are mechanistically similar to developmental pathways. In a study involving transgenic expression of the HC-Pro of *Turnip mosaic vi-rus* (TuMV) in *Arabidopsis thaliana*, this protein was shown to alter the accumulation of miRNAs and to prevent the endonucleolytic cleavage of a number of their cellular tar-gets. This effect coincided with the occurrence of morphological defects resembling those of Dicer-like partial mutants called dcl-1. Remarkably, similar defects were observed upon TuMV infection, providing support for the idea that some of the symptoms caused by this virus are actually the result of alteration of miRNA-guided functions by HC-Pro [53]. Analyses of *N. benthamiana* and *Nicotiana tabacum* plants expressing well characterized silencing suppressors derived from 6 different virus genera: P1 of *Rice yellow mottle virus* (RYMV) and *Cocksfoot mottle virus* (CfMV), P19 of TBSV, P25 of PVX, HC-Pro of PVY, 2b of CMV and AC2 of *African cassava mosaic virus* (ACMV) were performed. Interestingly, some of the silencing suppressors promoted specific phenotypic effects. HC-Pro caused a severely distorted growth habit in both *Nicotiana* spp., while the P25 protein of PVX caused a specific flower malformation and an early senescence phenotype in the *N. benthamiana* plants, although not affecting *N. tabacum*. Moreover, P19 expressing *N. benthamiana* plants had blistered leaf epidermis, hairy and serrated leaves in one of the lines and occasional bending of the flower stalks while in *N. tabacum* caused occasional malformation of flowers [54].

From almost all virus genus that infect plants, over 50 individual VSRs have been identified, strongly suggesting that successful virus infection requires their expression [47, 55]. The data available in the literature suggest that almost all viruses encode at least one suppressor, but in many cases, viruses encode more than one [47]. Virus-encoded suppressors seem to have primordial RNA-binding properties and often show preference for a specific RNA molecule [56, 57].

Studies comparing the activities of three distinct RNA silencing suppressors (P19, P21 and HC-Pro) *in vitro* and *in vivo* showed that all three silencing suppressors are dsRNA-binding proteins that interact physically with siRNA duplexes [57]. These three suppressors inhibit siRNA-directed target RNA cleavage in a *D. melanogaster in vitro* RNA silencing system. Moreover, P19, HC-Pro and P21 uniformly inhibit the siRNA-initiated RISC assembly pathway by preventing RNA silencing initiator complex formation through siRNA sequestration. None of these silencing suppressors inhibit pre-assembled RISC activity *in vitro* or *in vivo*.

Suppression can vary in degree and spatial detail ranging, for example, from suppression in all tissues of all infected leaves to suppression only in the veins of newly emerged leaves. This suggests that different suppressors might be targeted to different parts of the gene silencing mechanism [58] such as viral RNA recognition, dicing, RISC assembly, RNA targeting and amplification [20]. For example, DCL function was indirectly blocked by the *Cauliflower mosaic virus* encoded P6 protein (viral translational transactivator protein). Transgenic P6 expression in *A. thaliana* reduced levels of DCL4-dependent 21-nt siRNAs (DCL4 converts non-coding RNA precursors into 21-nt tasiRNAs controlling developmental timing and organ polarity), similar to the effect of inactivating *A. thaliana* DRB4 (dsRNA-binding protein that physically and specifically interact with Dicers). Moreover, immunoprecipitation assays demonstrated that P6 physically interacts with DRB4 and that the ability of P6 to move within cellular compartments (nucleus and cytoplasm) was important for its silencing suppression activity [59].

Direct interaction between the V2 protein of *Tomato yellow leaf curl virus* (TYLCV) with SlSGS3, the tomato functional homolog of the *A. thaliana* SGS3 protein (AtSGS3), which is a coiled-coil protein involved in siRNA signal amplification, interferes with RNA silencing. Furthermore, the fact that a V2 mutant is unable to bind SGS3 loses its ability to suppress silencing indicates that the V2–SGS3 interaction may represent one of the key events in V2-induced RNA-silencing suppression in TYLCV-infected plant cells [60]. The HC-Pro protein can also act in a different manner as a viral suppressor of RNA silencing and might additionally be involved in sequestration of RNA duplexes. It was demonstrated that the FRNK amino acid motif in the central domain of HC-Pro is a probable point of contact involved in siRNA and miRNA duplex sequestration [61]. Mutations of FRNK (severe strains) to FINK (attenuated strains) caused attenuation of symptoms in squash leaves upon infection by *Zucchini yellow mosaic virus* (ZYMV). A decrease in miRNA accumulation was also observed. This raises the hypothesis that interactions of the FRNK box with different plant miRNAs directly influences their accumulation and endogenous regulatory functions, thereby contributing to symptom development.

Viral suppressors can interfere with the activity of methyltransferase HEN1. Studies performed with transgenic *A. thaliana* expressing P21 of BYV, P19 of tombusviruses, or P1/HC-Pro of TuMV demonstrated that, in addition to affect miRNA duplexes, these proteins interfere with short RNA stabilisation by blocking HEN1 methylation. Because miRNA precursors are supposedly cleaved in the nucleus, the fact that they are blocked for methylation by cytoplasmic viral suppressors could be explained in three ways: first, they may compete with HEN1 for substrate miRNA/miRNA* duplexes (sequestration by the suppressors could prevent HEN1 from interacting with duplexes or prevent HEN1 access to the 2′ OH of the 3′ terminal nucleotide); second, the viral silencing suppressors may bind directly to HEN1 and inhibit its activity, or interact with other factors required for HEN1 function; and third, viral suppressors may affect the subcellular localisation of HEN1 [62].

Diverse VSRs have been shown to bind AGO proteins. The first protein identified that binds AGO1 and AGO4 *in vivo* was 2b protein encoded by CMV. AGO1 is the major effector in both miRNA-directed and virus-induced RNA silencing. The 2b protein co-localises with AGO1 both in the cytoplasm and in the nucleolus [63]. The direct interaction of 2b protein with the PAZ and PIWI domains of AGO1, leading to the inhibition of its slicer activity, was verified by bimolecular fluorescence complementation (BiFC) and co-immunoprecipitation (Co-IP) assays [64]. The 2b protein also interacts directly with AGO4 in the nucleolus [65]. AGO4 binds to 24-nt long repeat-associated siRNA (ra-siRNA) to participate in RNA-directed DNA methylation (RdDM) [66]. In this case, 2b competes with AGO4 for binding to 24-nt ra-siRNA, suppressing the DNA methylation mediated by AGO4 [65]. However, the effects of inhibition by 2b of the RNA-dependent DNA methylation phenomena on virus replication and spread remain to be investigated.

RISC activity also undergoes interference by viral suppressors. In agro-infiltrated leaves of *N. benthamiana* containing a GFP transgene, the P0 protein encoded by *Beet western yellows virus* (BWYV) was identified as having strong silencing suppressor activity [67]. Further studies on two *A. thaliana* infecting poleroviruses revealed that P0 contains a conserved minimal F-box motif that interacts with homologues of S-phase kinase related protein 1 (SKP1), a core subunit of the multi-component SCF family of ubiquitin E3 ligases. Mutations in the F-box motif interrupt the interaction between P0 and a SKP1 homolog in *N. benthamiana*, causing a decrease in virus pathogenicity. In transgenic *A. thaliana* plants, expression of P0 caused severe developmental defects similar to those observed in mutants affected in miRNA pathways. Downregulation of a SKP1 homolog in *N. benthamiana* resulted in plant resistance to polerovirus infection. These results support a model in which P0 acts as an F-box protein that targets an essential component of the host post-transcriptional gene silencing machinery [68]. The results of subsequent investigation of the molecular mechanism by which P0 impairs PTGS showed that P0 expression does not affect the biogenesis of primary siRNAs, but it does affect their activity. Furthermore, in transformed *A. thaliana* plants P0 expression leads to various developmental abnormalities reminiscent of mutants affected in miRNA pathways. In this system, P0 expression is accompanied by enhanced levels of several miRNA-target transcripts, suggesting that P0 acts at the level of RISC. It was also revealed that P0 physically interacts with AGO1 to trigger AGO1 protein decay in plants [69].

There are also viral suppressors with unspecified function, such as the triple gene block protein 1 (TGBp1) of PVX, cysteine-rich proteins encoded by hordeiviruses, tobraviruses, furoviruses, pecluviruses and carlaviruses, and the b protein of *Barley stripe mosaic virus* (BSMV) [52].

A number of assays have been established to verify the silencing suppression activity of a given viral gene and/or to identify multiple VSRs encoded by a single virus. In the case of plants, two assays have been widely used. The first is based on the transient, mixed expression of two transgenes in leaves co-infiltrated with two *Agrobacterium tumefaciens* strains. One strain induces RNA silencing of a reporter gene such as the green fluorescent protein (GFP) in the infiltrated leaf (local silencing); the subsequent spread of silencing into upper non-infiltrated tissues in transgenic plants that carry a homologous, integrated transgene (systemic silencing) is measured. The potential silencing suppressor is identified by the ability of the transiently expressed viral gene to enhance and/or sustain visibly higher levels of expression of the reporter gene. However, this assay was not capable of identifying several viral suppressors, including the CMV 2b gene, P25 of PVX, and the coat protein (CP) of *Citrus tristeza virus* (CTV) because they display very low suppression activity in agro-infiltration assays. Thus, their suppressor activities were confirmed by alternative approaches [40, 51]. The second type of assay, which is based on the use of grafting experiments, enables the identification of VSRs that are active against systemic silencing but not local silencing. In these experiments, selected transgenic plants stably expressing a candidate VSR are genetically crossed with a transgenic plant line that carries an autonomously silencing reporter transgene, such as 35S-GUS (β-glucoronidase) in tobacco line such as 6b5. Spreading expression of the viral protein suppresses reporter transgene silencing in the resulting F1 progeny (6b5xVSR) and can be determined by grafting reporter scions onto rootstocks made from the F1 plants. The reporter scions are from another transgenic plant line that expresses the reporter GUS transgene at high levels. The reporter gene becomes silenced a few weeks after grafting onto 6b5 rootstocks owing to the importation of a sequence-specific silencing signal from the silencing rootstock. Silencing does not occur in the scions if the VSR can inhibit either the synthesis of the mobile silencing signal in the F1 rootstocks or its export from rootstock to scion. Analysis of expression of the reporter transgene in the F1 progeny can also reveal whether the VSR suppresses local silencing, DNA methylation of the reporter transgene, or both [40].

Viruses are able to spread through infected plant cells using two ways of movement: cell-to-cell movement and long-distance movement. To combat this distribution, plants emit a silencing signal that spreads between cells. Because the effect of spreading is nt-sequence specific, the nature of the signal is likely to incorporate a siRNA or other RNA species [70]. Long-distance spreading depends on an RNA-dependent RNA polymerase (RDR), whereas short-distance movement of the signal does not [70]. A large number of suppressors of RNA silencing, including some effectors of long-distance virus movement through the phloem, are involved in these movements. For example, the P19 of tombusvirus is a suppressor of silencing that is not required for virus replication in isolated cells but is required for extensive viral invasion of systemic leaves [46]. P19 blocks the intercellular movement of the silencing signal by binding DCL4-dependent 21-nt siRNA [40]. Likewise, the potyviral HC-Pro and cucumoviral 2b proteins are suppressors [24] required for systemic virus infection [71]. *Cucumovirus* 2b protein inhibits the systemic movement of RNA silencing by either binding dsRNA/siRNA or inhibiting the slicer activity of AGO1 [23]. A site-directed mutation strategy involving the HC-Pro protein of TEV showed a correlation between silencing suppression and the ability to mediate long-distance virus movement [72]. Reevaluation of the role of *Tomato bushy stunt virus* (TBSV) P19 in the systemic invasion of *N. benthamiana* by the virus revealed a silencing suppression role for the structural protein (CP) of *Turnip crinkle virus* (TCV). The authors showed that a TBSV P19 deletion mutant, while capable of systemic movement in the plants, accumulated progressively less viral RNA in the systemic leaves due to loss of silencing suppressor ability. When the TBSV structural protein was replaced with TCV CP to create a chimeric virus, it restored close to wild type levels of virus accumulation in systemic leaves. This result shows that both of these genes participate in efficient systemic TBSV infection and suggests that TCV CP not only provides structural protein but also complements the silencing suppressor function of TBSV P19. Moreover, it is also suggestive that assembled virions are likely important for the effective unloading of viruses from the vascular system into the leaf mesophyll. This work provides direct evidence that P19 primarily enhances systemic invasion by suppressing the host PTGS responsible for eliminating viral RNAs in the infected plants [51]. The P25 protein of three 'triple gene block' (TGB) proteins of potexviruses is another example of suppressor that is required for cell-to-cell movement of the virus; it is an RNA helicase that moves cell-to-cell and modifies plasmodesmata [73, 74].

As previously mentioned, in many cases viruses encode more than one VSR. A good example is the closterovirus CTV, which encodes three different silencing suppressors [P23, coat protein (CP) and P20] and exhibits distinctive features related to silencing suppression. CTV has a plus-strand RNA genome of approximately 20 kilobases (kb) in length. Its P20 and P23 proteins, but not CP, suppressed RNA silencing in an agro-infiltration assay and were able to reverse transgene silencing. In addition, P20 and CP, but not P23, prevented intercellular silencing spread. It was suggested that P23 appears similar to HC-Pro because, although both are potent suppressors of intracellular silencing, neither prevents intercellular silencing or DNA methylation of the target transgene. On the other hand, the suppressor activity of P20 shares features with silencing suppression mediated by CMV 2b, i.e., both are potent but incomplete suppressors of intercellular silencing, and suppression of intercellular silencing is not associated with reduced DNA methylation of a target GUS transgene. In the case of CP, suppression of intercellular silencing spread is not associated with suppression of intracellular silencing, unlike P20, CMV 2b and P25 of PVX, which are known to interfere with intercellular silencing [75].

Silencing suppressors may confer biased protection against viral RNA and subviral parasites. It was shown that the P1/HC-Pro proteins of TEV caused an increase in the accumulation of the negative strand viral RNA of PVX [44], suggesting that negative-strand viral RNAs are more susceptible to the RNA silencing based host defence while positive-strand viral RNAs are better protected.

Viral infection is greatly influenced by changes in environmental temperature. A general explanation for this phenomenon is that RNA silencing-mediated plant defence is temperature dependent. Generally, at low temperature (15°C), both virus- and transgene-induced RNA silencing is inhibited; the level of virus- or transgene-derived siRNAs is dramatically reduced, leading to enhanced host susceptibility to virus infection and loss of silencing-mediated transgenic phenotypes. In contrast, with increasing temperature (27°C), RNA silencing is activated, and the amount of siRNA gradually increases. However, accumulation of miRNAs, which play a critical role in developmental regulation is temperature independent [76]. Because the replication of viruses does not appear to be disproportionately inhibited by higher temperature, one can assume that the activity of viral silencing suppressors is relatively constant over the temperature range that permits viral systemic infection. Thus, the level of silencing suppression activity should be relatively constant over this temperature range and therefore more readily overcome at higher temperature due to enhancement of the RNA silencing pathway. Conversely, it can be predicted that at low temperatures, the weakened RNA silencing would be more readily overcome by viral silencing suppressors [51].

3.1. Mechanisms of suppression

3.1.1. Viral suppressors

HC-Pro

As described above, HC-Pro was one of the first viral proteins to be identified as a suppressor of transgene- and virus-induced RNA silencing [77, 78]. HC-Pro is produced by plant viruses of the Potyvirus genus, family Potyviridae, the most important group of plant pathogenic viruses. HC-Pro has attracted renewed attention in recent years due to its multifunctionality and involvement in different steps of the potyvirus life cycle [79, 80]. Potyviruses, like the majority of plant viruses, have a single-stranded, positive-sense RNA genome that consists of approximately 10.080 nucleotides and is polyadenylated on its 3 'end and surrounded by a capsid [81]. The genomic RNA has a single ORF located between two noncoding regions, which are called 5'NTR and 3'NTR (non-translated region). Translation of the single ORF produces a polyprotein with molecular weight between 340 and 370 kilodaltons (kDa). This polyprotein is cleaved into functional proteins of the virus through the proteolytic activity of three proteases of viral origin (P1, HC-Pro, and NIa), resulting in 8-10 final products. HC-Pro and P1 act in *cis*, each carrying out its own cleavage, and NIa catalyses its own cleavage and that of six other polypeptides [80, 82].

A typical potyviral HC-Pro consists of approximately 460 amino acids and has a molecular weight of approximately 52 kDa, it performs a surprisingly large number of functions; in fact, among proteins produced by potyviruses, it is the protein for which the greatest number of features has been described [80]. Apart its role as silencing suppressor, HC-Pro plays several other roles as a proteinase, participates in aphid transmission, acts as an auxiliary viral replication factor and participates in virus cell-to-cell and long distance movement [44, 80, 83-88].

HC-Pro can be divided into three functional regions, a N-terminal region that is essential for transmission, a C-terminal region that is responsible for its proteolytic activity and a central region involved in all other functions described. However, recent studies show that most functions overlap along its primary amino acid sequence [89].

Concerning its ability to suppress silencing, HC-Pro was shown to restore GFP expression in both old and new leaves of post-transcriptionally silenced transgenic plants (reviewed in [90]). HC-Pro suppresses PTGS via interaction with one or more cellular proteins that are either components of the silencing machinery or regulators of the silencing pathway. Studies have shown that HC-Pro interferes with the accumulation of the small RNAs associated with silencing. These small RNAs derive from the cleavage of dsRNA by Dicer and HC-Pro may target the process at this step [91, 92]. Dicer could be blocked by HC-Pro in several ways: HC-Pro can prevent the small RNAs from being produced by preventing the enzyme from binding to the dsRNA template, thus blocking the cleavage step; alternatively, it could block at a step downstream of cleavage, preventing incorporation of siRNAs and making the silencing unstable. A model in which HC-Pro suppression of PTGS occurs upstream of accumulation of small RNAs has been proposed [93]. Furthermore, HC-Pro has been shown to transactivate the replication, and enhance the pathogenicity, of a broad range of heterologous plant viruses [44].

Since HC-Pro prevents accumulation of siRNAs of silenced genes, it prevents silencing in a universal manner; however, in tobacco HC-Pro was shown to increase the *in vivo* accumulation of several miRNAs, namely, *miR167, miR164* and *miR156* [94]. In addition, HC-Pro is not able to inhibit the systemic silencing signal, suggesting that HC-Pro works downstream from production of the systemic signal. It was suggested that HC-Pro works at the point of RISC assembly and that it most likely unwinds miRNA duplexes [95]. The specificity of HC-Pro binding to small RNAs was tested by the use of synthetic 21-nt or 24-nt siRNA duplexes and 19-nt or 21-nt blunt-ended RNA duplexes. The results showed that HC-Pro binds with specificity to 21-nt siRNA duplexes. Moreover, it has higher binding affinity for duplexes with 2-nt overhangs than for small single-stranded RNAs or blunt-ended small RNA duplexes [57].

HC-Pro is often mentioned in conflicting reports in the literature that address the relationship between PTGS and DNA methylation. In some instances, a good correlation between HC-Pro suppression of PTGS and the decrease of DNA methylation is observed. When introduced in a GUS-silenced tobacco line, for example, HC-Pro affected the accumulation of small RNAs of the PTGS pathway and reduced methylation of the corresponding GUS locus [96], suggesting that silencing is directly related to DNA methylation. In contrast, another study showed that HC-Pro increased DNA methylation of the promoter sequence of a silenced DNA target gene when silencing was induced by dsRNA directed against the promoter region [97]. In the same study, it was shown that the amount of promoter-derived siRNA molecules increased five-fold in the presence of HC-Pro.

P25

P25, the product of the first gene of the "triple gene block" (also known as TGBp1) encoded by PVX, is an RNA helicase that induces plasmodesmal gating. P25 promotes cell-to-cell movement of the virus and is also associated with suppression of RNA silencing [98]. P25 was one of the first VSRs to be identified and shown to inhibit transgene sense- or dsRNA-induced RNA silencing. The mechanism of action of P25 contrasts with that of HC-Pro, which acts at a downstream cellular signalling step. For this reason, mixed infections of PVX with any other potyvirus (which encode HC-Pro) normally result in synergistic disease [86]. Such diseases are common and often occur in plants as a result of the interaction between viruses that suppress silencing at various points of the silencing pathway [99, 100]. P25 is the only suppressor so far described that affects gene silencing but fails to recover silenced GFP expression post-transcriptionally. Surprisingly, P25 does not interfere with silencing of viral-induced sites [101].

Antiviral silencing suppression by P25 is required for cell-to-cell movement of the virus but has no apparent effect on viral accumulation in protoplasts, unlike most known VSRs such as cucumoviral 2b, tombusviral P19, and carmoviral P38. The analysis of a variety of random mutants of P25 showed that all produced defects in the suppression of silencing and in cell-to-cell movement. Some P25 mutants, defective in suppression activity, could be supplemented by heterologous viral suppressors. However, other mutants showed silencing suppression activity but were not functional as movement proteins. These results demonstrate a crucial role for P25 in cell-to-cell movement of the virus and also suggest the importance of an additional function of P25 in these activities [98].

P25 exhibits strong activity against silencing produced by both sense and inverted repeat transgenes in leaves of *N. benthamiana* and in transgenic *A. thaliana*. These observations indicate that P25 targets a downstream step in the synthesis of dsRNA [98, 101, 102]. As reported above, P25 inhibits systemic silencing but does not inhibit gene silencing induced by viruses in locally infected leaves. Moreover, it reduces the accumulation of both primary and secondary siRNAs but has no effect on the accumulation of endogenous miRNA and siRNA. It has been speculated that P25 does not interfere with programmed RISC [101, 102].

Co-immunoprecipitation assays indicate that P25 interacts with various members of the AGO family, including AGO1, AGO2, AGO3 and AGO4, but not AGO5 or AGO7. Furthermore, P25 promotes the proteasome-dependent degradation of AGO1 [103], indicating that its suppressor activity is dependent on AGO1 degradation. It is not currently known whether P25 inhibits the local motion silencing signal targeting AGO1 to promote movement of the virus [104].

P19

The tombusviral 19 kDa protein, P19, is one of the best studied viral silencing suppressors. The hypothesis that P19 is a viral suppressor arose in 1995 when Scholthof and colleagues reported that the 19 K protein of TBSV is a pathogenicity determinant. TBSV is a virus with a broad host range that induces a variety of symptoms in different hosts [105]. This virus contains a single copy of a positive-sense single-stranded RNA genome of 4800 nucleotides [106]. Five major ORFs are encoded by the TBSV genome. Two small nested genes located near the 3' terminus of the genome are expressed via a second subgenomic mRNA that directs synthesis of a 22 kDa protein (P22) and a 19 kDa protein (P19) [107]. P19 can act both as an elicitor of the HR response in *N. tabacum* or as an inductor of systemic necrosis in *N. benthamiana* [108]. Due to its activity as a host-specific symptom determinant, the P19 was suggested to play a role in overcoming host defence systems [108, 109]. This hypothesis was confirmed by the inoculation of silenced GFP tobacco plants with a recombinant PVX carrying the 19K coding region [48]. In these assays, plants infected with PVX-19K showed severe symptoms two weeks after inoculation while those already infected with PVX-m19K (with a nontranslatable P19 RNA) showed mild mosaic symptoms. Suppression of silencing occurred in PVX-19K infected plants but was manifested only in new emerging tissues and was most pronounced in the veins. However, symptoms of PVX-19K were visible on all areas of the leaves. Interestingly, P19 restores GFP expression in PTGS inactivated transgenic plants only around the veins of new emerging leaves [46], even though TBSV accumulates to a high concentration in the whole leaf [108].

Several recent studies report a breakthrough in understanding the molecular mechanism of the suppressor activity of P19. This suppressor prevents incorporation of siRNAs into effectors such as the RISC complex by binding specifically to 21-nt siRNAs *in vitro* and *in vivo* [110, 111]. This model was confirmed by three-dimensional structural resolution of the P19-siRNA complex showing that P19 acts as a clamp for dsRNA binding to the ends of the siRNA duplex [112, 113]. However, it was also reported that after the RISC complex is formed, P19 is no longer effective, being unable to bind to siRNA and miRNA [57].

P19 inhibits the onset of transgene-induced local and systemic silencing [110]. It does not interfere with the location of virus-induced silencing, but it can prevent systemic silencing. It was suggested that P19 depletes PTGS generated 21-25-nt dsRNAs, thus inhibiting the development of transgene-induced silencing and preventing the production of active signal complex. Interestingly, transgenic plants expressing biologically active P19 showed an altered phenotype, suggesting that the P19-targeted PTGS pathway might also have a role in developmental regulation. Low level expression of P19 altered leaf morphology in transgenic plants. In addition to leaf curling, some severely affected plants also showed delayed appearance of developed secondary stems. Although it is possible that developmental abnormalities in transgenic plants are not related to the silencing suppressor activity of P19, these findings are suggestive that the P19-targeted PTGS pathway plays a role in plant development.

The silencing suppressor activity of P19 is also observed in other hosts. Since siRNA binding by P19 does not require host factors *in vitro*, and that these short RNAs are specificity determinants of silencing effector complexes, P19 could be used to inhibit RNA silencing in heterologous systems, including *D. melanogaster*, worms and mammals [110].

The P19 protein of *Cymbidium ringspot virus* (CymRSV), a relative of the TBSV P19 protein, specifically binds to siRNAs *in vitro*, and two reports show co-crystallisation of P19 homo-dimers with siRNA [112, 113]. P19 also binds RNA duplexes with a blunt end and with a 2-nt 3′ overhanging end. In animals, Dicer digests from the ends of long dsRNAs [114] and therefore might produce long dsRNAs with 2-nt 3′ overhangs. Although it is possible that P19 competes with Dicer-related proteins for the 2-nt 3′ overhanging ends of long dsRNAs, the high level of 21-25-nt RNAs in CymRSV infected cells suggests that P19 fails to suppress Dicer-like activity.

A study using mutants of CymRSV demonstrated that lack of P19 suppressor did not affect most basic viral functions, including genome replication, cell-to-cell movement and phloem long-distance transport [109]. In contrast, the systemic infection of plants inoculated with a silencing suppressor mutant of CymRSV was seriously compromised and led to the development of a recovery phenotype, suggesting that P19 suppressor targets a non-cell-autonomous step of RNA silencing [110].

In CymRSV infected plants, siRNAs are present in P19–siRNA complexes, while in plants infected with the P19-defective mutant in which P19 was inactivated (termed Cym19stop), siRNAs were found as free molecules. P19 apparently does not affect virus-induced cell-autonomous silencing because CymRSV and Cym19stop viral RNAs, as well as siRNAs derived from these viruses, accumulate to the same levels in transfected single cells [110]. In addition, the P19 protein was shown to repress the accumulation of all size classes of siRNA produced in agroinfiltration assays [110, 115]. While CymRSV infects *N. benthamiana* system-ically and typically kills the host within two weeks, infection with the mutant virus results in a recovery-like phenotype showing mild symptoms and low virus levels in the upper leaves [116]. Moreover, P19-deficient and wild type CymRSV accumulate at similar levels in both protoplasts and inoculated leaves, indicating that this protein does not prevent RISC from degrading viral RNAs by sequestering viral derived siRNAs (vsiRNA) [37]. In systemic leaves, P19-deficient CymRSV accumulates only in the vascular bundles and exhibits defects in invading the surrounding tissues suggesting that blocking the local movement of RNA silencing by P19 is essential for systemic virus infection [117].

In this context, studies have shown that P19 specifically sequesters the DCL4-dependent 21-nt siRNAs derived from transgene RNAs; these siRNAs normally move into the neighbouring recipient cells and act as a silencing signal [118]. These results imply that P19 promotes systemic virus infection by sequestering vsiRNA, thus preventing the signal for RNA silencing from spreading out of vascular bundles into neighbouring cells [104]. Therefore, when P19 is absent, the systemic signal moves faster than the virus in the infected plant, thereby estab-lishing antiviral silencing in cells ahead of the infection front. As a result, any virus entering these cells is immediately controlled by silencing-mediated RNA degradation. In conclusion, the presence of the silencing suppressor is essential for the development of systemic virus infection [37].

3.1.2. Endogenous suppressors

In addition to the numerous viral suppressors of RNA silencing, endogenous RNA silencing suppressors have also been reported in eukaryotes. The endogenous suppression of RNA silencing negatively controls the presence of siRNAs and miRNAs in different ways. The generation and control of such siRNAs and miRNAs are essential for normal development of plants and animals [119-121].

The first endogenous RNA silencing suppressor was identified in *N. tabacum* and was named rgs-CaM (regulator of gene silencing CaM). This protein was found in a screen for proteins interacting with the viral suppressor HC-Pro. Expression of rgs-CaM can be induced in leaves of *N. tabacum* when HC-Pro is expressed either from a transgene or from infection with a virus that encodes HC-Pro. When expressed at high levels in *N. benthamiana*, rgs-CaM suppresses both PVX–induced gene silencing and sense transgene–mediated PTGS (S-PTGS) [49]. A recent study, however, demonstrated that rgs-CaM is not an endogenous suppressor of silencing [122]. In fact, this protein acts as an endogenous pattern recognition receptor able to bind to several viral silencing suppressors through their RNA-binding domains. Thus, rgs-CaM activity confers a countermeasure against viral suppressors.

In addition, an inhibitor protein of *A. thaliana* RNase L activity, called RLI2, was also described as having a silencing suppressor activity when expressed at high levels in transgenic *N. benthamiana* [123]. Another known endogenous suppressor, the *A. thaliana* exoribonuclease XRN4, suppresses silencing by promoting the degradation of aberrant, uncapped RNAs that constitute possible templates for an RNA dependent RNA polimerase (RdRP) pathway involved in silencing. These aberrant molecules represent important activators of silencing, serving as templates for the production of new dsRNAs by the action of the RdRP. Indeed, mutations in the gene *xrn4* promote RdRP-dependent silencing [124] and lead to over-accumulation of miRNA-generated cleavage products [125]. Three other suppressor proteins, the exoribonucleases XRN2, XRN3 and FRY1, were identified in *A. thaliana*, thus complementing existing knowledge of the suppression of silencing involving XRN4 [126]. While XRN4 is cytoplasmic, XRN2 and XRN3 are nuclear exoribonucleases [127]. XRN2 and XRN3 contribute to the suppression of RNA silencing by degrading miRNA-derived loops excised during miRNA maturation in the nucleus. In contrast, XRN4 acts exclusively in the cytoplasm, promoting degradation of uncapped messages such as miRNA target cleavage products [124-126]. Fry1 acts as a fine-tuning modulator of the activities of XRN2, XRN3 and XRN4 [126]. Interestingly a family of exoribonucleases known as small RNA degrading nucleases (SDN) degrades mature miRNA molecules in *A. thaliana*, acting specifically on single-stranded miRNAs [121].

3.1.3. Modifications of the host transcriptome

Viruses can counterattack RNA silencing immunity not only by acting directly on gene products that are required for silencing *per se* but also by inducing stress and plant defence responses that interfere with antiviral silencing [128, 129]. An interestingly example include the RAV2/EDF2 protein, which belongs to the RAV/EDF family of transcription factors. This protein is required for suppression of silencing by potyvirus HC-Pro and carmovirus P38, two viruses that belong to unrelated families. RAV2 is required for suppression of silencing in a direct way that involves blocking the activity of primary siRNAs as well as indirectly by its effects on upregulation of some stress and defence response genes [128].

The induction of biotic or abiotic stress activates other defence responses that can divert the host from antiviral silencing [129]. Therefore, RAV2 is a critical control factor for carmovirus and potyvirus suppressors [128].

Other viruses make use of alternative mechanisms for suppression of silencing. The TrAP geminiviral protein AC2 upregulates a gene coding for the cold- and abscisic acid-inducible protein KIN1 as well as five additional known or putative cold-regulated genes [50]. As already mentioned, the efficiency of RNA silencing is dependent on temperature; at low temperatures, inhibition of silencing occurs and the plant becomes susceptible to viral infection [76]. The inhibition of silencing at low temperature is a pathway used by AC2 to accomplish the suppression of silencing [50]. Another strategy exploited by geminivirus is up-regulation of an endogenous RNA silencing suppressor, Werner exonuclease-like 1 (WEL1), which is mediated by AC2. Interestingly, the related proteins MUT-7 (mutate 7) and Werner syndrome-like exonuclease (WEX) have been identified as positive regulators of RNA silencing in *C. elegans* and *A. thaliana*, respectively [130, 131]. Thus, AC2 up-regulation of Wel-1 results in interference with, or competition for, factors that are required for normal WEX function. Transient expression of a WEL-1 transcription unit is sufficient to suppress RNA silencing in *N. benthamiana* [50].

4. Applications

4.1. Virus-Induced Gene Silencing (VIGS)

Virus-induced gene silencing (VIGS) is a technique derived from the knowledge of RNA silencing. It uses recombinant viruses to specifically reduce or knock-down endogenous gene activity; it is based on post-transcriptional gene silencing (PTGS) [132]. When used to infect plants, recombinant viral vectors carrying segments of host genes produce siRNAs that are specific to host mRNA. The RISC complex mediates the degradation of target host mRNAs, leading to downregulation of gene expression. Thus, the infected plant has a phenotype similar to a loss-of-function mutant of the gene of interest [133].

VIGS is used as a tool for turning down host gene expression, especially in plants. In principle, a plant gene of interest can be silenced by infecting the plant with a viral vector that has been modified to express a nucleic acid sequence homologous to the host gene. As a proof of concept, several endogenous genes have been silenced using VIGS. The plant gene *phytoene desaturase* (PDS), a regulator of carotenoid biosynthesis, was silenced in *N. benthamiana* plants by the use of a recombinant TMV vector. As a result, degradation of the host PDS mRNA and resultant alterations in the pigment synthesis pathway were observed [133].

There are four main reasons for the popularity of VIGS. First, the methodology is simple, often involving agroinfiltration or biolistic inoculation of plants. Second, the results are obtained rapidly, typically within two to three weeks of inoculation. Third, the technology bypasses transformation steps and hence is applicable to a number of plant species that are recalcitrant to transformation. Fourth, the method has the potential to silence multi-copy genes [134].

Efficient silencing depends mainly on the choice of VIGS vector. There are many factors to be considered when choosing the virus to be used for VIGS. Among the factors to be considered are (1) the virus must produce few or no symptoms during infection, thereby facilitating easy visualisation and interpretation of the mutant phenotype; (2) it must induce persistent silencing, thus viruses with strong silencing suppressors are to be avoided because they can interfere with the establishment of silencing; (3) it is advantageous to have infectious cDNA clones of the virus for cloning purposes; and (4) the virus must retain infectivity after insertion of foreign DNA. The virus should also show uniform spread, infect most cell types including the meristem, and preferably show a broad host range [133].

Several RNA and DNA viruses have been modified to create VIGS vectors. The gene to be silenced is cloned in an infectious derivative of a viral DNA (DNA virus-based vectors) or cDNA (RNA virus-based vectors) derived from viral RNA. Plant inoculation with viral vectors is most commonly achieved via *A. tumefaciens* infection, but can also be achieved by mechanical inoculation of *in vitro* synthesized transcripts, or for DNA-based vectors, by biolistic delivery methods. During the course of viral infection, either double-stranded RNA or RNA with a high degree of secondary structure is often produced; both of these are efficient initiators of RNA silencing directed against the infecting viral RNA. Other factors that play an important role in gene silencing in VIGS are the orientation of the insert (inverted repeats are more efficient than antisense orientation, which, in turn is more efficient than same sense orientation) and systemic spread of the silencing effect (the silencing signal is believed to spread independently of the VIGS vector to other parts of the plant) [134]. More than 30 VIGS vectors have been developed, and these vectors have been widely used to study the functions of genes involved in basic cellular functions, metabolic pathways, development, plant-microbe interaction, and abiotic mechanisms [132].

The first viral vector used for VIGS was TMV. Shortly thereafter, another vector was produced based on another RNA virus, PVX carrying a cDNA fragment derived from the *PDS* gene [135]. However, although these first vectors were effective, they have intrinsic disadvantages. First, the VIGS phenotype is superimposed and sometimes complicated by chlorosis, leaf distortion and necrotic symptoms of virus infection. A second disadvantage of these viral vectors is their inability to invade every cell, such that cells in which the target gene is not silenced may obscure VIGS phenotypes [136]. A novel VIGS vector based on TRV was then established. TRV was shown to induce more efficient silencing of transgenes and endogenous genes. It could spread more vigorously throughout the entire plant, including meristem tissue, and the symptoms induced by TRV are much milder than those produced by other viruses [136].

A primary limitation of VIGS technology is that a viral vector can be used only in plants that are hosts of the virus used. The first VIGS vectors (e.g., PVX) do not infect the model plant *A. thaliana*. Therefore, new vectors such as the TRV-based vector [136] were developed to overcome this difficulty. TRV is also one of the few viruses that have been modified into a highly efficient cloning and expression system for use in large-scale functional genomics screening. TRV vectors can induce VIGS in a number of solanaceous hosts like *N. benthamiana*, tomato, potato, pepper, petunia, poppy (Eudicot species), and the model system *A.*

thaliana (family Brassicaceae) [133, 137]. VIGS vectors have been applied not only in dicotyledonous plants but also in monocotyledonous plants. For this, a modified VIGS vector based on *Brome mosaic virus* (BMV) was developed and validated in barley, rice and maize [138].

The VIGS system can be helpful in assessing gene function, especially for genes that cause zygotic/embryonic lethality when mutated and in species that are recalcitrant to genetic transformation. As aforementioned, it can be designed to silence multiple members of a gene family, thereby circumventing the problem of functional redundancy of genes [133].

4.2. Use of viral suppressors

The discovery of RNA silencing, and its derived technology (RNA interference; RNAi), has increased our knowledge of gene regulation and function. RNAi opened up novel avenues in biology, making it possible to develop fascinating strategies for application in genetic analysis, plant protection, and many other areas related to crop improvement [139]. In this context, a large number of silencing suppressor proteins have been described, and the discovery of the molecular basis of silencing suppression has inspired new concepts about the molecular basis of symptoms caused by viruses in plants [37].

Many biotechnological applications involving plants require high levels of protein expression. Generally, stably transformed plants are the preferred platform for large-scale production. To try to increase expression levels, transgenic lines that encode a replicating RNA virus vector carrying a gene of interest, a technology coined 'amplicon', have been exploited. The rationale of this method involves increasing the accumulation of the product of interest through transcription of an amplicon transgene that initiates viral RNA replication and gene expression. However, the strategy failed because the transformants consistently exhibited RNA silencing of the amplicon transgene [140]. The viral dsRNA replication intermediates produced in every cell of the transgenic plants were recognised as potent triggers of the silencing-based defence mechanism that is normally elicited in the course of natural infections. Based on those findings, it was subsequently reasoned that co-expression of viral suppressors might prevent this adverse response and permit the high levels of gene expression initially envisioned with the use of amplicons [141].

To test this idea, in reference [94] crossed transgenic tobacco plants expressing TEV HC-Pro with amplicon lines designed to express a GUS reporter gene from the PVX genome. Pairing the suppressor and the amplicon locus resulted in a dramatic increase in virus accumulation and gene expression such that the leaves of mature plants accumulated the GUS protein up to 3% of total soluble protein. Remarkably, in spite of high virus accumulation, the plants did not suffer from viral disease and remained symptomless.

As opposed to stable, transgenic expression, transient expression is of interest for achieving expression of useful proteins. In plants, recombinant strains of *A. tumefaciens* can be used for transient gene expression. In principle, this system could allow high levels of gene expression; however, its utility has thus far been limited because ectopic protein expression usually ceases after 2– 3 days [141]. RNA silencing is, in fact, a major cause of this lack of efficiency. It was therefore anticipated that co-delivery of *A. tumefaciens* cultures with silencing suppressors would enhance expression of the genes of interest [101]. Studies with the P19 protein of TBSV were among those that provided the best results. Expression of a range of proteins was enhanced 50-fold or more in the presence of this suppressor, and experiments with GFP indicated that the co-infiltrated tissues accumulated the protein up to 7% of total soluble protein [142]. Due to its simplicity and rapidity, the P19-enhanced expression system is currently used in industrial production as well as used as a research tool for the isolation and biochemical characterisation of a broad range of proteins without the need for the time-consuming regeneration of stably transformed plants [141].

5. Perspectives

The molecular basis of the silencing suppression of VSR proteins is quite complex and is currently incompletely understood. By the way, the discovery of the mode of action of different viral suppressors has demonstrated the existence of a complex interaction between VSR and plant silencing-regulated networks. For example, in addition to sequestering siRNA duplexes, the P19 protein of tombusviruses specifically controls antiviral AGO1 expression through enhanced miR168 expression, which arrests AGO1 translation [23]. It is likely that many other VSRs interact in diverse ways with RNA-silencing pathways. Many of these interactions remain to be discovered, and there are several gaps in our knowledge regarding the effectors of plant silencing machinery. Until very recently, the mechanisms of plant si/miRNA RISC assembly or the components of the plant RISC, which may also be potential targets of VSRs, were little known. The recently developed system of plant in vitro RISC [143] will likely accelerate the exploration of plant RISC assembly and RNA-targeting mechanisms mediated by this effector. This system will enable exploration of the mechanisms by which VSRs interact with one or more of the RISC components and prevent its assembly.

More information about the replication, subcellular localisation and regulation of the expression of viral genes, including VSRs, is required so that we may better understand the molecular mechanisms of VSR-mediated silencing suppression for the many plant viruses for which they are still not known. Because many VSRs have multiple functions in the virus life cycle, separate analysis of their silencing suppressor activities can lead to misinterpretations; thus, it is essential that VSRs be studied in their natural virus backgrounds [23].

Although common mechanisms of silencing suppression exist, there is also great variation in suppression mechanisms, likely driven by evolution and fitness, this variation has yielded viral strains with different properties. It is likely that additional differences will be found when plant viruses and their suppressors are tested in several plant species. This will provide us with a greater understanding of the parameters associated with the natural host range of a virus and may possibly lead to new strategies for crop protection [52].

Some of the already well described VSRs can be used as powerful tools for better understanding silencing pathways because they target specific steps of silencing machinery. Indeed, the P19 protein was recently used to demonstrate that siRNA duplexes function as mobile silencing signals between plant cells, in addition that P1, P38 and P0 proteins may prove to be powerful tools for studying the still unknown components of RISCs [118].

6. Conclusions

The discovery of the mechanisms involved in RNA silencing and in silencing suppression by virus has helped researchers to investigate several aspects of the plant-virus co-evolution. Our understanding of the underlying mechanisms served as a basis for the development of different technologies aiming plant manipulation to generate novel traits or pathogen resistance. Elucidation of the mode of action of different plant viral suppressors provided fundamental contributions to the comprehension of the RNA silencing phenomena. In this context, strategies exploiting viral suppressors to prevent the occurrence of RNA silencing in genetically engineered plants have been developed [144]. High levels of transgene expression should be expected through the implementation of such strategies. This review has addressed some important topics in the areas of RNA silencing and silencing suppression. Ongoing studies aimed at further clarification of the main points of these processes are currently being conducted by many different groups.

Acknowledgements

We are grateful to the Brazilian funding agencies CNPq, FAPESP and CAPES, for their financial support. A.T. Costa, R. K. Makiyama and A. V. Nunes are recipients of doctoral fellowships from CNPq (140299/2010-6), FAPESP (2010/03001-0), CAPES, respectively, and J.P. Bravo is the recipient of a CNPq/PNPD postdoctoral fellowship (558413/2008).

Author details

Alessandra Tenório Costa, Juliana Pereira Bravo, Rodrigo Kazuo Makiyama, Alessandra Vasconcellos Nunes and Ivan G. Maia

*Address all correspondence to: alebiologyt@hotmail.com

São Paulo State University "Júlio Mesquita Filho" Department of Genetics, Institute of Biosciences – UNESP, Botucatu Campus SP, Brazil

References

[1] Hull R. Matthew's Plant Virology (4ᵃ ed.). Academic Press; 2002.

[2] Laliberté J-F, Sanfaçon H. Cellular Remodeling During Plant Virus Infection. Annual Review of Phytopathology 2010;48:69-91.

[3] Rodrigues SP, Lindsey GG, Fernandes PMB. Biotechnological Approaches for Plant Viruses Resistance: From General to the Modern RNA Silencing Pathway. Brazilian Archives of Biology and Technology 2009;52(4):795-808.

[4] Collmer CW, Marston MF, Taylor JC, Jahn M. The I Gene of Bean: A Dosage-Dependent Allele Conferring Extreme Resistance, Hypersensitive Resistance, or Spreading Vascular Necrosis in Response to the Potyvirus Bean common mosaic virus. Phytopathology 2000;13:1266-1270.

[5] Krause-Sakate R, Redondo E, Richard-Forget F, Jadao AS, Houvenaghel MC, German-Retana S, Pavan MA, Candresse T, Zerbini FM, Le Gall O. Molecular Mapping of the Viral Determinants of Systemic Wilting Induced by a Lettuce mosaic virus (LMV) Isolate in Some Lettuce Cultivars. Virus Research 2005;109:175-180.

[6] Whitham S, Wang Y. Roles for Host Factors in Plant Viral Pathogenicity. Current Opinion in Plant Biology 2004;7:365-371.

[7] Baker B, Zambryski P, Staskawicz B, Dineshkumar SP. Signaling in Plant-Microbe Interactions. Science 1997;276:726-733.

[8] Durrant WE, Dong X. Systemic Acquired Resistance. Annual Review of Phytopathology 2004;42:185-209.

[9] Seo YS, Gepts P, Gilbertson RL. Genetics of Resistance to the Geminivirus, Bean dwarf mosaic virus, and the Role of the Hypersensitive Response in Common Bean. Theoretical and Applied Genetics 2004;108:786-793.

[10] Palukaitis P, Carr JP. Plant Resistance Responses to Viruses. Journal of Plant Pathology 2008;90 (2):153-171.

[11] Cooper B. Collateral Gene Expression Changes Induced by Distinct Plant Viruses During the Hypersensitive Resistance Reaction in Chenopodium amaranticolor. The Plant Journal 2001;26(3):339-349.

[12] Vance V, Vaucheret H. RNA Silencing in Plants-Defense and Counterdefense. Science 2001;292:2277-2280.

[13] Dielen A-S, Badaoui S, Candresse T, Germa-Retana S. The Ubiquitin/26S Proteasome System in Plant–Pathogen Interactions: A Never-Ending Hide-and-Seek Game. Molecular Plant Pathology 2010;11:293-308.

[14] Ballut L, Drucker M, Pugniere M, Cambon F, Blanc S, Roquet F, Candresse T, Schmid HP, Nicolas P, Gall OL, Badaoui S. HcPro a Multifunctional Protein Encoded by a Plant RNA Virus, Targets the 20S Proteasome and Affects Its Enzymic Activities. Journal of General Virology 2005;86:2595-2603.

[15] Zeng lR, Vega-Sanchez ME, Zhu T, Wang, GL. Ubiquitination-Mediated Protein Degradation and Modification: An Emerging Theme in Plant-Microbe Interactions. Cell Research 2006;16:413-426.

[16] Francki,RIB. Responses of plant cells to virus infection with special references to the sites of RNA. In: Positive strand RNA viruses (M.A. Brinton and R.R. Rueckert, eds.) Plan R. Liss, New York, 1987:423-436.

[17] El-Hoseny ME El-Fallal A A, A.K.El-S;,A.D, A.S.S. Biology Cytopathology and Molecular Identification of an Egyptian Isolate of Zucchini yellow mosaic Potyvirus (ZYMV-EG). Pakistan Journal of Biotechnology 2010;7(1-2):75-80.

[18] Hatta T, Bullivant S,Matthews REF. Fine Structure o f Vesicles Induced in Chloroplasts of Chinese Cabbage Leaves by Infection with Turnip yellow mosaic virus. Journal of General Virology,1973; 20, 37-50.

[19] Afrren B, Gulfishan M, Baghel G, Fatma M, Khan AA, Naqvi QA. Molecular Detection of a Virus Infecting Carrot and its Effect on Some Cytological and Physiological Parameters. African Journal of Plant Science 2011;5(7):407-411.

[20] Oakenfull EA, Riou-Khamlichi C, Murray JA. Plant D-type Cyclins and the Control of G1 Progression. Philosophical Transactions of the Royal Society B 2002;357:749-760.

[21] Chapa-Oliver AM, Guevara-González RG, González-Chavira MM, Ocampo-Velázquez RV, Feregrino-Pérez AA, Mejía-Teniente L, Herrera-Ruíz G, Torres-Pacheco I. Analogies between Geminivirus and Oncovirus: Cell Cycle Regulation. African Journal of Biotechnology 2011;10(55):11327-11332.

[22] Gutierrez C. DNA Replication and Cell Cycle in Plants: Learning from Geminiviruses. The EMBO Journal 2000;19(5):792-799.

[23] Burgyán J, Havelda Z. Viral Suppressors of RNA Silencing. Trends in Plant Science 2011;16(5):265-272.

[24] Anandalakshmi R, Pruss GJ, Ge X, Marathe R, Mallory AC, Smith TH, Vance VB. A Viral Suppressor of Gene Silencing in Plants. Proceedings of the National Academy of Sciences of the United States of America 1998;95:13079-13084.

[25] Fire A, Xu S, Montgomery MK, Kostas AS, Driver SE, Mello CC. Potent and Specific Genetic Interference by Double-stranded RNA in Caenorhabditis elegans. Nature 1998;391:806-811.

[26] Mlotshwa S, Pruss GJ, Peragine A, Endres MW, Li J, Chen X, Scott PR, Bowman LH, Vance V. DICER-LIKE2 Plays a Primary Role in Transitive Silencing of Transgenes in Arabidopsis. PLos ONE 2008;3(3):e1755. doi:10.1371/journal. Pone.0001755.

[27] Hannon G J. RNA interference. Nature 2002;418(11):244-251.

[28] Song L, Gao S, Jiang W, Chen S, Liu Y, Zhou L, Huang W. Silencing Suppressors: Viral Weapons for Countering Host Cell Defenses. Protein Cell 2011;2(4):273-281.

[29] Hohn T, Vazquez F. RNA Silencing Pathways of Plants: Silencing and its Suppression by Plant DNA Viruses. Biochimica et Biophysica Acta 2011;1809:588-600.

[30] Fagard M, Vaucheret H. (Trans) gene Silencing in Plants: How Many Mechanisms? Annual Review of Plant Physiology and Plant Molecular 2000;51:167-194.

[31] Hoffer P, Ivashuta S, Pontes O, Vitins A, Pikaard C, Mroczka A, Wagner N, Voelker T. Posttranscriptional Gene Silencing in Nuclei. PNAS 2011;108(1):409-414.

[32] Lin J, Chun-Hong W, Yi L. Viral Suppression of RNA Silencing. Science China 2012;55(2):109-118.

[33] Paszkowski J, Whitham SA. Gene Silencing and DNA Methylation Processes Current Opinion in Plant Biology 2001;4:123-129.

[34] Avramova, Z. Epigenetic Regulatory Mechanisms in Plants Handbook of Epigenetics: The New Molecular and Medical Genetics. Chapter 16 Elsevier 2011.

[35] Melnyk CW, Molnar A, Baulcombe DC. Intercellular and Systemic Movement of RNA Silencing Signals. The EMBO Journal 2011;30:3553-3563.

[36] Napoli C, Lemieux C, Jorgensen R. Introduction of a Chimeric chalcone synthase Gene into Petunia Results in Reversible Co-suppression of Homologous Genes in trans. Plant Cell 1990;2:279-289.

[37] Burgyán J. Virus Induced RNA Silencing and Suppression: Defence and Counter-Defence. Journal of Plant Pathology 2006;88(3):233-244.

[38] Ratcliff FG, MacFarlane SG, Baulcombe DC. Gene Silencing without DNA: RNA-Mediated Cross-Protection between Viruses. The Plant Cell 1999;11:1207-1215.

[39] Lindbo JA, Silva-Rosales L, Proebsting WM, Dougherty WG. Induction of a Highly Specific Antiviral State in Transgenic Plants: Implications for Regulation of Gene Expression and Virus Resistance. Plant Cell 1993;5:1749-1759.

[40] Li F, Ding S. Virus Counter Defense: Diverse Strategies for Evading the RNA-Silencing Immunity. Annual Review Microbiology 2006;60:503-531.

[41] Hamilton AJ, Baulcombe DC. A Species of Small Antisense RNA in Posttranscriptional Gene Silencing in Plants. Science 1999;286: 950-952.

[42] Sijen T, Vijn I, Rebocho A, van Blokland R, Roelofs D, Joseph NM, Kooter JM. Transcriptional and Posttranscriptional Gene Silencing are Mechanistically Related. Current Biology 2001;11:436-440.

[43] Shimura H, Pantaleo V. Viral Induction and Suppression of RNA Silencing in Plants. Biochimica et Biophysica Acta 2011;1809:601-612.

[44] Pruss G, Ge X, Shi X M, Carrington JC, Vance VB. Plant Viral Synergism: The Potyviral Genome Encodes a Broad-Range Pathogenicity Enhancer that Transactivates Replication of Heterologous Viruses. The Plant Cell 1997;9:859-868.

[45] Brigneti G, Voinnet O, Li W-X, Ji L-H, Ding S-W, Baulcombe DC. Viral Pathogenicity Determinants are Suppressors of Transgene Silencing in Nicotiana benthamiana. The EMBO Journal 1998;17(22):6739-6746.

[46] Voinnet O, Pinto YM, Baulcombe DC. Suppression of Gene Silencing: A general Strategy Used by Diverse DNA and RNA Viruses of Plants. PNAS 1999;96(24): 14147-14152.

[47] Ding S-W, Voinnet O. Antiviral Immunity Directed by Small RNAs. Cell 2007;130:413-426.

[48] Voinnet O. Induction and Suppression of RNA Silencing: Insights from Viral Infections. Nature Reviews 2005;6:206-221.

[49] Anandalakshmi R, Marathe R, Ge X, Herr JM, Jr Mau C, Mallory A, Pruss G, Bowman L, Vance VB. A Calmodulin-Related Protein that Suppresses Post-Transcriptional Gene Silencing in Plants. Science 2000;290:142-144.

[50] Trinks D, Rajeswaran R, Shivaprasad PV, Akbergenov R, Oakeley EJ, Veluthambi K, Hohn T, Pooggin MM. Suppression of RNA Silencing by a Geminivirus Nuclear Protein, AC2, Correlates with Transactivation of Host Genes. Journal of Virology 2005;79(4):2517-252750

[51] Qu F, Morris T J. Suppressors of RNA Silencing Encoded by Plant Viruses and Their Role in Viral Infections. FEBS Letters 2005;579:5958-5964.

[52] Alvarado V, Scholthof HB. Plant Responses against Invasive Nucleic Acids: RNA Silencing and Its Suppression by Plant Viral Pathogens. Seminars in Cell & Developmental Biology 2009;20:1032-1040.

[53] Kasschau KD, Xie Z, Allen E, Llave, C, Chapman EJ, Krizan KA, Carrington JC. P1/HC-Pro, a Viral Suppressor of RNA Silencing, Interferes with Arabidopsis Development and miRNA Function. Developmental Cell 2003;4:205-217.

[54] Siddiqui SA Sarmiento C Truve E,Lehto H, Lehto K. Phenotypes and Functional Effects Caused by Various Viral RNA Silencing Suppressors in Transgenic N. benthamiana and N. tabacum. Molecular Plant-Microbe Interactions 2008; 21:178–187.

[55] Díaz-Pendón JA, Ding S-W. Direct and Indirect Roles of Viral Suppressors of RNA Silencing in Pathogenesis. Annual Review Phytopathology 2008;46:303-26.

[56] Mérai Z, Kerényi Z, Kertész S, Magna M, Lakatos L, Silhavy D. Double-Stranded RNA Binding May be a General Plant RNA Viral Strategy to Suppress RNA Silencing. Journal of Virology 2006;80(12):5747–5756.

[57] Lakatos L, Csorba T, Pantaleo V, Chapman, EJ, Carrington, JC, Liu, Y-P, Dolja, VV, Calvino, LF, López-Moya, JJ, Burgyán, J. Small RNA Binding is a Common Strategy

to Suppress RNA Silencing by Several Viral Suppressors. The EMBO Journal 2006;25(12):2768-2780.

[58] Kasschau KD, Carrington JC. A Counterdefensive Strategy of Plant Viruses:Suppression of Posttranscriptional Gene Silencing. Cell 1998;95:461–470.

[59] Haas G, Azevedo J, Moissiard G, Geldreich A, Himber C, Bureau M, Fukuhara T, Keller M, Voinnet O. Nuclear Import of CaMV P6 is Required for Infection and Suppression of the RNA Silencing Factor DRB4. The EMBO Journal 2008;27(15): 2102-2112.

[60] Glick E, Zrachya A, Levy Y, Mett A, Gidoni D, Belausov E, Citovsky V, Gafni Y. Interaction with Host SGS3 is Required for Suppression of RNA Silencing by Tomato yellow leaf curl virus V2 Protein. PNAS 2008;105(1):157-161.

[61] Shiboleth YM, Haronsky E, Leibman D, Arazi, T, Wassenegger M, Whitham SA, Gaba V, Gal-On A. The Conserved FRNK Box in HC-Pro, a Plant Viral Suppressor of Gene Silencing, Is Required for Small RNA Binding and Mediates Symptom Development. Journal of Virology 2007;81(23):13135-13148.

[62] Yu B, Chapman EJ, Yang Z, Carrington JC, Chen X. Transgenically Expressed Viral RNA Silencing Suppressors Interfere with microRNA Methylation in Arabidopsis. FEBS Letters 2006;580:3117-3120.

[63] González I, Martínez L, Rakitina DV, Lewsey MG, Atencio FA, Llave C, Kalinina NO, Carr J P, Palukaitis P, Canto T. Cucumber mosaic virus 2b Protein Subcellular Targets and Interactions: Their Significance to RNA Silencing Suppressor Activity. Molecular Plant-Microbe Interactions 2010;23(3):294-303.

[64] Zhang X, Yuan Y-R, Pei Y, Lin S-S, Tuschl T, Patel DJ, Chua N-H. Cucumber mosaic virus-encoded 2b Suppressor Inhibits Arabidopsis Argonaute1 Cleavage Activity to Counter Plant Defense. Genes & Development 2006;20:3255-3268.

[65] Hamera S, Song X, Su L, Chen X, Fang R. Cucumber mosaic virus Suppressor 2b Binds to AGO4-Related Small RNAs and Impairs AGO4 Activities. The Plant Journal 2012;69:104-115.

[66] Mi S, Cai T, Hu Y, Chen Y, Hodges E, Ni F, Wu L, Li S, Zhou H, Long C, Chen S, Hannon GJ, Qi Y. Sorting of Small RNAs into Arabidopsis Argonaute Complexes is Directed by the 5'Terminal Nucleotide. Cell 2008;133:116-127.

[67] Pfeffer S, Dunoyer P, Heim F, Richards KE, Jonard G, Ziegler-Graff V. P0 of Beet Western Yellows Virus is a Suppressor of Posttranscriptional Gene Silencing. Journal of Virology 2002;76(13):6815-6824.

[68] Pazhouhandeh M, Dieterle M, Marrocco K, Lechner E, Berry B, Brault V, Hemmer O, Kretsch T, Richards KE, Genschik P, Ziegler-Graff V. F-Box-Like Domain in the Polerovirus Protein P0 is Required for Silencing Suppressor Function. PNAS 2006;103(6): 1994-1999.

[69] Bortolamiol D, Pazhouhandeh M, Marrocco K, Genschik P, Ziegler-Graff V. The Polerovirus F Box Protein P0 Targets ARGONAUTE1 to Suppress RNA Silencing. Current Biology 2007;17:1615-1621.

[70] Himber C, Dunoyer P, Moissiard G, Ritzenthaler C, Voinnet O. Transitivity Dependent and Independent Cell-to-Cell Movement of RNA Silencing. The EMBO Journal 2003;22:4523-4533.

[71] Ding S-W, Li W-X, Symons RH. A Novel Naturally Occurring Hybrid Gene Encoded by a Plant RNA Virus Facilitates Long Distance Virus Movement. The EMBO Journal 1995;14(23):5762-5772.

[72] Kasschau KD, Carrington JC. Long-Distance Movement and Replication Maintenance Functions Correlate with Silencing Suppression Activity of Potyviral HC-Pro. Virology 2001;285:71-81.

[73] Angell SM, Davies C, Baulcombe DC. Cell-to-Cell Movement of Potato virus X is Associated with a Change in the Size Exclusion Limit of Plasmodesmata in Trichome Cells of Nicotiana clevelandii. Virology 1996;215:197-201.

[74] Kalinina NO, Rakitina DV, Solovyev AG, Schiemann J, Morozov SY. RNA Helicase Activity of the Plant Virus Movement Proteins Encoded by the First Gene of the Triple Gene Block. Virology 2002;296:321-329.

[75] Lu R, Folimonov A, Shintaku M, Li W-X, Falk BW, Dawson WO, ' S-W. Three Distinct Suppressors of RNA Silencing Encoded by a 20-kb Viral RNA Genome. PNAS 2004;101(44):15742-15747.

[76] Szittya G, Silhavy D, Molnár A, Havelda Z, Lovas Á, Lakatos L, Bánfalvi Z, Burgyán J. Low Temperature Inhibits RNA Silencing-Mediated Defence by the Control of siRNA Generation. The EMBO Journal 2003;22(3):633-640.

[77] Wezel RV, Liu H, Wu Z, Stanley J, Hong Y. Contribution of the Zinc Finger to Zinc and DNA Binding by a Suppressor of Posttranscriptional Gene Silencing. Journal of Virology 2003;77(1):696-700.

[78] Hartitz M D, Sunter G, Bisaro DM. The tomato golden mosaic virus transactivator (TrAP) is a Single-Stranded DNA and zinc-binding Phosphoprotein with an Acidic Activation Domain. Virology 1999; 263:1-14.

[79] Riechmann JL, Lain S, Garcia, JA. Highlights and Prospects of Potyvirus Molecular Biology. Journal of General Virology 1992;73:1-16.

[80] Maia IG, Haenni A, Bernardi F. Potyviral HC-Pro: A Multifunctional Protein. Journal Genetics Virology 1996;77(7):1335-1341.

[81] Fauquet CM, Mayo MA, Maniloff J, Desselberger V, Ball LA. Virus Taxonomy. Eigth Report of the International Commitee on Taxonomy of Viruses. Academic Press; 2005.p1259.

[82] Carrington JC, Freed DD. Cap-Independent Enhacement of Translation by a Plant Potyvirus 5′ Nontranslated Region. Journal of Virology 1990;64(4):1590-1597.

[83] Carrington JC, Freed DD, Sanders TC. Autocatalytic Processing of the Potyvirus Helper Component Proteinase in Escherichia coli and in vitro. Journal Virology 1989;63:4459-4463.

[84] Cronin S, Verchot J, Haldeman-Cahill R, Schaad MC, Carrington JC. Long-distance Movement Factor: a Transport Function of the potyvirus Helper Component Proteinase. Plant Cell 1995;7:549-559.

[85] Kasschau KD. Carrington JC. Requirement for HC-Pro Processing during Genome Amplification of Tobacco etch potyvirus. Virology 1995;209(1):268-273.

[86] Vance V B, Berger P H, Carrington J C, Hunt A G, Shi X M. 5′ Proximal Potyviral Sequences Mediate Potato virus X Potyviral Synergistic Disease in Transgenic Tobacco. Virology 1995;206: 583-590.

[87] Rojas MR, Zerbini FM, Allison RF, Gilbertson RL, Lucas WJ. Capsid Protein and Helper Component-Proteinase Function as Potyvirus Cell-to-Cell Movement Proteins. Virology 1997;237:283-295.

[88] Urcuqui-Inchima S, Maia IG, Arruda P, Haenni AL, Bernardi F. Deletion Mapping of the Potyviral Helper Component-Proteinase Reveals two Regions Involved in RNA Binding. Virology 2000;268:104-111.

[89] Plisson C, Drucker M, Blanc S, German-Retana S, Le Gall O, Thomas D, Bron P. Structural Characterization of HC-Pro, a Plant Virus Multifunctional Protein. The Journal of Biological Chemistry 2003;278:23753-23761.

[90] Li WX, Ding SW. Viral Suppressors of RNA Silencing. Current Opinion in Biotechnology 2001;12:150-154.

[91] Bass B. Double-Stranded RNA as a Template for Gene Silencing. Cell 2000;101:235-238.

[92] Zamore PD, Tuschl T, Sharp PA, Bartel DP. RNAi: Double-stranded RNA Directs the ATP-dependent Cleavage of mRNA at 21 to 23 Nucleotide Intervals. Cell 2000;101:25–33.

[93] Mallory AC, Ely L, Smith TH, Marathe R, Anandalakshmi R, Fagard M, Vaucheret H, Pruss G, Bowman L, Vance VB. HC-Pro Supression of Transgene Silencing Eliminates the Small RNAs but not Transgene Methylation or the Mobile Signal. Plant Cell 2001;13(3):571-583.

[94] Mallory AC, Parks G, Endres MW, Baulcombe D, Bowman LH, Pruss GJ, Vance VB. The Amplicon-Plus System for High-Level Expression of Transgenes in Plants. Nature Biotechnology 2002;20(6):622-625.

[95] Chapman EJ, Prokhnevsky AI, Gopinath K, Dolja VV, Carrington JC. Viral RNA Silencing Suppressors Inhibit the microRNA Pathway at an Intermediate Step. Genes Development 2004;18:1179-1186.

[96] Llave C, Kasschau KD, Carrington JC. Virus Encoded Suppressor of Post-Transcriptional Gene Silencing Targets a Maintenance Step in Silencing Pathway. Proceedings of the National Academy of Science 2000;97:13401-13406.

[97] Mette MF, Matzke AJ, Matzke MA. Resistance of RNA-mediated TGS to HC-Pro, a Viral Suppressor of PTGS, Suggests Alternate Pathways for dsRNA Processing. Current Biology 2001;11:1119-1123.

[98] Bayne EH, Rakitina DV, Morozov SY, Baulcombe DC. Cell-to-Cell Movement of Potato potexvirus X is Dependent on Suppression of RNA Silencing. Plant Journal 2005;44:471-82.

[99] Marathe R, Anandalakshmi R, Smith TH, Pruss GJ, Vance VB. RNA Viruses as Inducers, Suppressors and Targets of Post-Transcriptional Gene Silencing. Plant Molecular Biology 2000;43(2-3):295-306.

[100] Matzke M, Matzke A, Pruss GJ, Vance VB. RNA-Based Silencing Strategies in Plants. Current Opinion in Genetics & Development 2001;11(2):221-227.

[101] Voinnet O, Lederer C, Baulcombe DC. A Viral Movement Protein Prevents Spread of the Gene Silencing Signal in Nicotiana benthamiana. Cell 2000;103:157-167.

[102] Moissiard G, Parizotto EA, Himber C, Voinnet O. Transitivity in Arabidopsis can be Primed, Requires the Redundant Action of the Antiviral Dicer-like 4 and Dicer-like 2, and is Compromised by Viral-Encoded Suppressor Proteins. RNA 2007;13:1268-1278.

[103] Chiu MH, Chen IH, Baulcombe DC, Tsai CH. The Silencing Suppressor P25 of Potato virus X Interacts with Argonaute1 and Mediates Its Degradation through the Proteasome Pathway. Molecular Plant Pathology 2010;11(5):641-649.

[104] Jiang L, Wei CH, Li Y. Viral Suppression of RNA Silencing. Science China Life Sciences 2012;55:109-118.

[105] Martelli GP, Gallitelli D, Russo M. Tombusviruses. In: The Plant Viruses. (ed.) Koenig R 1998.13-72.

[106] Hearne PQ, Knorr DA, Hillman BI, Morris TJ. The Complete Structure and Synthesis of Infectious RNA from Clones of Tomato bushy stunt virus. Virology 1990;177(1): 141-151.

[107] Hayes RJ, Petty ITD, Coutts RHA, Buck KW. Gene Amplification and Expression in Plants by a Replicating Geminivirus Vector. Nature 1988;334:179-182.

[108] Scholthof HB, Scholthof K-BG, Jackson AO. Identification of Tomato bushy stunt virus Host-Specific Symptom Determinants by Expression of Individual Genes from a Potato virus X Vector. Plant Cell 1995;7(8):1157-1172.

[109] Dalmay T, Rubino L, Burgyán J, Kollar A, Russo M. Functional Analysis of Cymbidium ringspot virus Genome. Virology 1993;194(2):697-704.

[110] Silhavy D, Molnár A, Lucioli A, Szittya G, Hornyik C, Tavazza M, Burgyán J. A Viral Protein Suppresses RNA Silencing and Binds Silencing-Generated, 21- to 25-Nucleotide Double-Stranded RNAs. The EMBO Journal 2002;21, 3070-3080.

[111] Lakatos L, Szittya G, Silhavy D, Burgyán, J. Molecular Mechanism of RNA Silencing Suppression Mediated by p19 Protein of Tombusviruses. The EMBO Journal 2004;23:876-884.

[112] Vargason J, Szittya G, Burgyán J, Hall TM. Size Selective Recognition of siRNA by an RNA Silencing Suppressor. Cell 2003;115(7):799-811.

[113] Ye K, Malinina L, Patel DJ. Recognition of Small Interfering RNA by a Viral Suppressor of RNA Silencing. Nature 2003;426:874-878.

[114] Elbashir S, Lendeckel W, Tuschl T. RNA Interference is Mediated by 21- and 22-Nucleotide RNAs. Genes & Development 2001;15(2):188-200.

[115] Hamilton A, Voinnet O, Chappell L, Baulcombe DC. Two Classes of Short Interfering RNA in RNA Silencing. The EMBO Journal 2002;21(17):4671-4679.

[116] Szittya G, Molnar A, Silhavy D, Hornyik C, Burgyán J. Short Defective Interfering RNAs of Tombusviruses are not Targeted but Trigger Post-Transcriptional Gene Silencing Against Their Helper Virus. The Plant Cell 2002;14(2):359-372.

[117] Havelda Z, Hornyik C, Crescenzi A, Burgyán J. In situ Characterization of Cymbidium Ringspot Tombusvirus Infection-Induced Posttranscriptional Gene Silencing in Nicotiana benthamiana. Journal of Virology 2003;77:6082-6086.

[118] Dunoyer P, Schott G, Himber C, Meyer D, Takeda A, Carrington JC, Voinnet O. Small RNA Duplexes Function as Mobile Silencing Signals between Plant Cells. Science 2010;328:912-916.

[119] Carrington JC, Ambros V. Role of microRNAs in Plant and Animal Development. Science 2003;301(5631):336-338.

[120] Mallory AC, Vaucheret H. Functions of microRNAs and Related Small RNAs in Plants. Nature Genetics 2006; 38:S31-S36.

[121] Ramachandran V, Chen X. Degradation of microRNAs by a Family of Exoribonucleases in Arabidopsis. Science 2008;321:1490-1492.

[122] Nakahara KS, Masuta C, Yamada S, Shimura H, Kashihara Y, Wada TS, Meguro A, Goto K, Tadamura K, Sueda K, Sekiguchi T, Shao J, Itchoda N, Matsumura T, Igarashi M, Ito K, Carthew RW, Uyeda I. Tobacco Calmodulin-Like Protein Provides Secondary Defense by Binding to and Directing Degradation of Virus RNA Silencing Suppressors. PNAS 2012;doi:10.1073/pnas.1201628109.

[123] Sarmiento C, Nigul L, Kazantseva J, Buschmann M, Truve E. AtRLI2 is an Endogenous Suppressor of RNA Silencing. Plant Molecular Biology 2006;61:153-163.

[124] Gazzani S, Lawrenson T, Woodward C, Headon D, Sablowski, R. A Link between mRNA Turnover and RNA Interference in Arabidopsis. Science 2004;306(5698): 1046-1048.

[125] Souret FF, Kastenmayer JP, Green PJ. AtXRN4 Degrades mRNA in Arabidopsis and Its Substrates Include Selected miRNA Targets. Molecular Cell 2004;15:173-183.

[126] Gy I, Gasciolli V, Lauressergues D, Morel JB, Gombert J, Proux F, Proux C, Vaucheret H, Mallory AC. Arabidopsis FIERY1, XRN2, and XRN3 are Endogenous RNA Silencing Suppressors. The Plant Cell 2007;19:3451-3461.

[127] Kastenmayer JP, Green PJ. Novel Features of the XRN-Family in Arabidopsis: Evidence that AtXRN4, One of Several Orthologs of Nuclear Xrn2p/Rat1p, Functions in the Cytoplasm. Proceedings of the National Academy of Sciences USA 2000;97(25): 13985-13990.

[128] Endres MW, Gregory BD, Zhihuan G, Foreman AW, Sizolwenkosi M, Xin G, Pruss GJ, Ecker JR, Bowman LH, Vance VB. Two Plant Viral Suppressors of Silencing Require the Ethylene-Inducible Host Transcription Factor RAV2 to Block RNA Silencing. Plos Pathogens 2010;6(1):1-12.

[129] Taliansky M, Kim SH, Mayo MA, Kalinina NO, Fraser G, McGeachy KD, Barker H. Escape of a Plant Virus from Amplicon-Mediated RNA Silencing is Associated with Biotic or Abiotic Stress. Plant Journal 2004;39(2):194-205.

[130] Glazov E, Phillips K, Budziszewski GJ, Meins FJr, Levin JZ. A Gene Encoding an RNaseD Exonuclease-Like Protein is Required for Post-Transcriptional Silencing in Arabidopsis. Plant Journal 2003;35(3):342-349.

[131] Ketting RF, Haverkamp TH, Luenen HG, Plasterk RH. Mut-7 of C. elegans, Required for Transposon Silencing and RNA Interference, is a Homolog of Werner Syndrome Helicase and RNaseD. Cell 1999;99:133-141.

[132] Becker A, Lange M. VIGS – Genomics goes Functional. Trends Plant Science 2010;15(1):1-4.

[133] Padmanabhan MS, Dinesh-Kumar SP. Virus-Induced Gene Silencing (VIGS). Elsevier Ltd; 2008.375-380.

[134] Purkayastha A, Dasgupta I. Virus-Induced Gene Silencing: A Versatile Tool for Discovery of Gene Functions in Plants. Plant Physiology and Biochemistry 2009;47:967-976.

[135] Ruiz MT, Voinnet O, Baulcombe DC. Initiation and Maintenance of Virus-Induced Gene Silencing. The Plant Cell 1998;10:937-946.

[136] Ratcliff F, Martin-Hernandez AM, Baulcombe DC. Tobacco rattle virus as a Vector for Analysis of Gene Function by Silencing. The Plant Journal 2001;25(2):237-245.

[137] Huang C, Qian Y, Li Z, Zhou X. Virus-Induced Gene Silencing and Its Application in Plant Functional Genomics. Science China Life Sciences 2012;55(2):99-108.

[138] Ding XS, Schneider WL, Chaluvadi SR, Mian MAR, Nelson RS. Characterization of a Brome mosaic virus Strain and Its Use as a Vector for Gene Silencing in Monocotyledonous Hosts. Molecular Plant-Microbe Interactions 2006;19:1229-1239.

[139] Jagtap UB, Gurav RG, Bapat, VA. Role of RNA Interference in Plant Improvement. Naturwissenschaften 2011;98:473-492.

[140] Angell SM, Baulcombe DC. Consistent Gene Silencing in Transgenic Plants Expressing a Replicating Potato virus X RNA. The EMBO Journal 1997;16(12):3675-3684.

[141] Moissiard G, Voinnet O. Viral Suppression of RNA Silencing in Plants. Molecular Plant Pathology 2004;5(1):71-82.

[142] Voinnet O, Rivas S, Mestre P, Baulcombe DC. An Enhanced Transient Expression System in Plants Based on Suppression of Gene Silencing by the P19 Protein of Tomato bushy stunt virus. The Plant Journal 2003;33:949-956.

[143] Iki T, Yoshikawa M, Nishikiori M, Jaudal MC, Matsumoto-Yokoyama E, Mitsuhara I, Meshi T, Ishikawa M. In Vitro Assembly of Plant RNA-Induced Silencing Complexes Facilitated by Molecular Chaperone HSP90. Molecular Cell 2010;39:282-291.

[144] Scholthof HB. Heterologous Expression of Viral RNA Interference Suppressors: RISC Management. Plant Physiology 2007;145(4):1110-1117.